B. D. ROSS

# Perfusion Techniques in Biochemistry

A LABORATORY MANUAL IN THE USE
OF ISOLATED PERFUSED ORGANS IN
BIOCHEMICAL EXPERIMENTATION

CLARENDON PRESS · OXFORD

1972

*Oxford University Press, Ely House, London W.1*

GLASGOW    NEW YORK    TORONTO    MELBOURNE    WELLINGTON

CAPE TOWN    IBADAN    NAIROBI    DAR ES SALAAM    LUSAKA    ADDIS ABABA

DELHI    BOMBAY    CALCUTTA    MADRAS    KARACHI    LAHORE    DACCA

KUALA LUMPUR    SINGAPORE    HONG KONG    TOKYO

Text set in 11/13 pt. Monotype Modern, printed by letterpress,
and bound in Great Britain at The Pitman Press, Bath

323767

# Preface

THE importance of the technique of isolated organ perfusion in bio-chemistry does not require emphasis.

As the understanding of metabolic processes at a molecular level becomes more complete, biochemists are turning to complex systems in which the integration of such processes in an intact tissue may be studied. Between the isolated organelle preparation, the tissue homogen-ate, and slice on the one hand, and the intact animal on the other, lies the technique of organ perfusion. A single organ or tissue may be studied *in vitro*, isolated from other organs, and maintained by a supply of oxygen and nutrients, perfused through the vascular bed. Cells are maintained in the normal anatomical and physiological relationship and are not fragmented or dispersed. There is no dilution of intra-cellular cofactors as occurs in homogenate or incubation experiments; there are no transected cells leaking their contents into the medium, as occurs with a tissue slice. Nevertheless, the experimental preparation must be closely defined if such perfused organs are to yield reproducible results. The existing literature is scattered and is deficient in two main areas. The details given of the apparatus, the perfusion medium, and the operative procedure are often insufficient, and the criteria by which the preparation may be evaluated are rarely discussed. These are of importance in guiding the new investigator to select the technique most appropriate to his needs, and in establishing an objective basis for comparison of results obtained with different preparations. These considerations suggested the need for a technical manual.

The technical information is offered here in more detail than can usually be given in a journal. The apparatus, the preparation of ani-mals, and the operative details for perfusion of all those organs for which techniques are available are given; in all, 71 methods of perfu-sion are described, in 22 organs and tissues, and the descriptions are introduced by a brief discussion of the anatomy, histology and physio-logy of the organ to be perfused. The anatomical and operative descrip-tions I think demonstrate that it is not necessary to have a formal training in either anatomy, physiology, or surgery to apply the tech-nique of organ perfusion to a biochemical problem.

The generalization of the technique is discussed in Chapters 1 and 2, and an assessment of the methods available follows each subsequent chapter. The methods described have been selected because there is some objective evidence of the viability and reproducibility of the preparation. Methods have been rejected as of limited value to the biochemist either because the technique is too complex or the preparation is not reproducible. Wherever possible the criteria used for such an assessment are discussed and a series of tests of viability and function is presented.

This review will have achieved some success if it encourages experimentalists in biochemistry and medical sciences to record and publish such information about new techniques of perfusion of isolated organs.

Finally, the techniques of organ perfusion established for biochemical use are increasingly being applied in medical and related fields of research. It is hoped that this account might be useful in suggesting applications of the method and will provide some guidance to a reader who wishes either to assemble such a preparation or to assess the results obtained by others.

B.D.R.

*Pall Mall*
*May* 1971

# Acknowledgements

WORK of this kind does not occur in isolation. I am grateful to my teachers, and to colleagues and friends who have allowed their brains to be picked for the purpose of making this a comprehensive account. In particular I am grateful to Professor Sir Hans Krebs, who introduced me to liver perfusion, with Mr. Reg Hems and Dr. Michael Berry, and who encouraged the setting-up of a system of kidney perfusion, with Dr. Margaret Nishiitsutsuji–Uwo. This work was done in Oxford, in the Medical Research Council for Metabolic Research, in the University Department of Biochemistry.

In addition, I have had useful and stimulating discussions with Professor Otto Wieland, in whose laboratory in Munich I had the privilege to work. Other work was done at Imperial College, London, with the support of a Fellowship from the British Heart Foundation. Most of this text was compiled while an honorary member of the Nuffield Department of Surgery, Oxford, and I am grateful to Professor P. R. Allison for the use of laboratory facilities in his Department. The work was completed in the Nuffield Department of Clinical Biochemistry, and I am grateful to the Director, Mr. J. R. P. O'Brien and, for the liberal interpretation of my duties which made this possible, to the Chemical Pathologist, Dr. R. H. Wilkinson.

For demonstrations of special techniques I am grateful to Dr. G. Powis (colon), Dr. N. B. Ruderman (hind-limb), Dr. K. Rookledge (diaphragm), and Dr. K. G. M. M. Alberti (pancreatic perfusion; Sussman technique), Dr. J. Honor and Dr. L. Beilin (tail), and for discussions on the technique of splenic perfusion with Dr. W. L. Ford, kidney perfusion with Dr. H. Bahlmann and Dr. D. A. Hems and liver perfusion with Professor O. Wieland, Dr. J. Biebuyck, Dr. L. A. Menahan, Professor H. Schimassek, and Dr. W. Söling.

Various sections of the text were kindly read and still more kindly criticized in preparation by Mr. R. Hems, Dr. J. R. Henderson, Dr. L. H. Opie, and Dr. P. F. M. Wrigley. Dr. Patricia Lund read much of the general material of Chapters 1–3.

In assembling the necessary literature, I acknowledge the help of Miss E. Heaton of the Radcliffe Science Library, University of Oxford, and the Chemical Society Research Unit in Information Dissemination and Retrieval. Miss Frances Blakemore patiently and expertly typed all of the text, much of it several times.

Finally I am grateful for the persistence and skill of the staff of the Clarendon Press, Oxford, in converting the script into a book.

## Acknowledgments for Illustrations

The figures and tables listed below are reproduced by permission of the authors, editors, and publishers concerned, to whom my thanks are due. Full references are given in the respective chapters.

*Acta Biologica et Medica Germanica.* Rabitzsch G., **20**, 33 (1968): Fig. 5.6.

*American Journal of Physiology.* Bahlmann *et al.* **212**, 77 (1967): Fig. 1.4(d); Neely *et al.* **212**, 804 (1967): Fig. 5.7; Neely *et al.* **212**, 815 (1967): Fig. 2.4; Flock and Owen, **209**, 1039 (1965): Fig. 3.15; Weiss Ch. *et al.* **196**, 1115 (1959): Fig. 4.8; Windmueller, Spaeth, and Ganote, **218**, 197 (1970): Fig. 7.3, Table 7.1; Robert and Scow, **205**, 405 (1963): Fig. 8.1; Hinke and Wilson, **203**, 1161 (1962): Fig. 8.9.

*Archives Internationales de Pharmacodynamie et de Thérapie.* Fulgraff *et al.* **172**, 49 (1968): Fig. 4.12.

*Biochemical Journal.* Hems, Ross, Berry, and Krebs, **101**, 284 (1966): Fig. 3.9.

*Biological Chemistry.* Exton and Park, **242**, 2622 (1967): Fig. 1.6; Morgan, Henderson, Regen, and Park, **236**, 253: Fig. 5.4.

*British Journal of Pharmacology.* Wade and Beilin, **38**, 20 (1970): Fig. 8.8.

*Cell and Tissue Kinetics.* Ford, **2**, 171 (1969): Fig. 8.6.

*Experimental Medicine.* Miller *et al.* **94**, 43, 1 (1951): Fig. 3.6.

*Journal of Applied Physiology.* Turner, Neely, and Pitts, **24**, 102: Fig. 2.6; Baumann, Clarkson, and Miles, **18**, 1239 (1963): Fig. 4.10; Kavin, Levin, and Stanley, **22**, 604 (1967): Fig. 7.4; Thompson, Robertson, and Bauer, **24**, 407 (1968): Fig. 8.4; Gerber, **20**, 159 (1965): Fig. 1.7.

*Journal of Physiology.* Bleehen and Fisher, **123**, 260 (1954): Fig. 1.3; Fisher and Kerly, **174**, 273 (1964): Fig. 1.4(c); Burgen *et al.* **109**, 10 (1949): Fig. 5.16; Parsons and Robinson, **194**, 59P (1967): Fig. 8.7.

Konferenz der Gesellschaft für Biologischechemia. *Stoffwechsel der isolierte perfundierten Leber* (Ed. Staib and Scholz) (1968): (i) Forthe: Figs. 1.2, 7.6; (ii) Kessler and Schubotz: Fig. 2.7; (iii) Scholz: Figs. 1.8(a), 2.8, 2; (iv) Schimassek: Figs. 3.7, 3.16.

*Laboratory and Clinical Medicine.* Bollman *et al.* **33**, 1349: Fig. 7.2.

*Lipid Research.* Ho and Meng, **5**, 203 (1964): Figs. 8.2, 8.3.

*Metabolism.* Sussmann, Vaughan, and Timmer, **15**, 466 (1966): Fig. 6.3.

Oxford University Press. *Data for biochemical research* (Ed. Dawson, Elliott, Elliott, and Jones) (1969): tables of physiological media (section 20) (i) Krebs mammalian Ringer solutions; (ii) Other Ringer solutions: Tables 1.5, 1.6.

*Pflügers Archives.* Hoff and Hayes, **305**, 292 (1969): Fig. 8.10; Röskenbleck *et al.* **294**, 88 (1967): Fig. 1.9(b); Niesel *et al.* **294**, 79 (1967): Fig. 5.5; Basar *et al.* **304**, 189 (1968): Fig. 4.9; Basar *et al.* **299**, 191 (1968): Fig. 5.15.

*Proceedings of the Royal Society.* Ford and Gowans, **161B**, 244 (1967): Fig. 8.5; Powis, **174B**, 503 (1970): Fig. 3.14.

Springer Verlag. Schulze, in *Pathologie des Laboratoriumstiere* (Ed. Cohrs, Jaffe, and Meesen), vol. 1, p. 99 (1958): Fig. 7.1.

*Transplantation.* Semb *et al.* **6,** 977 (1968): Fig. 1.9(a).

Additional text-figures were prepared in part by Dr. N. B. Ruderman (Fig. 5.17), Dr. M. Nishiitsutsuji–Uwo (4.6), Mr. N. Vincent (2.3), and Miss Sylvia Barker (5.9), the last from a photograph by Mr. D. R. Floyd, who also prepared the text-figure in Appendix II. The drawings of the perfusion cabinet in Appendix I were prepared by Mr. M. Pritchard in consultation with Mr. R. Hems and with Mr. J. Cox, of the University Department of Biochemistry Workshop, Oxford. Mr. E. G. Lee drafted designs for Figs. 3.1–2 and 5.10–14.

# Contents

# Part 1

# Introduction

THE number of organ perfusion techniques applicable to biochemistry is increasing rapidly. It is essential at an early stage to standardize their use to prevent the increase, in turn, of observational differences that might be attributable to differences in method alone. The idea of organ perfusion is not new; standard biochemical texts in the early years of this century included technical reviews of the subject and discussed its applications. Baglioni (1910) and Müller (1910), in particular, saw the method as an instrument for routine use, and outlined the principles of metabolic studies in such systems. While their discussion stands, the methods upon which their conclusions were based were far from reproducible and were easily and totally displaced by the advent of the tissue-slice technique, coupled with manometry. Krebs and Henseleit (1932) introduced their paper on the study of urea synthesis thus:

Two methods are available in the first instance for the study of metabolism in animal tissues: the perfusion technique of Ludwig (1867) and the tissue slice method of O. Warburg (1923). We have preferred the tissue slice, which has many advantages. . . .

(Zur Untersuchung des Stoffwechsels tierischer Organe kommen in erster Linie 2 Methoden in Frage; die Durchströmungsmethode von C. Ludwig (1867) und die Gewebeschnittmethode von O. Warburg (1923). Wir haben die Gewebeschittmethode vorgezogen, Die Methode hat vor alteren Verfahren viele Vorzüge.)

In the meantime there has been an abundant multiplication of methods of organ perfusion for biochemical experiment, and their decision might well be reversed. As yet, no serious attempt has been made at standardization in its application to biochemistry and no general appraisal of the technique has appeared in print. To many, the idea of perfusion still conjures up the image of a virtuoso surgeon maintaining a precarious preparation with complex and unique machinery. But while some biochemical studies have been carried out with these 'unique' preparations, the modern development that is of interest to biochemists is the appearance of highly reproducible preparations, which are nevertheless simple to establish and maintain under defined experimental conditions. In addition, the apparatus has been so simplified by comparison with that discussed by earlier reviewers that multiple perfusion experiments may be conducted by a single investigator, much as a battery of manometers or of incubating vessels might be used. As a result, organ perfusion has become a standard experimental tool in many biochemistry laboratories.

Perfusion is defined as the passage of fluid or blood through the vascular bed of an organ. For the present purpose it may be more closely delineated as the maintenance of an organ in a viable state, isolated from the animal by means of the mechanically assisted circulation of an artificial medium through its vascular bed.

By the very nature of a blood supply, the medium has in the course of its passage through the organ been in intimate contact with a high proportion of the cells of a tissue, providing nutrients by the physiological route and permitting the extraction of metabolites by the same means. When compared with the standard techniques used *in vitro*—tissue slices, homogenates, and incubations of subcellular fractions—it is a complex method, and its increasing introduction into biochemistry must be explained and justified.

### Reasons for perfusion

Biochemistry, in metabolism and in other branches, has reached a stage where many anabolic and catabolic pathways are defined in detail, and the actions of cofactors, coenzymes, and effectors of reactions have largely been localized. With this knowledge and the information that metabolic processes may be controlled genetically, the factors that control the amount and activity of key enzymes become of interest. Investigators are now turning to the study, not only of the control of individual enzymes, but of entire metabolic pathways and of their interactions. It is not sufficient for the pathway under investigation to be enzymically present; the rates and direction of reactions depend upon the concentrations of available substrates, cofactors, and effectors, and 'regulation' as a homeostatic mechanism must therefore be assumed to involve all those pathways with common intermediates, with common cofactors, and with competing or complementary energy requirements. The existing techniques for the study of metabolism and its regulation, viz. enzyme isolation and purification, isolation of mitochondria, incubation of tissue extracts, homogenates, and slices, all have deficiencies in this respect, since the system is in each case quite artificial. Enzymes are diluted or absent, tissues are damaged, ions leak from slices, and so on. The need is therefore for an experimental preparation in which the physiological rates of as many reactions as possible are reproduced. *A priori*, this might be expected to occur if the tissue is intact, and in recent years the isolated perfused organ has proved itself suitable for such biochemical experiments. The liver preparation perfused *in situ* (see Chapter 3) may be taken as an example; the organ is maintained in the anatomical position within the rat but isolated from the rest of the animal by virtue of cannulation of its major blood supply. Full oxygenation is maintained by this route and the liver differs from the 'normal' only in lacking its hepatic arterial and nerve supply. The common heart preparations are truly isolated, in that the organ is contained in an organ chamber, with medium circulating through the coronary vessels to supply all the muscle of the organ. Again, it differs from the normal only in being detached from its nerve supply. The hind-limb preparation allows a largely muscular mass to be maintained in isolation from the animal and the muscle can be made to perform work; the kidney preparation preserves the organ intact, perfused through its major vessel, the renal artery. The special anatomical relationship in this organ between cortex and medulla is preserved, urine is formed and can be collected via the ureter. In this way, as many as possible of the 'normal' functions of the organ—muscular work, secretion, excretion, concentrating, and transport processes—are maintained, on the assumption that all these factors are influenced by, and may themselves influence, the metabolism of the tissue. The preparations offer very special advantages over the organ *in vivo*, in which biochemical study is bedevilled by the complexity of heterogeneous tissues and organs and the difficulty of defining the conditions of an experiment.

A number of examples may be cited in which the perfused organ demonstrates higher rates of reaction than analogous experiments conducted in alternative preparations *in vitro*. In particular, Hems, Ross, Berry, and Krebs (1966) have demonstrated that the rate of glucose synthesis from non-carbohydrate sources is very much closer to the expected maximum than is observed in liver slices, while liver homogenates are scarcely capable of the synthesis. Oxygen consumption, both in the perfused liver and the perfused kidney, is higher than that observed in tissue slices from these organs. These reaction rates are closer to the maximum rates observed *in vivo*.

However, while maximum rate of reaction is a useful general aim in a preparation, there must be unphysiological conditions that are manifest by rates of reaction in excess of that expected for the normal organ *in vivo*. The oxidation of malate by isolated liver mitochondria (Greville 1966) cannot be observed in liver slices or in the perfused liver (Ross, Hems, and Krebs 1967); the oxidation of succinate by mitochondria and thin liver slices (Rosenthal 1937) is not observed in more recent liver-slice experiments or in the perfused liver. It is probable that a permeability barrier to both these substrates exists in the intact cell (Hems, Stubbs, and Krebs 1968). Accordingly, an attempt should be made to assess the normality of a preparation by several methods in parallel. The special metobolic measurements available in the perfused organ are discussed in Chapter 2; but a number of further arguments in favour of this type of preparation should be mentioned here. Mechanical work, and the less obvious but similarly energy-requiring secretory processes found in many organs, notably the kidney, may significantly alter the pattern of metabolism, so that studies conducted in fragmented preparations lack this critical component in the overall control of metabolic pathways. Without describing the evidence in detail, it is clear that the concentration of AMP or of ATP within the cells, or in a specific compartment of the cell, may influence the activity of rate-limiting enzymes (Krebs 1964, Atkinson 1966). The perfused heart, for example, and in particular the newer 'working' heart preparation, offers unique possibilities of studying the effect of mechanical work on metabolism. Similarly, the perfused kidney allows the effect of the work of sodium reabsorption upon intermediary metabolism to be studied. The metabolism of muscular contraction may be studied in the hind-limb preparation which can be perfused 'at rest' or 'working' (Ruderman, Houghton, and Hems 1971).

Another special advantage of perfusion is the maintenance of the physiological division of the organ into 'compartments'; the vascular, interstitial, extra-cellular and intra-cellular, and secretory–excretory compartments. In the intestine, a further 'compartment', the lumen, is maintained. In most other preparations *in vitro* these distinctions are lost. Such divisions are almost certainly of 'regulatory' import, since the maintenance of gradients between them is a function of many homeostatic mechanisms.

Some discussion on the advantages and disadvantages of perfusion in relation to individual organs is included in the relevant chapters of this book; at the present stage, probably only the liver and heart of all the organs perfused for biochemical purposes may be said to be essential preparations. For many other organs, the technical and biochemical problems have still to be analysed and the preparations compared with existing methods used *in vitro*. Nevertheless, perfusion under controlled conditions has much to offer in the study of metabolism in an integrated tissue.

There are disadvantages in perfusion for biochemical experiment, and many of them survive the establishment of a simple, routine, and reproducible preparation, perfused with a defined medium. Thus, the very complexity of a whole organ deprives the biochemist of access to single reactions. Compartments and permeability barriers may prevent substrates and effectors from having effects that are known to be relevant when the agent is allowed direct access to the enzyme or organelle of interest. The size of a single experiment, and the time required to perform it, may make perfusion a far less efficient use of time than a preparation *in vitro* that will demonstrate the same effects. Schimmel and Knobil (1969) in particular have pointed out the greater efficiency of establishing an experimental fact with tissue slices than with isolated perfused organs. Finally, despite all efforts to perfect a technique of maintaining the organ *in vitro* in as near a normal state as possible, the resulting preparation may differ in some highly significant manner from the organ *in vivo*, limiting the interpretation of results obtained in the system.

## Historical aspects of organ perfusion

There is little point in a technical manual in discussing in detail the history of the efforts to perfuse organs. Interested readers are referred to the papers of Brodie (1903) and Embden and Glässner (1902); the state of organ perfusion in biochemistry is reviewed by Skutul (1908), Baglioni (1910), Müller (1910), and Kapfhammer (1927).

Table 0.1 summarizes the information that was available to Skutul (1908) on the development of special apparatus for this purpose; Baglioni discusses the aims and objectives of perfusion. There has been no new enunciation of the subject since 1910, although much has been written on individual techniques. It is therefore relevant to summarize Baglioni's views, which fairly represent the place of perfusion as an experimental tool in biochemistry, both then and now.

In Baglioni's view, the 'theory' of perfusion is as follows:
(i) It is desirable to separate individual organs from the whole animal to permit the study of one, unimpeded by the others.
(ii) The tissue must in fact be 'surviving'; so some signs of life are necessary. While this is easy with some organs, it is more difficult with others, especially the endocrine organs. (Clearly, this is no

TABLE 0.1

*Chronology of information on perfusion apparatus (after Skutul (1908))*

| | |
|---|---|
| 1849 | Loebell. First attempt at perfusion (pig-kidney). No experimental details. |
| 1862 | Bidder used similar method and apparatus. |
| 1867 | Schmidt applied temperature control to medium and organ. |
| 1885 | Frey and Grüber used continuous and pulsatile blood flow. Previous apparatus was either constant flow or, worse, interrupted to change reservoir. |
| 1890 | Jacobji. Apparatus allowing simultaneous blood and blood-gas analysis and a more or less physiological flow system. |
| 1892 | Jacobji and v. Sobieranski. Modification to include a lung in the circuit and thereby avoid direct mixing of gas with the medium. Alternating pressure. |
| 1895 | Langendorff. Heart perfusion apparatus (see Chapter 5). Modified apparatus. Controlled flow by reservoir bottle attached to water tap and outflow controlled with electromagnetic control of the water inflow. Double reservoir system, warm cabinet for medium and for the organ. (Essentially based upon Schmiedeberg-Bunge (1877).) |
| 1895 | Haldane (modified and used by Pfaff and Tyrode (1903)). To perfuse kidney. |
| 1903 | Brodie. Apparatus with pump to provide constant and controllable perfusion pressure. Small volume of blood only required. Applied to several organs (see below). |
| 1904 | Sakusow. Careful temperature control with 'thermo-regulator' with warmed coil. |
| ⎰1903 | Kurdinowski. Isolated perfused uterus, etc. with Lock's solution. |
| ⎱1904 | Siewert used heating coil *and* pre-warmed blood to reduce fluctuations in temperature. |
| | Skutul modified Siewert's apparatus and describes it in detail here. |

The references in this table are given in detail by Skutul (1908).

longer the case, since with modern assays hormones are among the easiest of parameters to assess.)

(iii) Perfusion attempts to imitate the natural circulation and the design of a multitude of apparatuses for the purpose aims to prevent the damaging effects of anoxia. Nevertheless, Baglioni concluded that perfusion is really a study of a dying organ, in which the results may represent only the function of the small fraction of the organ that survives.

To the general experimental layout, he suggested the following applications and methods of study:

(i) The composition of the medium may show changes, viz. large molecules may appear as a result of synthesis; substances may disappear from the medium and each allows of both qualitative and quantitative study.

(ii) Since not all products re-enter the medium, nor are all substrates entirely removed from it, analytical study of the organ itself may be undertaken.

(iii) Artificial means of stimulating function may magnify effects: for example, electrical stimulation, chemical stimuli (poisons, etc.); temperature of 38–40°C is suggested as optimal, but clearly some experiments could be conducted at room temperature.

(iv) The use of a small volume of recirculating medium may magnify effects; and the substitution of Ringer's solution for blood is advocated to reduce analytical complications.

(v) An attempt should be made to preserve cell structure and to study total function of the cell, rather than the effects of cellular enzymes shed into the medium.

The remainder of Baglioni's paper deals with the general methods of organ perfusion and with special details of individual preparations. Unfortunately, the technique then available did not allow consistent preparation of perfused organs in which to carry out the experiments suggested. In the interim, numerous preparations have appeared, some of them meeting Baglioni's required standard but most suffering from the same irreproducibility of the early preparations. There have been a number of modern developments which make a re-assessment of the subject possible, and suggest that organ perfusion now has a formal place in biochemical experiment.

**Modern developments**

The emphasis in recent years has been upon simplifying techniques in order that they may be reproducible. To this end, much greater attention has been paid to defining the experimental system, a point made by Baglioni, but often ignored in later perfusion methods. Thus, Miller, Brauer, and others produced a successful liver perfusion in the rat. Their success depended partly upon the choice of rat as experimental animal; the dog was a particularly unfortunate choice of earlier workers (see Kestens (1964) for review), since the canine liver has vasomotor responses that limit perfusion. In addition, the critical effect of perfusion with the correct medium was realized. Homologous blood could be used either fresh or after storage with the advent of suitable methods of anticoagulation and storage. Previously, defibrination was necessary, with the release of ill-defined vasomotor substances which may have been responsible for poor and patchy perfusion. Citrate–glucose solution (ACD) allows blood to be collected, stored, and centrifuged to provide red cells. The washed cells provide a very satisfactory oxygen carrier in semi-synthetic media. The use of heparin (especially) allowed fresh homologous blood to be used, and heparin may contribute to the success of rat-liver perfusions when injected into the donor animal.

In parallel with these successful perfusions with homologous blood, the use of semi-synthetic media has permitted a much closer definition of the experimental system. Hechter *et al.* (1951), but more especially Schimassek (1963), showed that perfusion could be successful even if blood was omitted from the medium. In Schimassek's hands, perfusion with a defined medium has become a standard procedure and has been applied to many organs apart from the liver. Probably the most important innovation in the preparation of defined perfusion media has been the use of plasma substitutues as 'expanders'.

Crystalline bovine albumin appears to provide the ideal colloid components in the medium and does not react with either the perfusing organ, or with other components of the medium. Other plasma expanders have been of use in different situations, and the general conclusion may be reached that 'physiological' perfusion can be conducted with either homologous blood or with one of the newer semi-synthetic media. The use of small laboratory animals, which has been so helpful in the rapid development of new techniques of perfusion, has been aided by the advent of micro-analytical procedures. Where several grammes of tissue were required for a single, and relatively non-specific, chemical, determination, the newer enzymic analyses, combined with spectrophotometry and fluorimetry, allow a battery of analyses to be performed upon a few milligrammes of tissue or a few millilitres of medium. This, in turn, allows multiple samples to be removed from the medium in the control of the perfusion, without so altering the total circulating volume as to alter the experimental conditions.

Finally, the acceptance of these new perfusion systems as useful laboratory methods has been accelerated by the careful definition of the function of the organ and tissue and their comparison with results obtained *in vivo* and *in vitro*. This approach, introduced by Schimassek, and extended into the field of tests of biosynthetic function of Hems *et al.* (1966) should allow new methods to be rapidly assessed and modified and will accelerate the important standardization of technique which must occur if reproducible results are to be obtained.

**Special requirements of biochemistry**

The newer methods of organ perfusion are much better suited to biochemical experiments than were those available to Embden and others. Because the rat has become the animal of choice, numbers of experiments may be performed where earlier only a few were possible owing to the physical difficulty of perfusing the liver of the dog. The effect of individual variation is thereby minimized; in-bred strains of rat are less different from one another than are dogs, even of the same breed.

The conditions of experiment are closely controlled: pH, temperature, flow rate, and pressure may be chosen to lie in the physiological range, and all may be equally easily modified. Experimental conditions, once chosen, may be so maintained for 1 or 2 h or more, long enough to conduct many different types of experiment. Even enzyme induction, a process which requires several hours to produce a measurable change, may be detected in such systems under the correct conditions.

The full extent of the investigations which may be carried out in perfused organs is discussed in Chapter 2.

**Other applications of organ perfusion techniques**

*Physiology*

Isolated organ perfusion techniques have been in use for the study of physiological function considerably longer than for biochemical purposes, and most of the techniques to be described in the following sections owe their origin to physiologists interested in observing the factors required to maintain an organ in isolation. Studies on blood-flow, the effects of nerve stimulation, and the formation of secretions are too numerous to detail. The aims of a technique for such a purpose must essentially be to mimic as far as possible the conditions of the organ *in vivo*: hence the common use of homologous blood or even of auto-perfusion techniques for this purpose. The presence of the animal's own lung or heart or both, as a substitute for a pump and oxygenator, or the use of a supporting animal is quite acceptable in this context, whereas it has been eschewed in the development of perfusion techniques in biochemical research.

*Pharmacology*

'Once-through' perfusion techniques, with saline media of the simplest kind, have been used to demonstrate the effects of drugs on an organ in isolation, or the uptake of a drug, its binding to receptors, its rate of metabolism, and so on. While these methods are often suitable for biochemical experiments (a particular example is the Langendorff perfusion of heart, as conducted by Loewi and Navratil (1926)), their very simplicity prevents long-term survival, and usually little attempt is made to ensure that the preparation is adequately provided with oxygen or that its function is stable over a period of time.

*Histological fixation*

Access to the greatest possible surface area of a tissue is central to satisfactory fixation for histological purposes, and vascular perfusion has been often used for this. Clearly the requirements of such a perfusion are of the simplest, but histological fixation has also been used as an adjunct to perfusion experiments of a biochemical nature. The method can then be as elaborate as may be required for the metabolic study on hand and, as a final step, the standard perfusion medium is replaced by a suitable fixative. In this way very satisfactory histology may be obtained and may be more nearly representative of the tissue *in vivo*.

*Organ survival for preservation or transplantation* (see Norman (1968))

A great deal of current work is concerned with organ perfusion in larger animals. Its purpose is to establish the conditions necessary for prolonged organ survival, with the aim of transplanting organs in medical practice. In addition, methods have been sought whereby the organ may be perfused for many days or weeks. Some such work has also been conducted in smaller laboratory animals and is referred to in the appropriate sections. However, the conditions required may be very complex and do not lend themselves primarily to biochemical experiment. The techniques involved, although similar in principle to those to be discussed here, are different in practice. Because of the size of the organ and the volume of medium, quite different mechanical problems are presented. For this reason, this aspect of organ perfusion is not considered further in this book.

One important application of the methods of organ perfusion developed in smaller laboratory animals, and with simplified conditions for relatively short-term survival, is the development of tests of viability. Such tests, once established, may be applied to more complex perfusions, or to large animals, or ultimately to human organs for transplantation to assess their state of activity. It should not be necessary to transplant an organ into a recipient to establish whether or not it is alive.

REFERENCES

ATKINSON, D. E. (1966) Regulation of enzyme activity. *A. Rev. Biochem.* **35**, 85.

BAGLIONI, S. (1910) Stoffwechseluntersuchungen an überlebenden Organen. *Handb. biol. ArbMeth.* **3**, 364.

BRODIE, T. G. (1903) The perfusion of surviving organs. *J. Physiol., Lond.* **29**, 266.

EMBDEN, G. and GLÄSSNER, K. (1902) Über den Ort der Atherschwefelsäure-bildung im Tierkörper. *Hoffmeisters Beitrag* 2, Chem. Phys. Band 1, p. 310.

GREVILLE, J. B. (1966) *Regulation of metabolic processes in mitochondria* (eds. J. M. Tager, S. Papa, E. Quagliariello, and E. C. Slater), p. 86. Elsevier, Amsterdam.

HECHTER, O., ZAFFARONI, A., JACOBSEN, R. P., LEVY, H., JEANLOZ, R. W., SCHENKER, V., and PINCUS, G. (1951) The nature and biogenesis of the adrenal secretory product. *Rec. Prog. Horm. Res.* **6**, 215.

HEMS, R., ROSS, B. D., BERRY, M. N., and KREBS, H. A. (1966) Gluconeo-genesis in the perfused rat liver. *Biochem. J.* **101**, 284.

—— STUBBS, M., and KREBS, H. A. (1968) Restricted permeability of rat liver for glutamate and succinate. *Biochem. J.* **107**, 807.

KAPFHAMMER, J. (1927) Die Leber im Stoffwechsel. *Handbuch der Biochemie,* 2nd edn, Vol. 9 (ed. C. Oppenheimer), p. 98. Jena.

KESTENS, P. J. (1964) La perfusion du foie isolé. Edition Arscia, Bruxelles.

KREBS, H. A. (1964) Croonian lecture 1963: gluconeogenesis. *Proc. R. Soc.* **B159**, 545.

—— and HENSELEIT, K. (1932) Untersuchungen über die Harnstoffbildung im Tierkörper. *Hoppe-Seyler's Z. physiol. Chem.* **210**, 33.

LOEWI, O. and NAVRATIL, E. (1926) Über humorale Übertragbarkeit der Herznervenwirkung. *Pflügers Arch. ges. Physiol.* **214**, 678.

MÜLLER, F. (1910) Die künstliche Durchblutung resp. Durchspülung von Organen. *Handb. biol. ArbMeth.* **3**, 327.

NORMAN, J. C. (ed.) (1968) *Organ perfusion and preservation.* Appleton Century Crofts (Meredith Corp.), New York.

ROSENTHAL, O. (1937) The intensity of succinate oxidation in surviving liver tissue. *Biochem. J.* **31**, 1710.

ROSS, B. D., HEMS, R., and KREBS, H. A. (1967) The rate of gluconeogenesis from various precursors in the perfused rat liver. *Biochem. J.* **102**, 942.

RUDERMAN, N. B., HOUGHTON, C. R. S., and HEMS, R. (1971) Evaluation of the isolated perfused rat hind-quarter for the study of muscle metabolism. *Biochem. J.* **124**, 639.

SCHIMASSEK, H. (1963) Metabolite des Kohlenhydratstoffwechsel der isoliert perfundierten Rattenleber. *Biochem. Z.* **336**, 460.

SCHIMMEL, R. J. and KNOBIL, E. (1969) Role of free fatty acid in stimulation of gluconeogenesis during fasting. *Am. J. Physiol.* **217**, 1803.

SKUTUL, K. (1908) Über Durchströmungsapparate. *Pflügers Arch. ges. Physiol.* **123**, 249.

WARBURG, O. (1923) Versuche an überlebendem Carcinomgewebe (Methoden: II. Die Herstellung der Gewebeschnitte). *Biochem. Z.* **142**, 317.

# 1. General principles of organ perfusion

ORGAN perfusion, for the present purposes, may be defined as the maintenance of an organ in isolation in a viable state, by the mechanically assisted circulation of a suitable fluid through its vascular bed.

Except for the most primitive studies, special apparatus is required and each worker has usually adopted an individual approach to solving the technical problems associated with one organ or one aspect of biochemistry. The resulting literature is very scattered. It is, in consequence, difficult for a newcomer to the field of organ perfusion to see the advantages or applications of perfusion to his special problem, or to assess the merits of the various methods available. In the face of such complexity, there is in some circles a reluctance to embark upon the general technique of organ perfusion just in those areas in which it may offer special access to biochemical problems.

Unavoidably, the details of special apparatus, operative techniques, and conditions of perfusion must be discussed for each organ separately, since the anatomical and physiological differences between organs are the very features that are preserved in whole-organ perfusion. However, generalizations may be made, and the method of organ perfusion in biochemistry should be considered as a unified technique with general applications.

This chapter attempts to treat the component parts of a perfusion system individually. For this purpose, the features of a perfusion may be considered in terms of the *organ*, particularly the arrangement of its vascular bed, the *perfusion medium*, and the *apparatus*. The apparatus comprises the means of supply and circulation of medium (usually a *pump*), its *oxygenation*, *filtration*, and *temperature control*.

*The organ.* The organ to be perfused should be capable of isolation from neighbouring tissues, although this isolation need not be physical; a separate vascular bed is sufficient to ensure that only one tissue or organ is perfused. The organ may be perfused in isolation within an organ chamber, or may remain *in situ* but isolated from the vascular beds of adjacent tissues. In addition, the organ, to be suitable for perfusion, should have a blood supply that is both reasonably compact and anatomically constant. A single artery of supply is the ideal and, equally, a single effluent vein. Descriptions are given of exceptions and perfusion systems devised to cope with the differences; for example pancreas, which has two principal arteries, and diaphragm, in which the arterial supply is so diffuse that perfusion is conducted in a retrograde direction, through the vein.

*The perfusion medium.* A detailed discussion of the nature of fluids which may be used as perfusion media is presented on p. 17. The medium is chosen as a carrier of oxygen in sufficient concentration to supply the requirements of the respiring organ, and is in every way analogous to the incubation medium of other, *in vitro*, experimental systems. The important difference is that it remains always separable from the perfused organ, in a way that is not so for, say, enzyme or homogenate incubations.

*Means of supply of medium.* Medium is moved continuously through the vascular bed in the manner of the circulation in life, with the assistance of either gravity or some pumping device. The alternative methods are discussed on p. 63. Medium may pass once through the organ, to be collected as it emerges, or, in a more sophisticated application of the method, a finite volume of perfusion medium may be continuously recirculated for the duration of an experiment. The majority of methods to be described use this recirculation-perfusion system, the many advantages of which in biochemical experiments are discussed on p. 61.

*Oxygenation and temperature control.* The maintenance of the organ demands an oxygen supply in most cases, and controlled experimental conditions demand a means of maintaining its temperature. The techniques and apparatus used for this purpose are discussed on pp. 41 and 70 respectively.

*Path of medium through the organ.* Central to the method of perfusion is a knowledge of the details of blood supply, the vascular and capillary beds, and the venous drainage of the organ under investigation. Details are included in the 'anatomical' section that introduces each chapter.

The mammalian vascular system comprises the blood and lymphatic systems, which are intimately associated in the organism as a whole but are components of separate compartments within the tissue or organ. Arteries enter organs and tissues, where they branch according to a specific pattern of arterioles. These in turn are continuous with a close-meshed network of capillaries. The blood from the latter is collected into minute vessels, the venules, which join to form veins.

The systemic and pulmonary circulations are those supplied by the left and right sides of the heart respectively (see Chapter 5).

A portal circulation, of which the relevant example in the present context is that of the liver, arises as follows. Blood circulating through the spleen, pancreas, stomach, and intestines is not drained directly to the heart but is conveyed by the portal vein to the liver. In the liver this vein divides like an artery. The portal blood traverses two capillary systems, those of the viscera and the sinusoids of the liver, draining into the hepatic veins.

Blood-vessels may be perfused in their own right, and methods are discussed in Chapter 8.

Although anatomically blood flows into an organ through arteries to emerge from the veins, there are well established examples of adequate maintenance of an organ by retrograde perfusion (see pp. 175, 302, and 347) so that it is not essential to mimic the physiological direction of blood flow for biochemical perfusion.

Under normal circumstances *in vivo* and in the majority of perfusion techniques (the exceptions are those mentioned in the previous paragraph), a simple pressure gradient from artery to vein obtains and the relationship determined by Starling's law may be applied to the passage of fluid through the capillary bed (see p. 37). Starling's law (1895) relates capillary pressure (CP), tissue pressure (TP), and colloid osmotic pressure of the perfusing fluid ($COP_B$), and successful perfusion usually depends upon this relationship being maintained near to the situation *in vivo*.

$$CP-TP = COP_B-COP_T,$$

where $COP_T$ = colloid osmotic pressure of extravascular fluid. The simplest explanation for oedema (the increase in fluid content) of a perfusing organ is failure to observe these conditions.

### THE RAT AS AN EXPERIMENTAL ANIMAL

Most of the methods to be described here are devised for use with the laboratory rat, *Mus norvegicus albinus*. The suitability of the rat for biochemical experimentation is largely accepted, and much data on mammalian metabolism is derived from this source. In particular, the rat liver has been studied in detail.

Organ perfusion techniques have been devised for many other species, and in earlier methods the dog, cat, and larger domestic animals were used, even for predominantly biochemical experiments (see p. 8). The reasons were largely technical; operative techniques in larger animals were more familiar (this is still an important argument in the surgical application of organ perfusion), and chemical methods of assay required large amounts of tissue. The ability to apply the general technique to a small and readily available animal such as the rat has enabled organ perfusion to become important in a field long dominated by incubation techniques with homogenates and tissue slices. Micro-chemical assays remove the requirement for large amounts of tissue. For normal laboratory purposes a small animal such as the rat is very convenient and wherever a suitable technique has been described for this animal it is included here, in preference to even a well-established method in a larger animal. This general rule has been waived to include descriptions of techniques thought to be appropriate for biochemical use but where the rat has for some reason proved unsatisfactory. Examples include thyroid gland (rabbit) and adrenal gland (ox); in the special case of the heart, the specific insensitivity of the rat heart to digoxin and other cardiac glycosides has made the use of guinea pig a common alternative (see Arnold and Lochner (1965) and Chapter 5). Other examples are cited in the text.

It is tempting to apply techniques developed in one species to others, but the same rigorous criteria are required and each technique must be tested objectively in a new species, as has been suggested for the rat. *A priori*, transfer of general methods between species should be possible and this is a further argument in favour of the rat. All the preliminary work may be conducted in this species, which may save very considerable and expensive developmental work, especially when larger animals are to be used. It is beyond the scope of the present account to discuss organ perfusion in large animals.

### PERFUSION MEDIUM

The perfusion medium corresponds to the incubation medium used in simpler biochemical systems, and similar criteria should be applied to its composition. Thus, it should provide an environment of constant temperature, constant pH, and an oxygen and carbon dioxide tension

3

suitable to the study in progress. Both the initial pH and the total buffer capacity of the medium are important, particularly when the standard 'recirculation-perfusion' (see p. 61) is used.

Largely as a result of the history of organ perfusion, and a desire to maintain a situation near to the state existing *in vivo*, blood, or at least erythrocytes, is the main component of most perfusion media. This is dictated by the high oxygen consumption of many tissues and the relatively low proportion of the medium which is in contact with the tissue at any one time. Haemoglobin is the obvious oxygen carrier to use although more recently inert fluocarbon carriers, which bind oxygen and function very much like erythrocytes when suspended as an emulsion (Sloviter and Kamimoto 1967), have been suggested as an alternative. However, as discussed on p. 11 it is one of the principal aims of organ perfusion in biochemical use to adopt a perfusion medium which is defined and of reproducible composition. This may differ from the physiological 'ideal'; Hechter *et al.* (1951) in discussing the use of homologous versus heterologous blood as perfusate comment that 'it should not be taken to indicate that physiological media are necessary to obtain maximal activity of a particular metabolic system', an observation which has been amply borne out by studies with the perfused liver (see Chapter 3). They go on to suggest that a medium should be 'chosen for convenience'. But this view is extreme. Many physiologists would not accept that blood should be omitted from such a system. However, a most important recent change in this field has been the realization that perfusion media, like incubation media (Warburg 1923, Krebs and Henseleit 1932) are susceptible to systematic study. This has been done for the liver (see Chapter 3) and to some extent for aortic perfusion of heart (Morgan, Henderson, Regen, and Park 1961, and Chapter 5).

So many different media have been introduced into organ perfusion that it is important to establish which are the most suitable. It is suggested that the *simplest* medium which gives reproducible perfusion of an organ for the length of time required and fulfils the criteria of function of the organ concerned should be employed.

The requirements of a perfusion medium may be summarized as:

(1) An oxygen-carrying capacity sufficient for the requirements of the respiring organ and with sufficient reserve to allow of extra oxygen consumption, due to added substrates or other stimuli, or to reduced flow rate.

(2) A physiological pH, usually 7·4, with adequate buffer capacity at a suitable p$K$, for example bicarbonate, p$K$ 6·1.

(3) A physiological concentration of the principal ions. This well-established feature of incubation media for tissue slices (see Umbreit, Burris, and Stauffer (1964) and Krebs and Henseleit (1932) for examples) should be followed in organ perfusion, and 'essential' ions omitted only after objective test.

(4) In the absence of blood, an adequate colloid osmotic pressure must usually be established with plasma or a plasma substitute, adequate to balance the hydrostatic and tissue pressures.

(5) Substrate(s) of respiration may need to be added to maintain optimal function (for example, glucose in heart perfusion).

All these requirements are, of course, met by homologous blood, when it is adequately gassed and warmed to 37°C, and this is commonly the medium recommended (see, for example, Miller, Bly, Watson, and Bale (1951); liver perfusion medium of heparinized rat blood, diluted 30 per cent with Ringer's solution). However, numerous alternatives have been proposed. An attempt has been made here to classify media currently in use in biochemical organ perfusion techniques, giving detailed recipes where appropriate. Their main uses, advantages, and disadvantages are discussed, and some general proposals are made on which to base a choice of perfusion medium.

**Some standard perfusion media**

A medium may be modified in any or all of its major constituents: erythrocytes, buffer, salts, colloid or suspending fluid. The following media are in common use.

Medium I.  Homologous (rat) blood (diluted 30 per cent with Ringer's solution (Miller *et al.* 1951)).

Medium II.  Semi-synthetic medium; original (Schimassek 1962).

Medium III.  Semi-synthetic medium; modified (Hems *et al.* 1966).

Medium IV.  Semi-synthetic medium, without red cells, supplemented (Schnitger, Scholz, Bücher, and Lübbers 1965).

Medium V.  Simple salt medium, without colloid (Morgan *et al.* 1961).

With the exception of the last, all these media are used primarily for liver perfusion, where they have been exhaustively tested; medium V is used in a standard method of heart perfusion (see Chapter 5).

These media will be described in detail, together with details of balanced salt media and buffers in common use in organ perfusion. The range of media available has been tabulated for convenience. Details are available in the text, or in the references cited in the text.

*Medium I (Miller et al. 1951) for liver* (see also p. 151)

Fresh rat blood, obtained by cardiac puncture in ether-anaesthetized animals after 18 h starvation is used. To 130 ml blood is added 12 ml of a heparin solution containing 5 mg heparin/ml of Ringer's solution; this is then diluted by one-third with Ringer's solution (p. 26). The resultant haematocrit varies between 25 and 40 per cent when the medium is prepared in this way.

This medium, or minor modifications of it (see p. 28) has also been used in perfusion of lung (Levey and Gast 1966), kidney (Bauman, Clarkson, and Miles 1963), brain (Thompson, Robertson, and Bauer 1968) and about one in three of all liver perfusion techniques currently published in the rat employ this medium of Miller *et al.*

*Note.* For technical details of obtaining rat blood see p. 32.

*Medium II. Semi-synthetic medium; original (Schimassek* 1963)

This medium, and the following Medium III (Hems *et al.*) which is based upon it, are the most commonly used semi-synthetic perfusion media and provide the pattern for many other media in use in this field (see Table 1.7 for some of its applications).

Pure bovine albumin, Fraction V (Behringwerke) is dissolved to a concentration of 2·5 g % in Tyrode solution, and the final volume is made up to 100 ml by the addition of washed bovine erythrocytes.

TABLE 1.1

*Composition of Schimassek perfusion medium*

| | | | |
|---|---|---|---|
| NaCl | 137 mM | L (+) lactate | 1·33 mM |
| KCl | 5·9 mM | Pyruvate | 0·09 mM |
| $CaCl_2$ | 1·80 mM | Bovine albumin (puriss) | |
| | | (Behringwerke) | 2·5 g% |
| $MgCl_2.6H_2O$ | 0·49 mM | Bovine erythrocytes (twice washed) | |
| | | | 10 g%Hb |
| $NaHCO_3$ | 11·9 mM | Terramycin (tetracycline) | |
| $NaH_2PO_4.H_2O$ | 1·22 mM | | 1·5 mg% |
| D-glucose | 5·45 mM | Total volume | 100 ml    Temp. 37°C |
| | pH 7·1 | (gassed with 5% $CO_2$) | |

The final haemoglobin concentration is $10 \text{ g} \%$, and in addition, substrates (lactate, pyruvate, and glucose) and an antibiotic (tetra-cycline) are added. The pH of the medium is controlled by gassing with $95\% \text{ O}_2$ with $5\% \text{ CO}_2$, which with $11 \cdot 9$ mM bicarbonate gives pH $7 \cdot 12$. The final composition of the medium is given in Table 1.1, and details on the washing of red cells are included in this section (p. 34).

*Medium III. Semi-synthetic medium; modified (Hems et al. 1966)*

The preparation of this medium, essentially similar to that described above (Medium II), is described in Chapter 3, p. 169. It differs from Medium II in using:

(a) Aged human red cells, in place of bovine cells. Originally intro-duced because the glycolysis of fresh red cells interfered with the interpretation of experiments on gluconeogenesis, their use may now be justified in routine experiment, on the basis of objective tests conducted in rat liver perfusion (see Chapter 3).

(b) Krebs–Henseleit buffer (25 mM bicarbonate) is substituted for Tyrode solution, the bicarbonate content of which may be calculated to be too low for a physiological pH of $7 \cdot 4$ to be achieved with $5\%$ $\text{CO}_2$.

(c) No substrates or antibiotics are present in the basic medium, although these are added as and when required. This medium, or minor modifications of it, has been widely applied in organ perfusion in the rat (see Table 1.7). Its composition is given in Table 1.2.

### TABLE 1.2

*Composition of Hems perfusion medium*

| | | |
|---|---|---|
| NaCl | 118 mM | Bovine albumin, crystalline #5 |
| KCl | $4 \cdot 75$ mM | (Armour Labs, or Pentex; and normally dialysed before use) |
| CaCl$_2$ | $2 \cdot 5$ mM | $2 \cdot 5 \text{ g}\%$ |
| KH$_2$PO$_4$ | $1 \cdot 18$ mM | Aged (4 weeks min.) human erythrocytes (twice washed) $2 \cdot 5\%$ Hb |
| MgSO$_4$.7H$_2$O | $1 \cdot 18$ mM | Total volume 150 ml |
| NaHCO$_3$ | $25 \cdot 0$ mM | |
| | pH $= 7 \cdot 4$ (gassed with $5\% \text{ CO}_2$) | |

Note that apart from the difference in bicarbonate concentration, this medium contains sulphate, which is lacking from Tyrode solution, and hence, from Medium II; the metabolic significance of this alteration has not been investigated in perfusion.

*Medium IV. Semi-synthetic medium, without haemoglobin (Schnitger et al. 1965)*

Haemoglobin-free medium may be based on Media II or III by simply omitting red cells (see, for example, Williamson, Kreisberg, and Felts 1966 and see Table 1.7 for others) or, alternatively, the medium elaborated by Schnitger *et al.* (1965) may be used. Apart from a balanced salt mixture, and 24 mM bicarbonate, to give pH 7·4 when gassed with 5% $CO_2$, this medium uses Dextran in place of albumin and adds an arbitrary amino-acid mixture (see Table 1.3 for composition).

TABLE 1.3

*Composition of Schnitger medium*

| Electrolytes | | L-Amino-acids | |
|---|---|---|---|
| NaCl | 137 mM | aspartic acid | 0·1 mM |
| KCl | 3·0 mM | threonine | 0·2 mM |
| $NaH_2PO_4$ | 0·7 mM | serine | 0·3 mM |
| $CaCl_2$ | 0·5 mM | glycine | 0·5 mM |
| $MgCl_2$ | 0·5 mM | alanine | 0·6 mM |
| $NaHCO_3$ | 24 mM | glutamic acid | 0·9 mM |
| | | glutamine | 0·9 mM |

Dextran (Rheomacrodex, salt-free, Knoll AG Ludwigshaven)
7·0 g%
pH = 7·4 (gassed with 5% $CO_2$)

This amino-acid mixture may serve as a model for media in which such supplementation is thought to be necessary, for example in protein synthesis and enzyme induction studies in perfused liver (see John and Miller (1969)).

*Medium V. Simple salt medium, without colloid (Morgan et al. 1961)*

Krebs–Henseleit medium (q.v.) with 5·0 mM glucose as the only addition has been successfully used in most studies with the perfused heart (Morgan *et al.* 1961 and see Chapter 5). This medium lacks a plasma expander and its use has been somewhat restricted in consequence. However, the diaphragm (Rowlands 1969), a preparation of isolated tail vessels (Hinke and Wilson 1962), and a partially successful perfusion of rat liver (Bloxham 1967) should be mentioned. In addition, Tyrode solution with 5·6 mM glucose has been used in a preparation of rabbit thyroid (Williams 1966). In this case the advantage of adding haemoglobin in solution has been clearly demonstrated by a doubling of the survival time of the functioning organ. The composition of the medium is essentially that given on p. 26, plus D-glucose 5·0 mM.

### Standard salt media

A few standard buffers form the basis of the commoner perfusion media: Ringer's solution, Krebs–Henseleit buffer, Tyrode solution, Locke's solution, and 'physiological' (0·9%) NaCl, occur frequently. In addition, a number of tissue culture media have found application in the more complex organ perfusion techniques, especially those concerned with prolonged survival (e.g. Folkman, Cole, and Zimmerman 1966 and Brauer, Pessotti, and Pizzolato 1951) and, for convenience, the recipes of these media are included here; original sources are as quoted. In addition, the recognized sources of such information are recommended (Umbreit *et al.* 1964, Dawson *et al.* 1969). The medium, often called Ringer (Umbreit) solution, is identical in composition with Krebs–Henseleit (1932) buffer.

*Krebs–Henseleit bicarbonate medium* (1932)

The following stock solutions are required:

(1) 0·9% NaCl (0·154 M)
(2) 1·15% KCl (0·154 M)
(3) $CaCl_2$ (0·11 M)
(4) 2·11% $KH_2PO_4$ (0·154 M)
(5) 3·82% $MgSO_4.7 H_2O$ (0·154 M)
(6) 1·3% $NaHCO_3$ (0·154 m)
(7) Gas mixture with 5% $CO_2$, with which 6. is vigorously gassed for 1 h before use.

To prepare:

mix 100 parts soln. 1;   1 part soln. 4
4 parts soln. 2;   1 part soln. 5
3 parts soln. 3;  21 parts soln. 6

and gas for 10 min. Any precipitation may be reversed by gassing once again.

*Modifications of Krebs–Henseleit medium*

*Low-bicarbonate buffer.* This buffer may be prepared in bulk, for storage, and made up to the composition above by addition of stock bicarbonate solution (Krebs–Henseleit 1932). To the original mixture are added only 3 parts of solution 6 (bicarbonate) instead of 21 parts, and the final 'high bicarbonate' medium is completed by the addition of 16 ml of stock solution 6 to each 100 ml of low-bicarbonate saline.

*Low-calcium medium.* Bleehen and Fisher (1954) and Zachariah (1961) recommend the medium above, with half the calcium content (i.e. 1·67 mM $CaCl_2$ instead of 3·3 mM) for Langendorff perfusion of the heart. This more closely approximates the 'ionized' calcium content of serum, and is suitable for media to which no albumin is added.

*Low-magnesium medium.* Opie (1965) recommends a medium low in both calcium and magnesium (1·67 mM and 3·85 mM respectively) for perfusion of the heart. The composition is otherwise as described for Krebs–Henseleit medium.

*Millipore filtration.* Hillier (1968) among others, has pointed out the hazards to perfusion studies of the crystallization or precipitation of calcium phosphate from Krebs–Henseleit medium. This should be avoidable, if the instructions of Krebs and Henseleit are followed, particularly the adequate and early gassing of the medium with $CO_2$. However, some authors advocate millipore filtration (0·5 $\mu$m pore size) of media before use in perfusion (for example, Mansford (1969)).

*Modified pH.* The pH of bicarbonate buffers may be modified by altering either the bicarbonate concentration or the partial pressure of $CO_2$ added to the gassing mixture. This relationship is expressed in the Henderson–Hasselbach equation, which may be used to calculate the pH under such conditions:

$$pH = pK' + \log \frac{[HCO_3]}{[CO_2]}.$$

*Notes.* (i) $pK'$ in Krebs–Henseleit medium or in serum $= 6\cdot1$ (Peters and Van Slyke 1932). In the absence of salts, the value is 6·33. Salts depress the $pK'$ in proportion to the ionic strength

$$pK'' = pK' - 0\cdot5\sqrt{\mu}.$$

Umbreit *et al.* (1964) give a nomogram for calculation of pH of bicarbonate buffers at 'zero' ionic strength, which should not be used for calculations in sera or for perfusion media composed of balanced salt solutions. That provided by Peters and Van Slyke (1932) is appropriate to this purpose.

(ii) $CO_2$. The value is usually given as % $CO_2$ in a gas mixture, which may be converted to moles as follows:

$$CO_2 \text{ in moles} = \frac{P\alpha CO_2 \times 1000}{760 \times 22400 \times 100},$$

where $P$ = atmospheric pressure, $\alpha$ = solubility of $CO_2$ in the medium, and $CO_2$ = % $CO_2$ at the pressure given.

(iii) The $CO_2$ content of gas mixtures may be confirmed by direct determination, manometrically by the method of Krebs (1930), or polarographically.

(iv) Bicarbonate. The bicarbonate added may not be the final concentration in the complex media used for perfusion. Bovine albumin displaces (Hems *et al.* 1966), and binds bicarbonate. For accuracy of pH, the final bicarbonate content of a medium should be determined by the method of Natelson (1951).

For the medium of Hems (q.v.) a selection of different ($HCO_3$) and % $CO_2$ values with the pH obtained is given in Table 1.4.

### TABLE 1.4

*pH obtained with some standard $CO_2$ and bicarbonate concentrations in Hems perfusion medium*

| Concentration of $HCO_3^-$ (mM) | 9·0 | 11·8 | 15·0 | 9·0 | 35·0 | 25·0 | 15·0 | 39·0 | 35·0 | 32·0 | 40·0 |
|---|---|---|---|---|---|---|---|---|---|---|---|
| $CO_2$ content of gas mixture (%) | 5·0 | 5·0 | 5·0 | 2·5 | 7·0 | 5·0 | 2·5 | 5·0 | 2·5 | 1·4 | 1·4 |
| Concentration of $CO_2$ in medium (mM) | 1·12 | 1·12 | 1·12 | 0·56 | 1·57 | 1·12 | 0·56 | 1·12 | 0·56 | 0·31 | 0·31 |
| pH (calculated) | 7·01 | 7·12 | 7·23 | 7·35 | 7·45 | 7·45 | 7·53 | 7·64 | 7·90 | 8·11 | 8·21 |

For intervening values, see nomogram in Peters and Van Slyke (1932).

### Other media

Ringer saline, Locke, Tyrode, and alternative media to the Krebs–Henseleit detailed above are contained in many perfusion media (see Table 1.7). Tables 1.5 and 1.6 are reproduced from Dawson *et al.* (1969, pp. 507–8). It must again be emphasized that too low a concentration of bicarbonate is present in most of these media for their routine use with 5% $CO_2$ as a buffer system. It is strongly advised to substitute Krebs–Henseleit medium in such cases (see, in particular, Schimassek, Medium II, above).

## TABLE 1.5

### *Krebs mammalian Ringer solutions (from Dawson et al. (1969))*

Parts by volume

| Solutions required (all approximately isotonic with serum) | Krebs–Henseleit original Ringer bicarbonate† | Krebs original Ringer phosphate | Krebs improved Ringer I | Krebs improved Ringer II | Krebs improved Ringer III | Krebs substrate fortified serum |
|---|---|---|---|---|---|---|
| | A | B | C | D | E | |
| 0·90% NaCl (0·154M) | 100 | 100 | 80 | 83 | 95 | |
| 1·15% KCl (0·154M) | 4 | 4 | 4 | 4 | 4 | |
| 1·22% CaCl$_2$ (0·11M) | 3‡ | 3‡ | 3‡ | | 3‡ | |
| 2·11% KH$_2$PO$_4$ (0·154M) | 1 | 1 | 1 | 1 | 1 | |
| 3·8% MgSO$_4$.7H$_2$O (0·154M) | 1 | 1 | 1 | 1 | 1 | |
| 1·3% NaHCO$_3$ | 21§ | | 21§ | 3 | 3 | |
| 0·1M-Phosphate buffer ph 7·4 (17·8 g Na$_2$HPO$_4$.2H$_2$O+20 ml N-HCl diluted to 11) | | 21 | | | | |
| 0·16M-Na pyruvate (or L-lactate) | | | 4 | 4 | 4 | 3 |
| 0·1M-Na fumarate | | | 7 | 7 | 7 | 6 |
| 0·16M-Na-L-glutamate | | | 4 | 4 | 4 | 3 |
| 0·3M-(5·4%) glucose | | | 5 | 5 | 5 | 5 |
| 0·1M-Na phosphate buffer (100 vol. 0·1M-Na$_2$HPO$_4$ (1·78% Na$_2$HPO$_4$.2H$_2$O)+25 vol. 0·1M-NaH$_2$PO$_4$(1·38% NaH$_2$PO$_4$−H$_2$O)) | | | | 18 | 3 | |

(Columns A–E: 100 vol. serum prepared from rapidly cooled blood.)

† Gassed with 5% $CO_2$ in gas phase.

‡ Twice the concentration of ionized Ca in serum (*Nature, Lond.* **184**, 1315 (1959)).

§ Gassed with 100% $CO_2$ for 1 h before mixing with other solutions.

TABLE 1.6

*Other Ringer solutions*

|  | Ringer (frog heart) (g) | Locke (mammalian heart) (g) | Tyrode (rabbit intestine) (g) | Amphibian Ringer (g) |
|---|---|---|---|---|
| NaCl | 0·65 | 0·9 | 0·8 | 0·65 |
| KCl | 0·014 | 0·042 | 0·02 | 0·025 |
| $CaCl_2$ (anhyd.) | 0·012 | 0·024 | 0·02 | 0·03 |
| $MgCl_2$ (anhyd.) |  |  | 0·01 |  |
| $NaHCO_3$ | 0·02 | 0·01–0·03 | 0·1 | 0·02 (pH 7·0–7·4) |
| $NaH_2PO_4$ (anhyd.) | 0·001 |  | 0·005 |  |
| Glucose | 0·2 | 0·1–0·25 | 0·1 |  |

These salts and glucose are dissolved in water to produce 100 ml solution.

## Complex media

A number of perfusion media have been based upon established media from the field of tissue culture, since many of the requirements for cell survival have already been covered by workers in that field (Waymouth 1965). Their uses in organ perfusion conducted for biochemical purposes is debatable; in particular Medium 199 has been criticized for including substances over and above those found necessary for tissue survival in culture, without prolonging survival (Paul 1970). Medium 199 forms the basis of a perfusion medium used for a study on RNA synthesis on the rat heart (Fanburg and Posner 1969). Other media used in special situations include White's medium (Brauer *et al.* 1951), for prolonged liver perfusion, Eagle's medium (Folkman, Winsey, and Moglus 1967), for a number of long-term organ survival studies (see Table 1.7); and Waymouth's medium 'MB 752/1' has been used in the special situation of liver perfusion in which the medium was required afterwards to support tissue culture (McConaghey and Sledge 1970). Full details of these media are included in Paul's book *Cell and tissue culture*. There are special precautions required in the preparation of each of these media, and the reader is referred to the sources for details (see Paul 1970 or Waymouth 1965).

## TABLE 1.7

*Perfusion media*

| Gas | Principal components of medium | Organ | Page | Authors |
|---|---|---|---|---|
| 5% $CO_2$ | K–H saline, 5 g% bovine albumin (dialysed) | Kidney | 232 | Nishiitsutsuji-Uwo *et al.* 1967 |
| 100% $O_2$ | 6% Dextran, acetate, phosphate, glucose, salts | Kidney | 241 | Weiss *et al.* 1959 |
| ? | Whole rat blood, insulin, PAH, glucose, heparin | Kidney | 246 | Baumann *et al.* 1963 |
| ?5% $CO_2$ | as Weiss *et al.*, albumin, Haemaccel, insulin, urea, pyruvate, lactate, glucose, $PO_4$, acetate, $HCO_3$ | | 250 | Fülgraff 1968 |
| 5% $CO_2$ | 'Schimassek', 5·5 g% albumin, urea | Kidney | 253 | Bahlmann *et al.* 1967 |
| | Summary of media used in kidney perfusion | | 254 | several |
| None (air) | 'Krebs–Ringer' (Umbreit) ½Ca conc. | Heart | 265 | Langendorff 1895 |
| 5% $CO_2$ | Krebs–Henseleit | Heart | 266 | Bleehen and Fisher 1954 |
| 5% $CO_2$ | Krebs–Henseleit, ½Ca and ½Mg content | Heart | 270 | Morgan *et al.* 1961 |
| 5% $CO_2$ | Tyrode solution | Heart | 273 | Opie 1965 |
| ?100% $O_2$ | Krebs–Ringer (as Morgan) | Heart | 275 | Arnold and Lochner 1965 |
| 5% $CO_2$ | Krebs–Henseleit plus glucose (filtered) | Heart | 277 | Rabitzsch 1968 |
| 5% $CO_2$ | K–H and 1% albumin, 10% rat erythrocytes, glucose, pyruvate, glutamate, and fumarate | Heart | 285 | Chain, Mansford, and Opie 1969; Morgan 1965 |
| 5% $CO_2$ | K–H, 3·5% Haemaccel, glucose, insulin, and papaverine | Heart | 293 | Oye 1964 |
| ?5% $CO_2$ | Fresh, heparinized rat blood, diluted 1:1 with 2% gelatin in 0·9% NaCl | Heart | 297 | Basar *et al.* 1969 |
| 5% $CO_2$ | 4% Dextran in K–R (Umbreit)? 29 mM $HCO_3$ OR | Pancreas | 327 | Grodsky 1963 |
| 5% $CO_2$ | 4% human albumin in K–R (Umbreit)? 29 mM $HCO_3$ | Pancreas | 327 | Grodsky 1963 |
| 5% $CO_2$ | Fresh rat plasma 1:1 with Ringer or 0·9% NaCl | Pancreas | 328 | Khayambashi and Lyman 1969 |
| $O_2$ | Whole, heparinized rat blood | Pancreas | 328 | Anderson and Long 1947 |
| ?5% $CO_2$ | Fresh, heparinized rat blood 1:5 with K–R (Umbreit) plus 2·5 g% bovine albumin | Pancreas | 334 | Sussman *et al.* 1966 |
| 7% $CO_2$ | 2 g/litre bovine serum alb. in K–R (Umbreit) | Pancreas | 335 | Loubatieres *et al.* 1967 |

TABLE 1.7 (*continued*)

| Gas | Principal components of medium | Organ | Page | Authors |
|---|---|---|---|---|
| 5% $CO_2$ | Heparinized rat blood, plus added heparin, penicillin, streptomycin—p.c.v. 40% | Intestine | 364 | Windmueller 1970 |
| | Defibrinated rat blood (with all additions above) | Intestine | 364 | Windmueller 1970 |
| | Medium I above, plus pentobarb. or propanalol | Intestine | 364 | Windmueller 1970 |
| ?5% $CO_2$ | Rat or ox cells (washed), K–H, Rheomacrodex | Intestine | 371 | Kavin et al. 1967 |
| Air | Washed ox cells in dialysed ox serum (Tyrode) plus glucose (100 mg%) | Intestine | 375 | Hestrin-Lerner 1954 |
| 5% $CO_2$ | 30–35 ml rat blood plus 10 ml Ringer+glucose | Intestine | 376 | Gerber and Remy-Defraigne 1966 |
| 5% $CO_2$ | Whole, heparinized rat blood, 12 mM glucose plus pentobarbitone | Intestine | 378 | Lee and Duncan 1968 |
| 5% $CO_2$ | citrated r.b.c.s (washed) in ?K–H saline, bovine albumin (2 g%), dextran (2 g%), glucose (4·4 mM) pluronic F68 (surfactant) pH 7·4 orthophosphoric acid. | Intestine | 379 | Dubois et al. 1968 |
| ? | 0·9% NaCl, heparin, papaverine, promethazine | Intestine | 383 | Forth 1968 |
| 5% $CO_2$ | washed human r.b.c.s, 5·6% P.V.P., 3·4 g% bovine alb. in a saline of which no detail (?Schnitger medium q.v.) | Intestine | 383 | Forth 1968 |
| 5% $CO_2$ | K–H (Umbreit) $Ca^{2+}$ 1·25 mM, glucose 0·2% and penicillin and streptomycin | Intestine | 387 | Powis 1970 |
| 5% $CO_2$ | K–H with low $Ca^{2+}$, 3% bovine alb., 7% p.c.v. washed, ox cells | Intestine | 388 | Powis 1970 |
| | Table of drug additions for intestinal perfusion | | 392 | |

*Media used for intestinal lumen*

| | | | | |
|---|---|---|---|---|
| | NaCl, 150 mM; glucose 220 mM; taurocholate 10 mM | | 364 | Windmueller et al. |
| | K–H and 27 mM glucose | | 378 | Lee and Duncan |
| | 0·9% NaCl | | 380 | Dubois et al. |
| | KH or bicarbonate saline (Na 145, Cl 120, $HCO_3$ 25 mM) | | 381 | Powis 1970 |

TABLE 1.7 (*continued*)

| Gas | Principal components of medium | Organ | Page | Authors |
|---|---|---|---|---|
| 5% $CO_2$ | Tyrode | Blood-vessels | 426 | MacGregor 1965 |
| | 'depolarizing' medium, $K_2SO_4$, 92 mM, KCl 138 mM in place of $Na^+$ | Blood-vessels | 426 | MacGregor 1965 |
| 5% $CO_2$ | Whole, heparinized rat blood | Blood-vessels | 426 | MacGregor 1965 |
| 5% $CO_2$ | 'balanced salt soln.' w. 15 mM $HCO_3$ | Blood-vessels | 427 | Hrdina |
| ? | Tyrode solution | Blood-vessels | 427 | Kovalčík |
| | 50 mM sucrose in ?Tyrode, (14·9 mM $HCO_3$) and EDTA | Blood-vessels | 427 | Uchida and Bohr |
| 5% $CO_2$ | Tyrode and 4 g% bovine alb. and glucose 0·7 mM | Blood-vessels | 430 | Wade and Beilin |
| ? | K–H saline, 11 mM glucose and 0·026 mM EDTA | Blood-vessels | 432 | Hinke and Wilson |
| 5% $CO_2$ | K–H saline and glucose 100 mg% | Diaphragm | 303 | Rowlands 1969 |
| 5% $CO_2$ | as Hems (1966), 4% bovine alb. 7 g% Hb | Hind limb | 310 | Ruderman et al. |
| 5% $CO_2$ | Fresh, heparinized rat blood and glucose (100 mg%) | Hind limb | 315 | Mahler et al. 1969 |
| 5% $CO_2$ | Heparinized rat blood and penicillin, strep. | Hind limb (single) | 315 | Kuna et al. 1959 |
| ? | Homologous blood (cat, dog, sheep) | Mammary gland | 423 | Linzell 1954 |
| 100% $O_2$ | Citrated, whole bovine blood and penicillin, strep. and tetracycline | Ovary | 349 | Romanoff and Pincus 1962 |
| 5% $CO_2$ | Tyrode plus 5·6 mM glucose | Thyroid | 343 | Williams 1966 (rabbit) |
| 5% $CO_2$ | Tyrode+haemoglobin in solution (as D'Silva and Neil (1954)) | Thyroid | 343 | Williams 1966 (rabbit) |
| Air? | Eagle's medium+human Hb in soln. glucose (400 mg%), penicillin, neomycin, mycostatin | Thyroid | 343 | Folkman et al. 1967 |
| ? | autologous (dog) plasma+hep. (2·5 mg%)-spun | Thyroid | 343 | Gimbroni et al. 1969 |
| ? | Ditto, but not spun = 'platelet rich' | Thyroid | 343 | Gimbroni et al. 1969 |
| Air | bovine blood-citrated (0·5–1%, stored 3 days) | Thyroid | 345 | Grimm, Greer, and Inoue 1967 |
| ? | citrated, bovine blood passed through rat liver, in reverse perfusion, before use | Adrenal | 347 | Hechter et al. 1948 |
| 5% $CO_2$ | Tyrode (11·6 mM 'HCO_3$), or Locke (6mM $HCO_3$) | Adrenal | 347 | Smith et al., Schneider et al. |
| ? | Locke's solution | Adrenal | 348 | Douglas and Rubin 1961 (cat) |
| 5% $CO_2$ | Tyrode solution | Adrenal | 348 | Cession–Fossion 1964 (rat) |

TABLE 1.7 (continued)

| Gas | Organ | Principal components of medium | Page | Authors |
|---|---|---|---|---|
| 5% $CO_2$ | Testis | defibrinated, homologous blood (rabbit)+penicillin, streptomycin and glucose (8·3 mM) | 351 | Van Demark and Ewing 1963 |
| ? | Prostate | homologous (rat) blood (heparinized) | 352 | Farnsworth et al. 1963 |
| ? | Adipose tissue (parametrial fat) | defibrinated rat blood (p.c.v. 4·5%), diluted 1/10th with Tyrode+4 g% alb.+pen.+strep+glucose 2·8 mM | 397 | Robert and Scow 1963 |
| ? | Adipose tissue (parametrial fat) | 4% bovine alb. in Tyrode; no r.b.c.s 8% bovine alb. in Tyrode; no r.b.c.s | 398 | Robert and Scow 1963 |
| 5% $CO_2$ | Epididymal fat | Krebs–Henseleit+5% bovine alb. | 399 | Ho and Meng 1964 |
| ? | Brain (rat head) | Fresh (heparin) rat blood, p.c.v. 12–16% with glucose, terramycin +5% bovine alb. (Pentex) | 406 | Thompson et al. 1968 |
| ? | | K–H+canine (washed) r.b.c.s to p.c.v. 20–23% plus 'brain' albumin 7–8% deionized 20% FX80 in buffer above | 408 | Andjus et al. (1967) based on Geiger (below) |
| | Brain (rat head) | | 408 | Sloviter and Kamimoto 1967 |
| | Brain (cat) | Complex blood-containing medium | 409 | Geiger and Magnes 1947 |
| ? Air | Thymus | as Thyroid I med. III, p. 343 | 343 | Folkman et al. 1968 |
| ? | Salivary gland | Locke's solution ±$Mg^{2+}$+$Ca^{2+}$ | 416 | Douglas and Poisner 1963 |
| ? none | Bone | Autologous blood (cat) | 419 | MacIntyre et al. 1967 |
| | Bone (half cat tibia) | Autologous blood (cat) | 420 | Parsons and Robinson 1967 |
| ? | Rat femur | Fresh, heparinized rat blood (filtered) | 421 | Dornfest et al. 1962 |
| ? | Rat femur | Fresh heparinized centrifuged (leucocyte-depleted) | 421 | Dornfest et al. 1962 |
| 5% $CO_2$ | Lung | K–H saline, albumin 2·5 g% | 444 | Leary and Ledingham 1969 |
| ? | Lung | Krebs–Ringer (Umbreit 1964) | 447 | Bakhle et al. 1969 |
| 8% $CO_2$ | Lung | I. Whole rat blood, heparin 100 units/10 ml | 448 | Hauge et al. 1968 |
| ? | Lung | II. Rat plasma (spin med. I) | 448 | Hauge et al. 1968 |
| 5% $CO_2$ | Lung | Heparinized rat blood with 0·9% saline-p.c.v. 20–25% glucose (9 mM) | 451 | Levey and Gast 1966 |

Note. ? implies lack of published information.
References are given at the end of the chapter that describes the individual technique.

In each situation in which such a complex medium has been applied to organ perfusion techniques, there is an adequate parallel with a semi-synthetic defined medium among those included in the introduction to this section. For biochemical purposes, the simpler alternative is recommended.

### Carriage of oxygen

Oxygen may be transported in a number of ways for the purpose of organ perfusion.

(a) In simple solution, for example, heart perfusion; see Chapter 5.

(b) As oxyhaemoglobin, bound to haemoglobin in solution, for example, thyroid (Williams 1966) and liver (D'Silva and Neil 1954).

(c) As oxyhaemoglobin, bound to haemoglobin within erythrocytes. The erythrocytes may be from autologous, homologous, or heterologous blood and may be fresh or stored before use. Examples are discussed below.

(d) Bound to an inert 'fluorocarbon' emulsion, for example, brain (Sloviter and Kamimoto 1967; Chapter 8).

(e) In gaseous form. Magnus (1902) demonstrated the maintenance of the heart when 'perfused' with gaseous oxygen through the coronary arteries. Lochner, Arnold, and Müller-Ruchholtz (1968) have revived the method and demonstrated that ATP levels are normal. No attempt has been made to use gases to perfuse other organs, and for biochemical purposes in which assays of change in the medium are central to many experiments it would appear to be only of academic interest.

The choice of method of oxygen transport depends upon the expected flow rate and the oxygen consumption of the tissue perfused. It is usually possible to omit red cells if this is useful for the experiment in hand (for example, Schnitger *et al.* (1965) for surface fluorimetry), but most workers prefer to include them, since the extra oxygen-carrying capacity gives a margin of safety. Thus, Medium II has red cells to a haemoglobin 10 g %, while Medium III uses a haemoglobin of 2·5 g %. When applied to liver perfusion, this difference does not contribute to differences in metabolism (see Chapter 3).

(i) *Whole rat blood*. It is usual to collect blood into containers with powdered or liquid heparin (6 units/ml suffices), unless the experimental design precludes the use of heparin (see below). Cardiac puncture or cannulation of the abdominal aorta are the methods of choice, the former method having the marginal advantage that animals may be allowed to survive for repeated small donations of blood.

In the method of cardiac puncture (Morland 1965), the donor rat is lightly anaesthetized with ether (or Nembutal), the cardiac apex is located with an index finger to the left of the mid-thorax, and a $\frac{3}{4}$-in needle attached to a 5- or 10-ml syringe which has been previously moistened with heparin in saline (6 units/ml) is inserted boldly between the ribs beneath this point. Burhoe (1940) writes that 5 ml may be taken in this way every 3 months, with survival of the animal.

Cannulation of the aorta is the method preferred for perfusion studies, since the number of donor animals is reduced to a minimum, an important consideration when planning for a large number of experiments.

Under ether anaesthesia, the abdomen is opened in the mid-line, by means of a scissor-cut through skin and through the linea alba. The intestine is deflected to the animal's right, exposing the aorta and inferior vena cava. The connective tissue covering of the aorta below the level of renal arteries is separated by blunt dissection (using the fingers) over sufficient of its length for a 'bulldog' clip to be applied proximally. With fine scissors, the lower aorta is incised, the vessel is grasped more distally in forceps, and a stiff nylon or polythene catheter (ideally 1·0 mm diam., cut obliquely as described for 'cannulae') some 15 cm in length is inserted as far as the 'bulldog' clip. The clip is now removed, and blood flows freely into the prepared container which stands in ice. In this way, a 300-g rat may yield 10–12 ml of blood, so that sufficient for a standard experimental method (for example, that of Miller *et al.* (1951)) may be obtained from six animals. Because collecting rat blood is inconvenient when applied to the already complex method of organ perfusion, in some media heterologous red cells have been substituted with satisfactory results (see Chapter 3). Nevertheless, perfusion with homologous blood remains important since 'unexpected' findings will always be suspected as 'unphysiological' unless confirmed in a perfusion more closely allied to the live state (Staib *et al.* 1969).

(ii) *Preparation of heterologous red cells.* Blood is usually collected into an anticoagulant mixture, 'ACD', and stored on ice or at 4°C until used. ACD = disodium citrate (monohydric) 2 g, dextrose (anhydrous) 3 g in water (120 ml/420 ml blood) (Loutit and Mollison 1943). Although most workers use the cells immediately, some allow their use up to 1 week after collection (Andjus, Suhara, and Sloviter 1967). Hospital blood-banks allow the use of human blood up to 4 weeks after collection; cells obtained from blood which has 'expired'

4

in this way are perfectly suitable for liver perfusion (see Medium III, Hems *et al.*, above).

Cells are washed free of citrate, serum, and accumulated lactate before use in semi-synthetic perfusion media. The method used by Schimassek (1963) is as follows. Approximately 100 cm³ of blood is spun in a cooled centrifuge for 20 min at 2000 rev/min. Using water-suction and a pasteur pipette, the serum, ACD, and the buffy coat are removed from the sedimented cells. Twice the volume of 0·9% NaCl or of the appropriate buffer in which the medium is finally prepared (Tyrode for Medium II, K–H for Medium III) is added. The cells are mixed and spun again at the same speed and for the same time. This is repeated once or twice, removing as much of the buffy coat as possible each time, and the cells are finally resuspended in approximately half the original volume of buffer. This suspension contains about 20 g % haemoglobin and may be used in preparing the final medium as described on p. 20. Full details for the preparation of Hems medium (Medium III above) are given in Chapter 3.

*Note.* Schimassek (1968) quotes Berry who suggests that the washing is best conducted at room temperature rather than at 4°C for more efficient removal of lactic acid from the cells.

### Determination of haemoglobin; cyanmethaemoglobin method (see Wintrobe 1967)

Drabkin's solution is prepared, i.e $NaHCO_3$ 1·0 g KCN 50 mg $K_3Fe(CN)_6$ 200 mg made up to a litre with water. Blood (0·02 ml) is added to 5 ml of this solution and read against a standard cyan-methaemoglobin solution (BDH Ltd.) at 540 m$\mu$m.

### Determination of the rate of glycolysis

Allowance must be made for the metabolic changes induced by red cells in the perfusion medium. Glycolysis in particular may proceed at such a rate as to obscure the metabolism of the perfused organ. In addition, enzymes may be released from red cells as they lyse. In recirculation perfusion, in which the circulating mass of red cells exceeds the mass of tissue perfused, this is of significance. The rate of glycolysis may be determined for each batch of red cells in a simple incubation test.

Hems *et al.* (1966) introduced the use of aged human red cells, stored for 4–6 weeks before use, in which oxygen carriage is unimpaired, while the rate of glycolysis is considerably reduced. This was of particular relevance to studies of gluconeogenesis but also has general application.

The appearance of glycolytic enzymes in the medium may in part be due to lysis of red cells, and authors especially interested in this group of enzymes advocate the omission of red cells from the medium (Schmidt 1968); alternatively, a control perfusion (for example, apparatus with medium but without the organ) may give some impression of the quantitative extent of such interference.

In liver perfusion (and possibly also in kidney perfusion), the rate of glycolysis in the medium may be more than balanced by the rate at which lactate is reconverted to glucose by gluconeogenesis in the liver. In examining the data of Exton and Park (1967) it is clear that no correction needs to be made for red-cell glycolysis when the sub-strate studied (for example alanine or glycerol) is converted to glucose in the liver at a rate lower than that obtained with lactate as substrate (see Ross, Hems, and Krebs (1967) for comparative rate of gluconeo-genesis in liver). Any glucose which is converted to lactate by the medium will be regenerated, provided that the liver's capacity for gluconeogenesis is not already saturated. The reverse is true for studies in which the substrate, for example dihydroxyacetone or fructose, is converted to glucose more rapidly than is lactate. In such cases, the rate of lactate accumulation in the medium may be determined and a correction should be applied on the basis of a test *in vitro* of red-cell glycolysis as described above.

The simplification attained by the use of 'non-glycolysing', aged red cells is apparent.

## Oxygen binding

Dissociation curves of $HbO_2$ are given by Valtis and Kennedy (1954). There are species differences and 'aged' cells also appear to be different from fresh cells. In calculating the oxygen content of the medium for studies on oxygen consumption, this information, together with oxygen solubility figures, is essential. The use of chemical methods of oxygen determination (for example the Van Slyke method) allows such variations to be circumvented. Information on oxygen solubility may be used for approximate calculation of the oxygen content of different media and allow suitable haemoglobin concentration and flow rate to be selected.

## Haemoglobin in solution

Haemoglobin in solution fulfils two functions; oxygen transport and colloid osmotic agent, and it has been used as a basis of perfusion media for liver by D'Silva and Neil (1954), and for thyroid gland by Williams (see Chapter 6). Haemoglobin may be prepared from defibrinated ox blood as follows.

(a) Red cells from 1 litre of blood are washed four times in 0·9% NaCl; ether is added in 10 ml aliquots to produce a free-flowing solution (Solution X, D'Silva and Neil).

(b) Alumina cream (300 ml, from 50 g ammonium alum), is mixed with 450 ml of solution X, which is allowed to stand for 15 min and is then centrifuged (3000 rev/min for 15 min).

(c) The supernatant is dialysed against running tap-water until $Na^+$ and $K^+$ content are negligible. Stock solutions prepared in this way have an $O_2$ capacity of 17–18 ml/100 ml, and a saline medium prepared to have an oxygen capacity of 7–8 ml 100 ml has a haemoglobin content 5·5–6·0 g Hb/100 ml and a colloid osmotic pressure close to that of serum.

In addition to binding oxygen, haemoglobin exerts a colloid osmotic effect, and this may be the mechanism of the improvement noted by Williams in rabbit thyroid gland perfused with Tyrode solution to which haemoglobin in solution was added (Chapter 6).

*Fluorocarbons as alternative oxygen carrier*

Only in the perfusion of brain has a comparative study been published in which the chemical oxygen carrier was substituted for erythrocytes. Sensitive parameters of function, in particular the EEG, are maintained, and there appears to be no difference between the effect of the two media. While there may be biochemical applications of this material, at present it is particularly of interest in perfusion for long-term organ survival, a subject that is reviewed by Norman (1968).

## Prevention of clotting

(a) *Defibrination*

The older method of preventing clotting, defibrination, is still in use in some perfusion systems (for example, Felts and Mayes 1966), particularly when the use of the alternative, anticoagulant heparin is contra-indicated by the experimental design. Methods are simple and mechanical: blood may be stirred with plastic rods, or with wooden spatulae, or shaken with glass beads.

(b) *Anticoagulants*

(i) *Heparin*. Heparin is the only anticoagulant to have found general use in perfusion studies and it may be injected intravenously into the experimental animal (see individual methods for dose, route, and timing of injection) or added to the perfusion medium. Little systematic work has been done to compare the two methods; there is a brief discussion of this aspect of the perfusion of liver in Chapter 3.

(ii) *Citrate*. Blood collected from large donor animals (ox, human) may be anticoagulated with citrate (ACD) (see p. 33) and the citrate later removed by washing or dialysis.

(iii) *Alternative anticoagulants*. The metabolic effects of heparin cannot be ignored and there is a need for an alternative rapidly-acting anticoagulant for use in biochemical perfusion studies.

## Maintenance of colloid osmotic pressure

The simple exposition of Starling's law (Starling 1895) is that blood circulating through a capillary bed loses fluid to the interstitial space when the hydrostatic pressure within the capillary exceeds the sum of the external tissue pressure and the colloid osmotic pressure within the lumen of the vessel. Since the hydrostatic pressure drops between the arterial and venous ends of the capillary, fluid is at first lost and then regained by the blood stream passing through it. In perfusion experiments, the artificial medium and perfusion pressure, as well as the venous pressure should be selected to achieve a similar balance, so that net fluid loss from the vascular compartment does not occur. To this end, a number of plasma substitutes or 'expanders' have been introduced and employed to maintain the colloid osmotic pressure in perfusion media. The commonest are:

*Albumin:* usually bovine albumin, Cohn Fraction V.
*Dextran:* the polysaccharide product of the action of Leuconostoc mesenteroides on crude sugar ('Zuckersaft'), with a mean molecular weight 40 000 (Rheomacrodex) or 70 000 (Macrodex).
*Polyvinylpyrrolidone (PVP):* a synthetic polymer from acetylene, ammonia and formaldehyde, M.W. 25 000–30 000. A 3·5% solution develops an osmotic pressure of about 400 mm water.
*Gelatin:* a protein degradation product of collagen (for example Haemaccel is a breakdown product of cellulose) which swells in water.

Work has been done in organ perfusion with each of these products, and there exists an objective basis upon which to make a choice. In addition, media to which no colloid is added (for example, Medium V above), or in which a concentrated solution of sucrose (for example, Uchida and Bohr (1969); Chapter 8) is substituted, are discussed.

### Albumin

Schimassek (1963) pioneered the use of bovine albumin, crystalline Fraction V in organ perfusion. This is the product of an alcohol fractionation of total plasma protein, described in detail by Cohn, Hughes, and Weare (1947).

Various commercial preparations are available and have found wide application in perfusion media: Armour; Pentex; Cutter Laboratories; Behringwerke appear to be interchangeable, with the reservations discussed below. In addition, a preparation of crystalline Fraction V human albumin has been used in pancreas perfusion (see Chapter 6; Grodsky *et al.* (1968)).

*Choice of concentration.* The normal serum albumin concentration is around 3–5 g %, but many perfusion systems use higher concentrations in high-pressure, arterial perfusion, to prevent the formation of oedema. The choice has usually been pragmatic—1 g/litre in some heart perfusions (Arnold and Lochner 1965), to 7 g % in a perfused hind-limb preparation (Ruderman *et al.*; see Chapter 5). In liver, no difference in the rate of glucose synthesis from lactate occurs in experiments conducted with albumin concentrations varying from 1 to 6 g albumin per cent.

*Problems in the use of bovine albumin.* Although albumin is probably the most satisfactory substitute for plasma for the present purpose, there are a number of disadvantages in its use.

(i) Batches differ considerably in an ill-definable way. Some are toxic to the preparation, which may fail completely; this has been reported for the perfused heart and for the liver (Schimassek 1963). Some authors advocate dialysis of albumin before its use in perfusion (see below for small molecular contaminants), but even this may not prevent the occasional toxicity of a batch of albumin. It is a wise precaution to note carefully the batch numbers and manufacturers' details of those batches of albumin which are satisfactory, and if possible to limit the use of one batch to any one series of experiments. Fortunately, unexplained 'toxicity' of albumin has not proved a frequent occurrence in liver perfusion, the method with which it is most commonly employed.

(ii) Binding properties of albumin, and the presence of biologically active contaminants may complicate the preparation of a reproducible medium, and interfere with parameters of function in the organ perfused. Binding of $H^+$ ions and of bicarbonate, together with the displacemement of bicarbonate when albumin is added to a buffer may influence the pH of the final medium in an unpredictable manner, and correction for this is discussed in Chapter 3 (see Hems *et al.* (1966)). Not surprisingly, albumin also affects the solubility of oxygen in buffers, and allowance must be made for this in calculations of theoretical oxygen content of a perfusion medium. Calcium is a special case: Krebs–Henseleit medium has twice the concentration of 'ionized' calcium. In the presence of albumin, some calcium is bound, and the concentration of ionized $Ca^{++}$ correspondingly reduced. This has not been observed to have critical effects in liver perfusion, but could in theory be of importance in heart perfusion studies (see Chapter 5).

(iii) Interference with assays performed on the medium, for example enzyme assays or intermediate assays by enzymic methods, is known to occur. Precipitation by simple methods, such as perchloric acid, may be used to remove albumin entirely from solution in the perfusion medium; this is one of its great advantages as a plasma expander in biochemical studies.

*Precautions in the use of bovine albumin.* For the reasons outlined above, some workers prefer to pre-treat commercially prepared bovine albumin, Fraction V by dialysis or 'defatting' or both. Dialysis tubing should be washed (or boiled in water) before use. A sealed vessel is required if Krebs–Henseleit buffer is used for dialysis. Since long-chain fatty acids are known to influence metabolism in a variety of ways, it may be wise to reduce the amount of these substances added initially to the medium. This has the further advantage that studies on the effects of fatty acid may be conducted against a true 'control' base line of 'nil' fatty acid content. Two methods are in common use.

In the charcoal method (Chen 1967) albumin (7·0 g) is dissolved in 70 ml of distilled water at 23°. Darco (3·5 g) is mixed into the solution and the pH lowered to 3·0 by the addition of 0·2 N HCl. The solution is placed in an ice bath and mixed magnetically for 1 h. Charcoal is then removed by centrifugation at $20\ 200 \times g$ for 20 min in a Sorvall RC 1 centrifuge with an SS 34 rotor at 2°. The clarified solution is then brought to pH 7·0 by the addition of 0·2 N NaOH.

Using the Dole assay for free-fatty acid, the following figures are given for the fatty acid content before and after treatment.

| *Source of albumin* | *FFA (mol/mol protein)* | |
| --- | --- | --- |
| Pentex (crystalline) | 1·05; 0·06 | 0·07; 0·00 |
| Armour (crystalline) | 0·81; 0·88 | 0·00; 0·01 |
| Armour (Fraction V) | 0·18; 0·48 | 0·00; 0·00 |

In the solvent method (Goodman 1957) a solution of Fraction V serum albumin is lyophilized and covered with a mixture of 5% glacial acetic acid in iso-octane (vol./vol.). (This mixture is pre-treated with anhydrous $Na_2SO_4$ to remove traces of water.) Extraction is carried out for 6 h or more; the mixture is then decanted and the albumin washed with iso-octane. The extraction is repeated, and the washing thereafter carried out twice. A vacuum is exerted over the albumin to remove acetic acid and iso-octane, and the product is dissolved in distilled water. This solution is cloudy, and is dialysed exhaustively against distilled water to remove the last trace of solvent. The product, when lyophilized for storage, contains less than 0·01 mol FFA/mol albumin, from an initial value of 1·8 mol/mol.

### Dextran

The use of Dextran in perfusion media is well established (see, for example, Bahlmann *et al.* (1967) for kidney, Schnitger *et al.* (1965) for liver, and Table 1.7 for other examples), but it has disadvantages.

(a) Aggregation of red cells is more likely to occur in the presence of Dextran, particularly Rheomacrodex (Whitmore 1968) so that its use in semi-synthetic media of which red cells are a component (Media II and III above) may cause failure of perfusion.

(b) Metabolic effects of Dextran, or perhaps due to the absence of albumin, are documented in the perfused liver. The rate of glucose synthesis from lactate, an acceptable parameter of function (Hems *et al.* 1966 and Chapter 3) in the liver occurs at a reduced and variable rate (see Table 3.6). Schimassek (1968) reports what are probably excessive losses of enzymes from the liver when Dextran is substituted for albumin (see Chapter 3) in the medium.

(c) Interference with assays, particularly of glucose and glycogen, may occur if Dextran remains in the extracts of perfusion medium; complete removal of Dextran from solution presentse som difficulties.

'*Haemaccel*'

A cellulose product that has found application in organ perfusion is Haemaccel. In the perfused liver it permits glucose synthesis at just less than 90 per cent of the rate which is observed when albumin is used in its place. It may not therefore be a totally adequate substitute.

*Perfusion without colloid*

Sucrose exerts an osmotic effect since it is not transported across cell membranes. Its use to date has been confined to a perfusion of the vessels of the tail (Hinke and Wilson 1962), and of liver (Claret and Coraboeuf 1968). Only in the perfused heart has perfusion without colloid been adopted with success (Morgan *et al*. 1961 and see Chapter 5). In the liver, attempts to perfuse with simple salt media result in oedema and poor metabolic function (Bloxham 1967), although for special purposes this may be acceptable. There is no information upon the effectiveness of colloid-free medium in supporting the bovine adrenal preparation of Hechter (Chapter 6) since comparative studies are lacking.

*Conclusions*

The use of plasma expanders has been satisfactory in many types of organ perfusion. Bovine albumin Fraction V is preferred on the basis of metabolic tests of organ function, but has disadvantages due to contaminants and due to its binding properties for small and large molecules. Other plasma substitutes appear not to be entirely satisfactory in that metabolic functions, where determined, show signs of tissue damage. The use of perfusion media without a plasma expander is not recommended, except in an organ such as the heart, in which oedema formation is not a problem.

## OXYGENATORS

### The need for an oxygenator

One of the prerequisites of an organ perfusion apparatus is that the medium reaching the organ contains sufficient oxygen to exceed the maximum requirements of the tissue under investigation (see Introduction). In other systems used *in vitro*, for example Warburg manometric flasks, the gas atmosphere above the tissue is determined at the beginning of the experiments, and is usually chosen to give the maximum oxygen tension, i.e. either 100% $O_2$ at 1 atm pressure. or 95% $O_2$, allowing for the need to have 5% $CO_2$ in the gas when the standard (25 mM) bicarbonate buffer system is used. In a closed system the oxygen available is limited, and the mass of tissue incubated may be such as to limit the duration of incubation by its rate of oxygen consumption. Conversely, $CO_2$ generated or metabolic changes contributing to a change in pH are cumulative, and the total buffer capacity may be the limiting factor to incubation.

A perfusion combines these elements of a 'closed system' with those of an open system. The volume of medium is predetermined and usually chosen for its total buffer capacity and for convenience in following the metabolic changes expected. The volume of gas, on the other hand, is unlimited and determined above all by the efficiency with which it can be brought into solution in the perfusion medium. To accelerate this process, some form of oxygenator is usually used.

## Characteristics of an oxygenator

### Animal lung

Investigators not especially concerned with the advantages of metabolic isolation offered in perfusion have used the animal's own lung, or that of another, supporting animal, as the most efficient oxygen:$CO_2$ exchanger. In the present context this is unacceptable as it reintroduces some of the variability of experiments *in vivo* as well as introducing a heterogeneous tissue.

### Artificial oxygenators

An oxygenator is required to saturate the medium with the gas or gases used in the shortest possible time. In a once-through perfusion system, this time is not important since the medium can be oxygenated in its reservoir before perfusion. When, alternatively, non-oxygenated medium is run through an oxygenator to the organ, saturation must be achieved in a single passage.

In a recirculation system, the medium can be saturated with oxygen without limit on time, before beginning perfusion, but once perfusion begins the system should be capable of replacing the oxygen uptake of the organ in one passage through the oxygenator. Some latitude is offered by the fact that most recirculation systems have an oxygenator with a bypass which allows a greater circulation of medium through the oxygenator than through the organ in the same period of time (see

Fig. 1.1). When this is not the case (Baggiolini and Dewald (1968) is an example) then oxygenation must be completed in a single passage. Clearly, system (b) requires only $x/2$ or 3 the time to saturate all the medium, assuming an oxygenator of maximal (100 per cent) efficiency. Oxygenation in this case includes the removal of excess $CO_2$ and the restoration of pH. The requirements of an oxygenator are therefore simply defined as an apparatus which will saturate with oxygen the

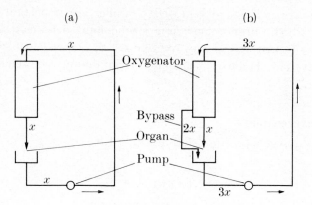

FIG. 1.1.    Diagram of a recirculation perfusion circuit. The use of a bypass to permit recirculation through an oxygenator. $x$ is the flow rate through the perfused organ. (a) Circuit without bypass—flow through oxygenator cannot exceed the flow through the perfused organ. (b) A bypass allows more effective use of an oxygenator with a flow several times that through the organ.

volume of perfusion medium passing through the organ per unit time, in the same time.

In addition, the following points are of importance in the design for any particular organ perfusion.

(a) Efficient gas exchange, the volume of gas brought into solution, or the volume of medium which can be saturated in a given time, in excess of the requirements of the metabolizing tissue.

(b) Total gas collection should be possible if required, and provision should be available for the determination of gas flow rate through the apparatus.

(c) Micro-bubbles, a potent cause of failure in perfusion due to embolism in small blood-vessels should not be formed during gassing, and if they are, as a result of the particular form of the oxygenator, then an efficient bubble trap must be incorporated. Coarser bubbling and frothing, which occur so readily with protein solutions as perfusion medium, should be avoided. Frothing has the mechanical effect of reducing flow rate through tubes and may lead to the loss of perfusion volume. In particular, the volume of medium in a reservoir or a pool supplying a pump may be so reduced as to make pumping inefficient. Further, frothing causes denaturation of protein progressively with time, and in a medium containing red cells haemolysis may become a serious problem.

(d) Ease and rapidity of cleaning, as well as sufficient robustness of construction to withstand repeated cleaning, are important. As discussed on p. 83, chromic acid or a strong ionic detergent should be used, and the material of any oxygenator to be used in biochemical experimentation should either fulfil this minimum requirement or be disposable.

## Types of oxygenator

Types of oxygenator are listed in Table 1.8.

TABLE 1.8

*Basic designs of oxygenator*

| | | |
|---|---|---|
| Open | (i) | Bubblers |
| | (ii) | Falling columns; gas-lift systems |
| | (iii) | Films |
| | (iv) | Rotating chamber |
| | (v) | Rotating discs: rotating cylinder |
| | (iv) | Others |
| Closed | | Membrane oxygenators |

## Open type

(i) *Bubblers.* Non-protein media may safely be gassed by means of the conventional sintered-glass bubbler (see, for example, Morgan *et al.*, heart perfusion). The use of such direct gassing with albumin solutions, however, leads to frothing, which may denature proteins with an uncontrollable formation of foam within the apparatus. When erythrocytes are present in the medium, direct bubbling produces an unacceptable degree of haemolysis.

Bubbling alone may be insufficient to oxygenate the medium, depending upon the nature of the circulation (i.e. once-through or recirculation) and upon the volume to be oxygenated. Morgan *et al.*

(1961) incorporated a sintered-glass bubbler which delivers gas below the level of fluid in the reservoir within a simple column oxygenator (see below). This provides a descending film of medium over which the escaping gasses ascend. Oxygen tensions of 550 mm Hg are reported for the aortic inflow in this system (Morgan 1961). A bubbler alone is used in the upper reservoir of the standard Langendorff perfusion by Opie *et al.* and by Forth, since there is no limit to the time required for the oxygenation of the medium. Forth, on perfusion of small intestine (Chapter 7), and Leichtweiss *et al.* on kidney (Chapter 4) use simple bubblers in the presence of protein solutions (plus other plasma expanders) but do not comment on frothing. In each case, there is a large

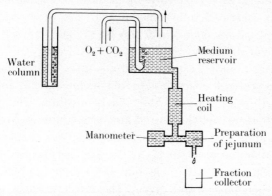

Water column — $O_2 + CO_2$ — Medium reservoir — Heating coil — Manometer — Preparation of jejunum — Fraction collector

Fig. 1.2. Perfusion by gas pressure. The apparatus of Forth (1968) for perfusion of jejunum combines oxygenation and a means of circulation by passing gas into a closed compartment (see Chapter 7, p. 383 for details).

static volume of fluid which is not as likely to foam as that, for example, in the Morgan heart-perfusion system. However, it is the presence of a high pressure of gas over the surface of the reservoir in the Leichtweiss apparatus (and others) which limits frothing. The inevitable bubbles are dealt with by an efficient bubble trap in the Leichtweiss apparatus, but not apparently in that of Forth (see Fig. 1.2). This may explain the relatively short periods of perfusion available in the latter preparation.

The use of large-bore tubes to introduce gas into a medium is obviously not as efficient as the use of the sintered-glass attachment described and is to be avoided. Only in the combined approach to oxygenation and circulation of fluid, as seen in the gas-lift (see Chapter 7) is the gas introduced as large bubbles. The efficiency of oxygenation in these systems is low and they should usually be supplemented by an oxygenator of another type, or even with a bubbler in a suitably placed reservoir.

(ii) *Falling column oxygenators, gas-lift systems.* The falling column oxygenator, in which no special provision is made to ensure the formation of a film of medium over the surface, has found little general application. However, it remains the oxygenator most widely used in heart perfusion apparatus based on the design of Neely, Liebermeister, Battersby, and Morgan (1967) such as that of Chain, Mansford, and Opie (1969). It is not clear to what extent oxygenation is achieved by means of the bubbler incorporated into the lower compartment of this apparatus (see Fig. 5.7). High rates of gas flow (1–2 l/min) are essential to maintain a saturated inflow medium. Under these circumstances, however, inflow gas tensions of 550 mm Hg (72 per cent saturation) are achieved without difficulty.

Anderson and Long (1947) described a simple 'falling film' oxygenator in which a straight-sided glass tube is inclined at an angle of some 80° to the vertical and receives the effluent medium. The gas is introduced at the bottom and the rate of flow of medium is slow enough to allow some filming and adequate gas exchange was claimed. A similar system is used by Grodsky in his pancreatic perfusions but has now apparently been abandoned in favour of a simple bubbler. No figures for inflow oxygen tension are given. The latter workers used no red cells in the medium.

No improvement on Morgan's working heart preparation has been introduced, so that this remains the only current 'falling film' oxygenator in common use. Thought should be given to improving this, for the likely developments in the field include the use of protein-containing media, red cells, other metabolites, with perhaps higher oxygen uptake, and possibly the use of the hearts of larger animals, for example guinea pig, with oxygen consumption beyond the capacity of the present oxygenator to replenish.

*Gas-lift systems.* Oxygenation and circulation through the lumen of the bowel is often achieved in perfusion with this device, introduced by Fisher and Parsons (1949). Circulation of perfusion medium is less generally achieved by this means. The heart perfusion apparatus of Bleehen and Fisher (Fig. 1.3) does so. The expense of a pump is spared but frothing and haemolysis precluded its use with albumin or with red-cell media.

(iii) *Multibulb glass oxygenator.* The basic design of the multibulb glass oxygenator (which appears to owe its introduction to Miller *et al.*) is of a vertical glass tube, narrow at its entry and expanding to form a series of approximately spherical bulbs before it ends in a small reservoir with an overflow. Essentially an idea derived from the design of condensers for chemical use, the constrictions between bulbs were considered to result in redistribution of gas within the column. A gas inlet is placed low down, and another orifice high up allows gas to leave

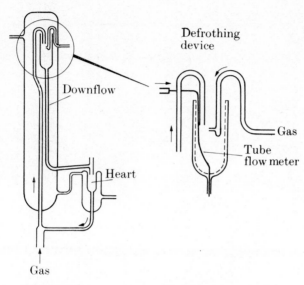

FIG. 1.3.   Bleehen and Fisher (1954) apparatus for perfusion of rat heart (see Chapter 5, p. 266 for details) incorporating gas-lift system of oxygenation and recirculation. A stream of large gas bubbles lifts the ascending column of medium. A second gas stream is introduced to prevent frothing of medium as it emerges from the top of the column (see inset).

the chamber. The critical features of design (not in fact met by all the oxygenators of this type in current use) are a smooth inner surface and a critical diameter at the inlet which does not allow the formation of bubbles in the continuously entering medium. The inlet size and shape of the first bulb, conical rather than spherical, is also critical in ensuring even delivery of medium to the whole circumference of the 'lung'. Then, the formation of a uniform film over the much expanded surface provided by a succession of bulbs, usually four or five in number without 'pooling' at the lower rim of each successive bulb, is the important feature. It allows a relatively short length of glass to oxygenate media to saturation in a limited number of passages. Fig. 1.4 illustrates the commoner types in use. The Miller design, as that used by Schimassek, Wieland, Söling, and others, can be manufactured by a competent glass-blower without special facilities. With special technical facilities it has been possible to improve on this design in detail

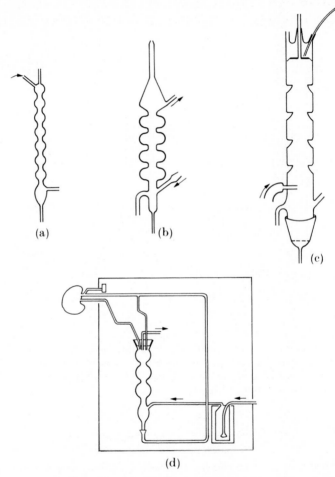

(a)          (b)          (c)

(d)

FIG. 1.4. A selection of multibulb glass oxygenators in use in perfusion. Arrows indicate the direction of flow of gas. (a) Original design of Miller *et al.* (1951). (b) Modification by Hems *et al.* (1966). Full details of dimensions are given in Fig. 1.5. (c) Oxygenator of Fisher and Kerly (1964). The formation of a film is facilitated by concentric guttering on the inner surface of the cylinder. (d) Perfusion system of Bahlmann *et al.* (1967) (for details see Chapter 4, p. 252). Here the oxygenator, which is below the level of the kidney, receives medium from the pump as well as outflowing from the perfused organ. It is not provided with an overflow.

(see Hems *et al.* (1966)) and its advantage over the less critical designs are as follows.

(i) There is no frothing, even at flow rates of 100–150 ml/min (this refers to the maximum output of the Watson–Marlow pump).
(ii) Provided the lung is wet immediately before the circulation of medium, the formation of a film is certain, even at low flow rates; it is quite difficult to obtain the annoying 'streaming' with this apparatus.
(iii) This, in turn, allows the use of low pump speeds, with a reduction in the rate of haemolysis.
(iv) No pooling of medium at the rim of the bulb occurs so that the volume of medium within the oxygenator at any one time is a minimum. (The volume of the oxygenator reservoir can be varied in manufacture.)

*Notes on the design of the multibulb glass oxygenator* (*Hems et al.* 1966)

The construction of the oxygenator of Hems *et al.* requires special glass-blowing techniques but the dimensions and details given allow it to be accurately reproduced. The oxygenator is constructed of Pyrex laboratory glass, which is borosilicate and highly reparable. Silica may be used, but soda glass is probably too weak for routine use. The starting material is heavy weight tubing, 22-mm external diameter (internal diameter 18 mm, i.e wall thickness = 1·8–2·0 mm). The dimensions are given, together with the suggested limits of error, in Fig. 1.5(a), (b). Additional points to be noted are as follows.

(i) Inlet diameter, its rate of widening to the introductory tube, and of this tube to the first, conical 'bulb' are critical. The dimensions given produce a film without frothing or breaking of the column of medium.
(ii) The gas outlet tube arises from the first 'bulb' at a true tangent to its lower border. This prevents 'spillage' of medium and does not break the flow of the film.
(iii) 'Throat' diameter and length are largely determined by the shape of the bulbs; however, the length is kept to the minimum compatible with machining, to maintain the continuity of the film.
(iv) 'Bulb' dimensions: the obtuse angle of entry to the bulb determines the formation of a film, while the angles of the lower half of the bulb prevents 'heaping' of medium. The 'squashed sphere' shape that

5

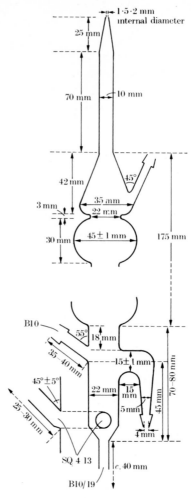

Fɪɢ. 1.5.    Details of Hems multibulb glass oxygenator. The dimensions of the oxygenator used in the liver perfusion apparatus of Hems *et al.* (1966, see Chapter 3), the kidney perfusion apparatus of Nishiitsutsuji–Uwo *et al.* (1967, see Chapter 4), and hind-limb preparation (Chapter 5) are given in detail. Measurements are critical to obtain a satisfactory preparation. All figures given are outside diameters in mm. Wall thickness = 1·8–2·0 mm. Bore of original column = 22 mm o.d. (18 mm i.d.). 'Heavyweight' borosilicate glass. (a) Upper part. In addition to the conical entry bulb, the complete oxygenator has four bulbs as shown (see Fig. 1.4. for comparison). (b) Lower part of oxygenator. The distance from the lower rim of the lowermost bulb to the overflow should be kept to a minimum. The 'reservoir' below the level of overflow contains 15–20 ml. To avoid frothing the relation between the levels of overflow and gas inflow is critical.

results is empirically the best form for the maintenance of a film of uniform and minimum thickness.

(v) Four bulbs of this shape, together with the inlet bulb are sufficient for routine use. For other purposes a shorter or a longer oxygenator may be required, but it is thought that this may be the minimum size suitable for use with the perfused liver of the rat.

(vi) The lower reservoir: the functions of the reservoir in this apparatus (and in other multibulb or falling film devices) are as follows. The overflow determines the constant head of perfusion pressure, used particularly in portal perfusion of the liver (Miller *et al.* 1951). A volume of 15–20 cm medium, and a height of some 70–80 mm damps out frothing and pressure variations which result from the arrival of the falling film in the reservoir. Sampling and adequate mixing of inflowing medium is possible, for example temperature, pH, or gas electrodes, or physical sampling. A safety factor is that the volume of medium (up to 20 ml) that is available to flow through the liver in the event of a mechanical failure elsewhere in the circuit allows a few extra seconds in which to save the preparation. In practice this has proved of great value. Modifications to the lower reservoir should be made only after reference to these suggested advantages to its present form.

(vii) The relative positions of the gas inlet, and the overflow tube shown diagrammatically here are critical, as are their dimensions.

The gas inlet must, of course, be above the lower rim of the overflow, but need not be any higher; the length of this part of the oxygenator may thereby be kept to a minimum. The 'weiring' of medium over the overflow tube is obtained by the critical shape of its lower margin, a design familiar to users of the Warburg flask with side-arm.

*Modifications of multibulb oxygenator design*

(i) *Water jacketing.* The basic design of multibulb oxygenator takes no account of temperature regulation and is usually used within a perfusion cabinet (see Chapter 3). Staib *et al.* (1968) have published a description of a similar liver-perfusion apparatus which incorporates a water jacket so that temperature control can be achieved with a simple water heater and recirculating pump. The disadvantage of this modification is difficulty in cleaning and consequently there is a greater chance of poor film formation. It is difficult to immerse the whole in chromic acid, and flushing with detergent solution, the usual alternative, may in time allow the accumulation of grease.

(ii) *Cascade system*. Fisher and Kerley (1964) illustrate an oxygenator with the same purpose as the multibulb oxygenator mentioned in (i). It is provided with a pronounced lip at the lower border of each 'bulb' so that medium descending in a film is delayed and pours over the rim to film over the next bulb. The advantages claimed for this arrangement by its designers is that passage through the oxygenator is delayed sufficiently for more effective oxygenation to occur. Originally used with blood media, it was later applied to experiments in which simple saline media were used. In the latter case it maintains a film despite lower viscosity of the perfusion medium and may thus have an advantage over the simple design discussed above. For the same reason, its use has been proposed if a falling-film oxygenator is preferred in a system which is to be siliconized (see p. 362). Cleaning and construction provide difficulties.

(iii) *Punctate form*. Biebuyck *et al.* (1970) found that the formation of a film in a multibulb oxygenator is improved if the surface of each bulb is interrupted with small punctate indentations.

(iv) *Inverted form*. Various modifications, including one in which the oxygenator is below the outflow from the perfused organ (Bahlmann *et al.* 1967) have been suggested but are less efficient than the simple design discussed above.

*Rotating chamber oxygenator*

Mortimore and Tietze (1959) introduced into small-organ perfusion an oxygenator which has special applications with blood as perfusion medium; the medium flows over the inner surface of a sloping glass cylinder 35 cm × 9 cm, which is rocked about its long axis (90° arc; 80 rev/min. Mortimore (1961) has since modified this basic design to give a larger surface area for filming. A 500 ml flask tilted at an angle of some 45° is rotated about its axis at 60 rev/min, so that medium forms a film over most of its inner surface. Gas is led in at a rate of 0·5 l/min by means of a tube which opens over the surface of the medium. It does not bubble beneath the surface and foaming is not a problem with this design of oxygenator. A delivery and pick-up tube from the perfused organ and to the pump respectively, enter the neck of the flask. A similar but simpler pattern has been used successfully in perfusion of spleen (cat), liver, and large intestine (see corresponding sections and Powis (1970*b*), Blakeley and Brown (1963), and Powis (1970*a*) (unpublished)).

A further modification has been introduced by Exton and Park in their application of Mortimore's liver perfusion technique. The oxygenator is illustrated in Fig. 1.6 and consists of a plexiglass chamber, now with its axis of rotation horizontal. This provides for a larger surface area for filming and can be driven by a simple motor (see, however, cautionary note below). The same three tubes—gas inflow, medium deliver, and medium pick-up—enter along the fixed axis of rotation of the oxygenation chamber. The authors recommend 200 ml/min gas flow and rotation at 60 rev/min; medium flow is at 6·8–7·2 ml/mm through

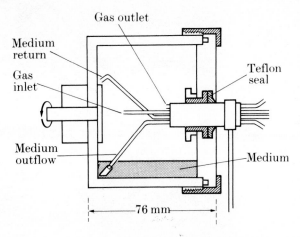

Fig. 1.6.    Rotating drum oxygenator (Exton and Park 1967). A cross-section of the oxygenator is shown. It is driven about the horizontal axis at 60 rev/min.

the liver; total volume of medium (for recirculation experiments) is 65 ml.

A cautionary note to the use of this oxygenator has been provided by Newsholme and Cox (unpublished) to the effect that in its original form the escape of gas may make electrical appliances within the cabinet a source of danger. Their suggested modification overcomes this objection and allows the driving motor for the oxygenator itself as well as the perfusion pump to be enclosed with the perfusion cabinet, with corresponding gain in compactness and a shortening of connecting tubes. Inflow and outflow tubes pass along the rotating shaft of the oxygenator, which is sealed by a spring-loaded glass ball-joint. The whole drum can be slid along its horizontal axis on to the drive shaft of a motor, and two oxygenators may be driven by the single motor.

An advantage of the closed oxygenator is that total gas collection is more efficient. Exton and Park report more than 95 per cent recovery of $C^{14}O_2$ from $NaHC^{14}O_3$ in the bubble traps.

### Rotating disc oxygenator

Gerber (1965) has designed an oxygenator that is suitable for small volume perfusion (40–50 ml). It consists of a spherical reservoir (Fig. 1.7) with inlet and outlet vents for gas and for medium. Mounted

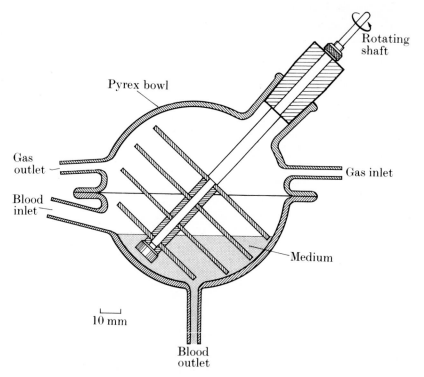

Fig. 1.7.   Rotating disc oxygenator of Gerber (1965). The medium is contained in two Pyrex bowls of 9-cm diameter, fixed together. The central spindle carries discs of Lucite or of stainless steel through the medium. (For details see text.)

within the reservoir, on an axis inclined at some 30° from the vertical, are a series of three Lucite discs, of a diameter suited to the circumference of the reservoir. The axis can be rotated at 40 rev/min by means of a heavy-duty high-torque motor, and the discs are thereby carried through the pool of medium. A continuous film of medium is in turn carried through the chosen gas atmosphere of the chamber.

The reservoir consists of two Pyrex bowls of 9 cm diameter, applied along their circumference to enclose the Lucite discs. This allows rapid disassembly for cleaning. Gerber (1965) gives the following flow rates for adequate oxygenation in liver and small intestinal perfusions: perfusion volume 10 or 40 ml; blood flow 3·5–12·5 ml/min; gas flow 10 or 50 cm$^3$/min. No gas tension measurements are available to confirm the claim quantitatively but the data included in the relevant sections (Chapter 7) suggest that oxygenation is complete. It is interesting to compare the rate of gas flow for the simple rotating chamber of Mortimore with this more sophisticated design. It appears to be possible to reduce it by 75–90 per cent; efficiency of oxygenation is imperative in the Gerber 'low volume' system, since in a liver perfusion, for example, the entire circulating volume passes through the organ every 4 min. Critical tests with liver perfusions in which maximal rates of oxygen uptake are induced (for example those of Hems *et al.*) have not been carried out.

*Rotating cylinder oxygenator*

Two types of rotating cylinder oxygenator (or 'screen' oxygenator) are in use in small organ perfusion: that of Scholz (1968) is a small and compact unit sealed within the heating block of the liver perfusion apparatus devised in Munich for haemoglobin-free perfusion. The apparatus is illustrated, but unfortunately this view shows nothing of the construction of the oxygenator itself (Fig. 1.8). Within the aluminium block lies a plastic mesh cylinder 29 cm long and 8 cm diameter. It has a surface area of almost 2000 cm$^2$, and when revolving within the similarly shaped chamber at 120 rev/min the circulating volume of 150 cm$^3$ of medium is spread over a surface of 20 m$^2$. The authors comment that foaming of the albumin solution which is used as perfusion medium in this system is avoided by freeing the medium of bubbles before it enters the chamber. Gas entering the oxygenator is pre-warmed by passage through small, warming cylinders.

The best test of the efficiency of gas exchange in this type of oxygenator is that on changing from one gas mixture to another, for example from oxygen:$CO_2$ to argon:$CO_2$, equilibrium with the new gas mixture is complete within 1 min. The rate of gas flow has not been published. In the experimental system for liver perfusion used by these authors high flow rates of medium are used and oxygen tension has been determined. Oxygenation appears to be adequate under all the experimental conditions tested (see Chapter 3).

The second 'screen' or rotating cylinder oxygenator in common use

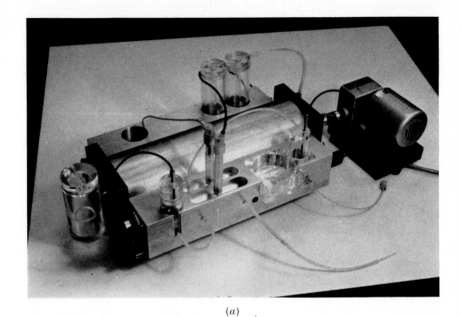

*(a)*

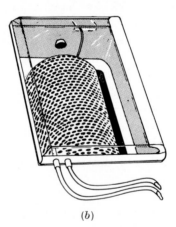

*(b)*

FIG. 1.8.   (a) Perfusion apparatus with rotating-cylinder oxygenator (Scholz 1968). The thermostat (a warmed aluminium block) contains the oxygenator, a filter with a contact thermometer, two flow-cells with oxygen electrodes, the liver chamber, and reservoir. Gases flow first through warming chambers (top right). (b) Perfusion apparatus with rotating cylinder oxygenator (Ambec, Beck Industries, Inc.). A plastic-covered cylinder of wire mesh revolves at about 40 rev/min about its long axis, within a rectangular chamber. Medium and gas are introduced through the leads illustrated, and the Perspex lid is gas-tight.

is that designed by Vaughan and incorporated in the commercially available 'Ambec' perfusion apparatus (see Chapters 6 and 7). A large rectangular reservoir houses the medium, together with a plastic-coated mesh cylinder which revolves about its long axis within the chamber, carrying a film of medium up into the gas-filled atmosphere above. A motor within the heating chamber drives the cylinder. A more-or-less tightly fitting lid carries the gas inflow and outflow lines as well as the medium inflow and outflow lines (Fig. 1.8). A defect in the design of this apparatus is that a guaranteed seal cannot be obtained at this point and, as a result, total gas collections are difficult to achieve. No doubt a minor modification would eliminate this difficulty. Cleaning is difficult. In addition, despite the very compact layout of the Ambec perfusion apparatus, a considerable length of polythene or tygon tubing is interposed between the oxygenator and the arterial inflow of the organ chamber. As discussed elsewhere (p. 82), excessive gas loss can occur and the inflow oxygen tension may not be the same as that of the medium leaving the oxygenator. Nevertheless, this oxygenator has been successfully employed in perfusion of the pancreas according to the Sussman technique (see Chapter 6) and of the small intestine (Dubois *et al.*; see Chapter 7).

*Other designs of open oxygenator*

A number of oxygenators has been designed for small-organ perfusion apparatuses, many of them too complex for the purpose. A common design, based on the original apparatus of Hooker (1915), consists of a vertically-positioned straight-sided cylinder with a disc near the upper end which is rotated about the vertical axis. Medium enters through a small orifice in the top of the cylinder, falls on to the disc, and is thrown outwards on to the walls of the cylinder, whence it films down to the reservoir at the bottom of the cylinder. Gas is admitted at the bottom and leaves at the top in the usual way. Such a design has been advocated by Kavin *et al.* (1967) for intestinal perfusion, and by Schreifer for his 'reversed' liver perfusion (see Chapter 3). In neither case has it any advantage over the multibulb oxygenator described on p. 50, this could supplant an oxygenator which has extra moving parts and is correspondingly difficult to clean and assemble.

Despopoulos (1966) describes an oxygenator constructed from a series of bowls over which the medium cascades. It would appear to have no advantages over the more conventional oxygenators discussed above.

**Closed type**

*Membrane oxygenators* (Fig. 1.9)

A common form of oxygenator in pump technology for bypass operation, organ preservation, etc., known as the membrane oxygenator, has been applied to small-animal organ perfusion. The principle of

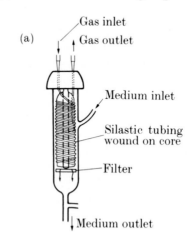

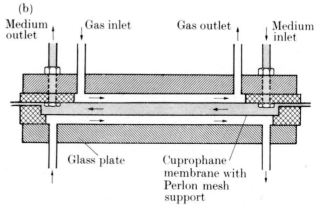

FIG. 1.9. Varieties of membrane oxygenator. (a) Silastic coil oxygenator (Folkman *et al.* 1968). A length of fine silastic tubing is wound onto a glass cylinder. This coil is contained within a cylindrical chamber through which medium flows. Gas is passed through the lumen of this coil and gas exchange occurs across the silastic membrane. Semb (1968) has used this oxygenator for kidney perfusion. (b) Counter-current dialyser modified for use as an oxygenator (Röskenbleck *et al.*). A cuprophane membrane separates two compartments; the outer receives the gas, and the inner conducts medium. For further details see Röskenbleck *et al.* (1967), and as an illustration of the use of the oxygenator in heart perfusion see Fig. 5.5.

this type of oxygenator is that gas and medium do not come into contact but are separated by a gas-permeable membrane, usually a plastic. Frothing, gas embolism, and bacterial contamination from the gas supply cannot occur. However, it is a more expensive form of oxygenator than the simple glass apparatuses already described, because its design is such that it may have to be made disposable. It has been applied in kidney perfusion (Semb, Melville-Williams, and Hume 1968) in which experiments with allogeneic lymphocytes were performed under sterile conditions, in liver perfusion of rat and mouse (Kvetina and Guaitani 1969) in which no gas uptake studies were contemplated, and in perfusion of thymus (Folkman, Winsey, Cole, and Hodes 1968) and of heart (Huhmann, Niesel, and Grote 1967) in which, again, no metabolic or gas exchange studies were undertaken.

(i) *Folkman oxygenator*. This consists of 6 m of silastic tubing, i.d. 0·012 inches; o.d. 0·025 inches wound round a central core, 2 cm in diameter. The whole is contained within another tube of 3 cm diameter, through which the perfusion medium is passed. Gas is passed through the silastic tubing, which is extremely permeable (see p. 82). No details of the efficiency of this oxygenator are given.

(ii) *Röskenbleck oxygenator*. This apparatus is a counter-current dialyser, adapted as an oxygenator by leading gas through the outer compartment. Although complex in construction, it has been demonstrated to be very efficient; graphic records show equilibration with new gas mixtures within 1 min (Röskenbleck *et al.* 1967).

## Collection of gas samples

The general discussion will be found in Chapter 2. With a multibulb oxygenator the facilities for total gas collection are not entirely satisfactory. It is an open system with a high gas flow.

To collect all outflowing gas, the upper vent of the oxygenator is connected by thick-walled tubing to two or more wash bottles containing caustic soda or another $CO_2$ trap. Beyond the wash bottles is a pump with low stroke volume. (In Hems apparatus (Chapter 3) the same Watson–Marlow pump may be used to draw gas through the oxygenator as is used for perfusion, by using a second 'track'.) The suction applied must be observed at regular intervals, so that there is no significant build-up of either gas or medium in the oxygenator.

*Likely causes of loss of gas*. The diffusion of gas through polythene and other plastics has been discussed (p. 82). The loss within the oxygenator, and between it and the collection is probably very small—some 20 cm of tubing is involved. Other lengths of tubing, particularly that between reservoir and oxygenator, may be more significant sources of loss. Finally, in the open liver perfusion system of Hems *et al.* there is possibly some loss from the liver surface. Total gas collections are more easily made in the fully isolated liver where the organ is within a sealed glass compartment (for example, that of Miller).

The only apparatus that offers error-proof gas collection for liver perfusion is that of Morris (1960) (discussed in Chapter 3) but its disadvantages in other respects make it likely that most workers will prefer to accept the small percentage loss of $CO_2$ found in more recent apparatuses.

### MEANS OF SUPPLYING MEDIUM

Organ perfusion implies the continuous exchange of medium within the vascular bed, which in turn requires a mechanical means of circulating fluid. In the early apparatuses, it was sufficient to allow an 'inexhaustible' supply of medium to flow once through the organ and then to be analysed and discarded. For this purpose, gravity or gravity assisted by an increased pressure such as is provided by a Marriot flask (see Fig. 5.3, Langendorff (1895)) sufficed. In more recent apparatus the device of Leichtweiss *et al.* (1967) for kidney perfusion and of Forth (1968) for the intestine are developments of this approach. A gas cylinder with a rotating valve exerts a continuous or discontinuous pressure over the surface of a reservoir which is elevated above the perfusing organ. This forces medium through the organ and perfusion is obtained without the use of a pump.

An alternative approach of special relevance to biochemical studies is recirculation perfusion, in which a defined volume of medium is perfused through the organ and constantly recirculated after mixing in a reservoir. This method is that most often meant by the term 'perfusion' and some discussion of the relative merits of the two approaches of once-through and recirculation perfusion is required.

### Once-through perfusion

Once-through perfusion has the advantage that simpler apparatus is required: in the simplest form of perfusion an oxygenator and pump may be dispensed with, without sacrificing the viability of the preparation (as, for example, in Langendorff's method). However the range of studies available in this system is limited.

(i) A large volume of perfusion medium is required, since the duration of an experiment is limited by the size of this volume. Medium may therefore need to be 'simplified', either because of consideration of cost or lack of materials, such as rat blood, substrates, or unusual effectors in short supply.

(ii) No 'equilibrium' is established in the way characteristic of incubation experiment, since the medium is constantly changing. This in itself is no disadvantage, since the composition of the medium is constant if it is delivered from a single well-mixed reservoir. It may, however, influence the length of time required for 'pre-perfusion'.

(iii) Metabolic changes in the medium must be established by determining arterio-venous concentration differences; this may be beyond the means of available micro-methods to determine. More important, low rates of 'uptake' from a high initial concentration of a substrate in the perfusion medium (such as that of glucose by the heart) cannot be accurately determined on the basis of small arterio-venous differences and it is here that the recirculation-perfusion method has special advantages.

(iv) As a corollary to the determination of uptake by arterio-venous difference, the rate of flow of medium through the organ must be accurately known; this is not necessary in recirculation perfusion. Thus, at the simplest:

rate of metabolism ($\mu$mol/min) = arterial conc. —

—venous conc. ($\mu$mol/ml) $\times$ flow rate (ml/min).

Further discussion of the kinetics of once-through and recirculation perfusion is given by Nagashima and Levy (1968), who deal especially with the problem of whether uptake is itself dependent upon flow rate (see p. 106).

*Recirculation perfusion*

A pre-selected volume of perfusion medium (the volume is chosen with reference to the nature of the study) is continuously recirculated through the organ for the duration of an experiment. In this respect, recirculation perfusion is analagous with incubation studies using tissue slices, homogenates, or extracts. The advantages of recirculation perfusion over the once-through technique are as follows.

(i) The volume of medium required is much smaller, and does not influence the duration of a perfusion. This is especially useful if rat blood is the perfusion medium. Similarly, the quantities of substrate required may be less than in a once-through perfusion.

(ii) An 'equilibrium' is established between the medium and the tissue, the relative volumes of which are fixed. This is especially illustrated by experiments such as those of Schimassek (1963) in which lactate and pyruvate concentrations in the perfusion medium achieve a plateau at a ratio lactate/pyruvate = 10/1, a figure that appears to be characteristics for the liver and to have parallels *in vivo*. Similarly, Schimassek and Gerok (1965) found that amino acids are released from the liver into the recirculating perfusion medium until something approaching the plasma concentrations existing *in vivo* are achieved, whereupon a plateau of amino-acid release is established.

(iii) Rates of uptake or production may be determined by serial analysis of the mixed medium in a reservoir. Changes are cumulative in the same way as in an incubation experiment, and quite small metabolic effects may be magnified by reducing the volume of medium or by prolonging the perfusion. Sampling errors are minimized compared with single determinations of arterial or venous concentration, and changes against a high background can be followed since they too are cumulative.

Recirculation perfusion has, however, the following disadvantages.

(i) Accumulation of products or exhaustion of substrates and other essential factors may influence the course of reactions in the perfused organs. Linearity of reactions with time (see Chapter 3, p. 190) or repeatability of a response at a later stage in a perfusion experiment (see, for example, Schnitger *et al.* (1965) and Scholz (1968)) are invaluable evidence that the organ is not influenced by these factors in a given experimental situation. But awareness of this risk has led Abraham *et al.* (1968) to incorporate a continuous dialysis into their recirculation liver perfusion apparatus (see Chapter 3).

(ii) A mechanical system of recirculation is essential, as well as a method of reoxygenating the medium. Thus, pumps (or gas-lift; see below) and oxygenators are an essential feature of recirculation-perfusion systems. Both may be omitted from once-through techniques. With the introduction of pumps comes the possibility of damage to the perfusion medium, such as haemolysis, frothing, and denaturation of protein, all of which may contribute to alterations in the organ during perfusion.

The advantages outweigh the disadvantages, and the majority of methods of perfusion devised for biochemical studies, and to be discussed in the following chapters, are essentially recirculation perfusions. The special case of the pancreatic–duodenal preparation may be mentioned (see p. 325) as an example of once-through perfusion by choice.

## Pumps

As discussed in the previous section, pumps are required in most perfusion apparatuses. Their use may be circumvented in once-through perfusion by the use of the Marriot flask (q.v.) or pressure from a gas cylinder. Alternatively, the gas-lift of Fisher and Parsons has been applied (for example, by Bleehen and Fisher 1954) in recirculation perfusion of the heart. Autoperfusion, the use of the donor's circulation (or that of a second animal) to perfuse the locally transplanted organ has only a limited place in biochemical studies. Müller (1910) illustrates such a system and it has been extensively exploited in cross-circulation studies (for example, by Heymans and Neil (1958)); more recently, thyroid perfusion and pancreas perfusion have been conducted in this way. The method has been of value in establishing the vasomotor characteristics of an organ for later perfusion, but it introduces many of the complications of whole-animal experiment and will therefore be omitted from further discussion.

Pumps are used in two differing ways, the requirements of which are (a) to supply medium directly to the afferent vessel or (b) to fill an elevated reservoir from which perfusion of the organ occurs under a constant hydrostatic pressure. In some situations, the two approaches are combined; Morgan's heart-perfusion apparatus (Morgan *et al.* 1961), described in Chapter 5, is an example

(i) *Pumped afferent supply.* In this case, the pulse and pressure characteristics of the pump are of central importance, since medium is transferred directly from the pump to the vascular bed of the perfused organ.

(ii) *Hydrostatic perfusion.* When the pump is required only to fill a reservoir, its pulse and other requirements are less exacting. However, the ideal pump is suited to either method of perfusion, and the characteristics of such a pump are discussed by many workers in this field. Sussman and Vaughan (1967) have designed a pump specifically for small-organ perfusion, but commercially available pumps are often well suited to the purpose. Some of these are discussed in the following paragraphs.

*Choice of a pump*

Prime consideration in the choice of a pump is its mechanical robustness. It should be waterproof and the medium chamber should be accessible but entirely separated from the mechanism of the pump in order to limit the possibility of contamination of the medium. The pump should not overheat, since temperature regulation of the medium is important. Flow and pressure should be well maintained and, in addition, it is useful if either or both of these functions can be varied at will. Pumps which require the operator to insert new cogs to vary the output are inconvenient for organ perfusion. The total output of the pump should be considerably in excess of the maximum flow rate through the organ to be perfused since, particularly in the 'hydrostatic' perfusion, the flow rate through the oxygenator may be two to three times this level (see for example, Hems *et al.* (1966); Chapter 3).

*Types of pumps*

Table 1.9 outlines the types of pumps in common use in small-animal organ perfusion apparatus. The terms 'open' and 'closed' are used to distinguish pumps in which the medium is exposed to the mechanism of the pump, usually in the form of valves ('open'), as opposed to those using a 'flexible impeller' (see *Pumping manual*, p. 144), in which the medium remains within a continuous circuit ('closed').

Brief descriptions of the pumps listed in Table 1.9 are as follows.

## TABLE 1.9

### *Types of pump used in organ perfusion*

(1) Open (valves), e.g. Dale–Schuster
(2) Closed (flexible 'impeller'), e.g. roller pump
  (a) Steady flow      (i) Constant, e.g. Vario-Perpex
                         (ii) Variable, e.g. Ambec
  (b) Pulsatile flow     (i) Sine-wave, e.g. Harvard finger-pump; Watson–Marlow
                         (ii) Square wave
                         (iii) Special wave form, e.g. Blakeley pump; with dicrotic notch
(3) Pumpless perfusion, e.g. Marriot flask, gas pressure, or gas-lift systems

(1) *Open (valve) pumps* are used by Rowlands (Chapter 5) for diaphragm and by Anderson and Long (Chapter 6) for pancreatic perfusion, and a commercial Dale–Schuster pump is available from Holter

and requires special preparation. Weiss *et al.* (1959) have designed a useful cannula with a side arm, incorporating a bubble trap and manometer (see Fig. 4.8). Brodie's cannula (1903), a glass tube shaped to hold a ligature and incorporating a 'bubble trap' is difficult to make sufficiently robust in the sizes required for the rat.

(c) *Metal cannulae.* Commercially available metal cannulae, such as the Frankis–Evans cannula have found a place in organ perfusion. A hollow trocar, with a bevelled and sharpened tip is enclosed within a cannula of suitable diameter which has a blunt square tip. After the cannula has been inserted, the trocar may be removed, leaving a cannula in place which has little tendency to 'cut-out' of the vessel. The hollow trocar is preferred to those needles which are provided with a solid trocar, since it may be necessary to fill the cannula from without before attaching the organ to the perfusion circuit. This may be done by attaching a full syringe to the trocar and injecting medium to the tip of the cannula. The Frankis–Evans cannula has been used for liver (Hems *et al.* 1966) and for hind-limb perfusion (Ruderman *et al.* Chapter 5) and may have more general uses. Its main disadvantage for small animals is its weight. Dubois, Vaughan, and Roy (1968) illustrate a special metal and plastic cannula with pressure-recording from a side arm, which has been applied to pancreas and small intestinal perfusion (see Chapters 6 and 7).

*Conclusion*

In certain situations, notably the kidney and the working heart, the design of the cannula is critical. More generally, a uniformity of size and material is advocated since this determines the effective perfusion pressure.

## MATERIALS

Most perfusion apparatus is built in the laboratory or workshop to the individual specifications of the investigator and the question of suitable materials arises. Here we are concerned with the materials of the perfusion apparatus proper; those that come directly into contact with the perfused organ or the perfusing medium. It is most important that such materials are not toxic to animal tissues. In addition, it should be possible to apply rigid standards of cleanliness, including bacterial sterility, if biochemical experiments are to be conducted under controlled conditions. To this end, translucent materials are preferable to opaque materials since gross appearance of cleanliness, the detection of bubbles, or simply changes in colour of the perfusion medium can be observed directly. Other factors, such as the strength, elasticity, and disposability are concerned. Plastics, silicone rubber, and similar synthetic materials are widely used in this field.

*Glass*

A skilled glass-blower can construct most of the apparatus described in the following chapters (see, for example, Oxygenators) and in perfusion as in other chemical methods, glass has ideal properties which allow chemical and microbiological standards of cleanliness to be applied—chromic acid baths, oven-drying, autoclaving, etc. The same cannot be said of many plastic substitutes. It has, however, the unfortunate capacity to react with or bind some possible constituents of perfusion media—platelets, angiotensin, and insulin are examples— and is in addition fragile and rigid.

The problem of reaction with components of the medium is only occasionally significant in biochemical experiment, where an effort is made to define the composition of the medium, but when agents thus affected are to be used the problem may be overcome by treating the glass with silicone.

*Metals*

Apart from its occasional use in cannulae (see Hems *et al.*; Morgan; Chapters 3 and 5), for which high-grade stainless steel is recommended, some perfusion apparatuses incorporate metal heating blocks (Bücher and Scholz, for example; Chapter 3) or wire-mesh oxygenators. In these cases it is more usual to coat the metal with a plastic material to avoid direct contact with the perfusion medium (Ambec oxygenator q.v. and Chapter 6) and since the techniques of plastic coating often leave much to be desired, metal is a somewhat hazardous material of which to construct an apparatus, the composition of which must be non-toxic to the perfused organ. Cleaning metals for biological purposes presents further problems.

*Plastics*

Plastics have many features that make them suitable for the present purpose, and plastics of all sorts have found wide application in the construction of perfusion apparatus. Two properties of plastics must be considered, however, which make them occasionally unsuitable—toxicity to biological material and permeability to gasses. One perfusion apparatus (Hems *et al.* (1966); see Chapter 3), apart from the reservoir and oxygenator with their immediate fittings, contains various plastics: cannulae (one metal)—nylon, filter—polypropylene, tubing—vinyl, connections—p.v.c., pump tubing—silastic. The heart perfusion apparatus of Morgan *et al.* (Chapter 5) has similar components, with the additional use of Teflon bungs. Despite the fact that most plastics appear to be reassuringly inert, each material must be considered on its merits.

*Toxicity*

Boyd and Pathak (1964) tested a variety of plastics in common use by perfusing a buffer in which they have been bathed for varying lengths of time through a standard preparation of frog heart. Semi-quantitative information was obtained on the basis of mechanical activity of the preparation. Oskoni (1968) presents evidence that polypropylene can inhibit glucose uptake under similar test conditions in the perfused rabbit heart, and Stein and Stein (1963) found that white silicone rubber released materials into a perfusion medium containing serum albumin, which gave a titratable acidity and showed up after methylation as several distinct peaks on gas–liquid chromatography.

There is no systematic study on the biochemical effects of such plastic materials in other experimental systems and the papers quoted form the basis of the usual precautions that are taken in the construction and cleaning of perfusion apparatus.

According to Boyd and Pathak, all types of tubing tested are toxic to some extent: Portex; standard p.v.c.; Xlon p.v.c.; Portex crystal vinyl; Esco water clear, of the plastics, and Esco TC156 and Dunlop DSR 557 of the silicone rubbers produced marked effects on the force of contraction, heart rate, or perfusion pressure of perfused frog heart, and these effects often survived, or were even enhanced, after standard washing procedures. It is reassuring that Morgan's 'working' heart preparation in the rat (see Chapter 5), which might be thought every bit as sensitive to 'toxins', contracts forcibly and has linear rates of glucose uptake as well as physiologically normal ATP concentration, etc. despite the presence of such plastics as are condemned by Boyd and Pathak.

A test of 'toxicity' *in vitro*, based upon the suppression of growth of

7

bone marrow in tissue culture (Boyse, Old, and Thomas 1962), is advocated by Atkins, Robinson, and Eiseman (1968; 1970) for the independent assessment of plastics in perfusion apparatus designed for organ-survival experiments. However the ultimate test remains the objective determination of a metabolic parameter of function in a perfused organ (for example, glucose uptake by heart or glucose synthesis by the liver). A source of data on toxicity and other physical characteristics of plastics is the *Pumping manual* (1968, p. 143).

*Permeability of plastics to oxygen and CO$_2$.* Widely known and yet almost equally widely ignored, is the permeability of many plastics to gases. The rate of transfer of oxygen across p.v.c. thin-walled tubing in a perfusion apparatus may be so high as to invalidate oxygen consumptions studies (Nishiitsutsuji–Uwo *et al.* 1967). Perfusion medium which is saturated with oxygen in an oxygenator may still arrive at the inlet cannula with a much lower oxygen tension, and the greater the length of plastic tubing between the two, the more oxygen is lost.

In experiments to test the effects of total anoxia on the perfused heart, Mansford and Opie (see Chapter 5) gassed the medium in the 'oxygenator' with 95% N$_2$:5% CO$_2$. Nevertheless, the medium contained appreciable concentrations of oxygen when it reached the oxygen electrode on the atrial side of the heart. This could be prevented by jacketing the polythene tubing with a second vinyl tube of three to four times larger diameter, and passing the same gas mixture down the intervening cavity. In the liver perfusion studies of Hems *et al.* (see Chapter 3) reliable oxygen uptake data were obtained only after the majority of the vinyl tubing between the bottom of the oxygenator and the portal vein cannula had been replaced by jointed glass. Similar precautions were taken by Nishiitsutsuji–Uwo *et al.* in a kidney perfusion apparatus which, because it relies on hydrostatic pressure, offers a length of some 100–120 cm between the oxygenator and the kidney. No such problems were encountered on the 'venous' side of either circuit, presumably because the gradients between inside and outside gas tensions are smaller.

A similar explanation may account for the fact that loss of CO$_2$ has not been considered a problem. $^{14}$CO$_2$ recovery experiments rarely show such losses as might be expected if massive leakage of gas through plastic tubing were occurring and, when it has been tested, pH in bicarbonate-buffered media is not abnormal, provided that an efficient oxygenator and accurate bicarbonate concentrations are used.

However, no direct study has been undertaken, and this potential hazard should be considered.

When tested as flat sheet the diffusion constants for oxygen and carbon dioxide were, in ml (s.t.p.d)/min/m$^2$/mil at 25°C,

|          | $D_{O_2}$ | $D_{CO_2}$ | $D_{CO_2}/D_{O_2}$ |
|----------|-----------|------------|---------------------|
| Teflon   | 30        | 78         | 2·6                 |
| Silastic | 1210      | 6310       | 5·2                 |

(Galletti, Snider, and Silbert-Aiden 1966) and, regardless of membrane thickness, the order of decreasing permeability was consistently $He > CO_2 > O_2 > N_2$ for Teflon, and $CO_2 > O_2 > He > N_2$ for Silastic. This is not the order of molecular diameter ($He > O_2 > N_2 > CO_2$) and some process other than simple diffusion through pores is thought to account for this finding. Gases may be soluble in the plastics studied. Technical data is contained in the *Pumping manual* (1968). This property of plastics is used to advantage in membrane oxygenators (Chapter 1) and oxygen electrodes.

## Methods of cleaning

The question of cleaning the complex apparatus used in perfusion is important on two counts. Chemical cleanliness is essential if micro-analytical work is to be accurately performed on medium or tissue. The added complication of toxic substances which leach from plastics, both during perfusion and during washing procedures, is discussed in the preceding section. Apart from chemical cleanliness, the problem of contamination with bacteria is magnified in perfusion, in which a partially open system is incubated at 37° for long periods. This aspect of perfusion is discussed on p. 84 and in Chapter 3. As mentioned there, many perfusion media incorporate antibiotics, but this is not the ideal solution in metabolic studies. Accordingly, the problem resolves itself into one of rendering the apparatus free of bacteria.

For most purposes, standard washing procedures are adequate. Kavin *et al.* (see Chapter 7) suggest rinsing with tap water, detergent, and distilled water and finally drying.

Hauge *et al.* (1966) suggest 1% benzalcon solution for plastic parts, rinsing 10 times with water and drying at 45°C, while Allison *et al.* (1961) advocate chromic acid for glass, and boiled rubber components (these have been mostly eliminated from perfusion apparatus for biochemical purposes) in 2% bicarbonate, rinsing them afterwards with water. Fisher and Kerly (1964) suggest washing glassware in 'boiling' 5% KOH in ethanol, and disposing of the plastic tubing after each perfusion. Windmueller, Spaeth, and Ganote (1970) advocate the additional use of steam to sterilize the apparatus once assembled.

The procedure followed by Hems *et al.* for the apparatus described in Chapter 3 is described below. After a perfusion, the medium is drained from the apparatus and its volume measured. If a second perfusion is to follow it is sufficient to wash out the apparatus as it stands with several changes of buffer and deionized water, and to leave buffer recirculating while a new batch of medium is prepared. At the end of the series (i.e. at the end of each day), water is pumped through the circuit to displace all perfusion medium and the apparatus is disassembled. Tubing is left to soak overnight in detergent (Pyroneg) and the oxygenator is washed and immersed in chromic acid. The remaining glassware too, should ideally be acid cleaned (with chromic acid).

Before reassembling the apparatus, the detergent-soaked components are washed through with tap water, with deionized water, and finally with distilled water. The filter is usually replaced, after three or four perfusions, since cleaning it completely is not feasible. Excess water is shaken out of the tubing, and the tubes blown through with compressed air. The apparatus is assembled, without the oxygenator, which is only washed (with copious water and distilled water) when the medium is ready to circulate.

Clearly, no formal sterilization is involved in this regime; nor have Boyd and Pathak's strictures over toxins released from plastics during detergent washing been observed. Bacterial contamination and toxic materials have, however, not been encountered by Hems *et al.* when using this regime.

In special circumstances other methods are required. For example, insulin added to the medium binds to glassware, which should then be cleaned with dilute HCl. Radioactive experiments require added care and attention to disposal of washings.

Bacterial contamination should be minimal if a simply designed apparatus is regularly cleaned in this way. The introduction of organisms is minimized if the medium is filtered through Millipore 0·45 $\mu$m before use (see Fanburg and Posner 1969). If truly sterile conditions are required in order to carry out prolonged perfusions with highly nutrient media, the apparatus of Carrel and Lindbergh (1938) is suitable, and these authors describe in great detail the assembly and sterilization of their apparatus.

*Test of sterility*

The effectiveness of sterilization in preventing interference from bacteria in metabolic experiment should be tested in each new perfusion system and, in addition, in the well-established methods if unexpected or irregular results appear. The simplest test is to plate some medium on agar before and after perfusion. The plates should be incubated both aerobically and anaerobically. In the special case of the liver, phagocytosis may obscure the appearance of organisms in the medium (Bonventre and Oxman 1965) and it is suggested that a simple saline homogenate of the perfused liver should be cultured in addition (see Chapter 3).

A more subtle test is suggested by the observations of Fanburg and Posner (1969) in the perfused heart. When antibiotics were omitted from the medium in their experiments, the form in which RNA appeared in the medium suggested a bacterial component. The finding of this component corresponded closely to the appearance of colonies on culture.

## METHODS OF ANAESTHESIA

Part of the operative procedure is the induction and maintenance of anaesthesia. The alternative, killing the animal by decapitation, has been used in heart perfusion (Morgan *et al.* 1961) where the operation takes only a few seconds. In any prolonged operation (liver, for example, requires 15 min and pancreas about 30 min), anoxia and ischaemia are considered to affect the state of the perfused organ and anaesthesia is the only suitable method of avoiding them.

While the metabolic effects of each agent *in vivo* are partly documented, little study has been made of any persistence of their effect in the perfused organ. Some discussion of this point is included in the section on liver (p. 212), but no definite conclusion is reached in this, the only organ in which sufficient data is available for any comparison to be made.

### (a) *Ether*

The simplest method (and the most variable in the state of anaesthesia produced) is to use ether, in a sealed box. The animal, placed in a box within which a pad soaked in ether is enclosed, falls asleep within 3–4 min, with varying degress of excitement during the induction, and anaesthesia is maintained during operation by a discontinuous administration of ether via a 'face' mask. Biebuyck, Saunders, Harrison, and Bull (1970) advocate a more systematic use of ether to reduce the degree of anoxia, and to provide an easily controlled level of anaesthesia. To this end an EMO vaporizer is included in the circuit. Liver perfusion experiments conducted with this form of anaesthesia in Miller's system gave parameters of function which did not differ significantly from those of Miller *et al.*

### (b) *Fifty per cent $CO_2$ in air*

Introduced by Miller *et al.* as an alternative to ether, this method gave identical results in the perfusion of liver. It has not found general application.

### (c) *Pentobarbitone (Nembutal)*

The commonest barbiturate anaesthetic in use in the rat, this agent gives satisfactory anaesthesia when injected in doses of 5–6 mg/100 g body weight, or somewhat less in alloxan diabetic rats. At this dose, a rat will remain asleep for the following hour at least. Note that commercial preparations of Nembutal (for example Abbot Laboratories) contain ethanol, in which barbiturates are more readily soluble. The metabolic effects of ethanol, even in small amounts, cannot be ignored. Thus the acetate production of the perfused rat liver was found by Keene, Williamson, *et al.* (unpublished) to be significantly increased from this source and these authors recommend that powder of Nembutal be made up in normal saline for intra-peritoneal injection.

### (d) *Halothane*

Advocates of halothane anaesthesia for small animals (Luschei and Mehaffey 1967) point out that unlike ether, it produces minimal mucous secretion and causes less depression of respiration than does pentobarbitone. It is administered as an inhalation anaesthetic, much as ether (see above). At present there are no reports of its use in organ perfusion experiments.

### (e) *Alternative agents*

Bloxham (1967) suggested the use of chloralose (60 mg/kg, intravenously) plus phenobarbitone Na (50 mg/kg intravenously) for liver perfusion. Other agents are listed in the text, with their appropriate preparation. It must be stressed that no objective comparison of the metabolic effects of these agents on the animal or on the isolated perfused organ has been undertaken, and in view of the recent observation by Alberti and Biebuyck, that Nembutal and ether both cause the release of insulin, such studies should be undertaken with other anaesthetic agents before they are unreservedly recommended for use in metabolic perfusion experiments.

A detailed introduction to operative techniques in the rat is provided by Lambert (1965) who discusses many basic techniques, including tying knots with instruments, methods of suture, cannulation, and so on. Supplementary information is available if required in any surgical text (see, for example, Rob and Smith 1968). However, most perfusion methods require only the most rudimentary surgical skills. A routine operation proceeds as follows.

(a)  Anaesthesia.

(b)  Immobilization of the animal.

(c)  Incision(s).

(d)  Ligation and section of vessels, isolation of the organ, and general dissection.

(e)  Cannulation techniques.

Any technique that requires more elaborate surgery cannot be recommended for routine use in biochemistry, where other, well-tried techniques of incubation are available to perform similar studies. Some techniques that are considered too difficult in the rat may be feasible for routine use in larger animals, for example, in testis of rabbit and adrenal gland of ox. None of the operations to be described in the succeeding chapters is any more difficult than many of the technical procedures to which biochemists are accustomed and the descriptions do not presuppose any knowledge of anatomy or of surgery.

(a) *Anaesthesia*

A number of simple methods exist, and their various merits are discussed above (p. 85). Most commonly, simple ether anaesthesia or intra-peritoneal Nembutal are used. The former method has the disadvantage that once induced, ether anaesthesia needs to be maintained by discontinuous administration of ether with a mask (for example, a funnel containing ether-soaked cotton wool over the animal's snout) with the danger of the animal dying during the operation if this is carelessly done. Nembutal in the correct dose keeps the animal asleep.

(b) *Immobilization of the animal*

Once asleep, the animal is best immobilized in a position suitable
to the operation. This is usually with the animal on its back and limbs
extended. Hems *et al.* use an animal tray for this purpose, which
serves in addition as a support for the animal during the perfusion
*in situ* of the liver. Adhesive tape is the simplest material for the
purpose. The tray itself may be of Perspex, since in the perfusion technique used *in situ* it may come into contact with the perfusion medium
(see Fig. 1.13). When heparin is to be injected into the femoral vein as,

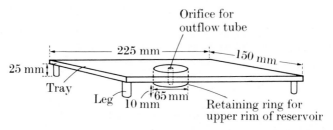

FIG. 1.13.    Animal tray of Hems *et al.* (1966) for liver perfusion *in situ.* A
Perspex tray designed to fit over a reservoir with air-tight joint. The animal lies
on its back on the tray; outflowing medium is carried to the reservoir through a
hole placed eccentrically in the tray. Legs allow the tray to be used for operation,
before transfer to the cabinet.

for example, by Hems *et al.* (1966) and Chain, Mansford and Opie (1969),
this is best done before the animal is fixed to the tray.

(c) *Incision*

Scissors, rather than a scalpel, are used for the incision through skin
and muscle layers, to expose the relevant viscera. To open the abdomen, a fold of skin low down in the mid-line is lifted between forceps
and a small V-shaped cut is made. Through this cut the blade of the
scissors may be inserted to direct the incision in the required direction,
i.e. towards the head, in the mid-line. The linea alba is bloodless and is
incised in the same way to expose the abdominal contents. The two
layers should be opened individually to avoid damage to organs,
particularly to the liver, which lies very superficially in the upper right
quadrant. When wider exposure is required, lateral incisions at the
level of mid-abdomen are made, but the major vessels that run either
side of the linea alba (the superior and inferior epigastric arteries and
veins) should be clamped above the intended incision. The cut is then
directed between adjacent circumferential vessels which are easily

seen coursing through the muscle of the abdominal wall. If the clamps are left in place, they serve as retractors to expose the abdominal viscera widely. Other exposing incisions are described as relevant.

### (d) *Ligation of vessels*

Vessels may be ligated immediately or a loose ligature may be passed round the vessel, which is occluded later in the operation. In some cases two ligatures (a double ligature) are applied to the vessel at a suitable interval, so that the vessel may be cut between.

A simple ligature is positioned by passing a curved forceps behind a segment of the vessel which has been previously cleared from neighbouring structures. The single (or doubled) end of a ligature some 10–15 cm long (e.g. 3/0 Mersilk) is grasped and drawn behind the vessel as the forceps are withdrawn. If the doubled loop is cut across, two ligatures are simultaneously available. A more complex technique of tying side-branches which are not clearly visible is discussed in Chapter 7, p. 388.

Tying ligatures by hand is adequate but may be clumsy. A reef knot is always required for safety. An alternative method, which is of value for small and fragile vessels, is to tie the ligature with two pairs of forceps. Rob and Smith (1968) illustrate the method, which is very simple and, with practice, considerably faster than the 'hand' method.

### (e) *Cannulation techniques*

Cannulae are described on p. 78 and individual methods of inserting cannulae into vessels are discussed as they arise. The general method most suited to the small vessels used in the present techniques is as follows.

The vessel is cleared of connective tissue sufficiently to place loose ligatures. Two or three ligatures are required, one distally to close the afferent vessel and two proximally to hold the cannula in the vessel. The distal ligature has the additional function of holding the cannula parallel to the vessel when it is tied round the cannula after the first knot to occlude the vessel.

To insert a cannula the vessel is grasped in curved non-toothed forceps and stretched distally. Fine pointed scissors are used to make a small incision across no more than half the diameter of the vessel, and the point of the cannula is inserted. This step is facilitated if the axis of the cannula is held at first somewhat perpendicular to the vessel, and only brought into the parallel axis after the wall of the vessel is passed. In special situations discussed in the text, a sharp cannula may be used to penetrate the vessel (for example, portal vein and right atrium in liver technique of Hems *et al.*, Chapter 3) and no incision is needed.

In arterial cannulation the danger of air-embolism, with even small bubbles occluding blood supply to large areas of tissue, is avoided by filling the cannula to the tip with medium or with heparinized saline (as, for example, by Vaughan *et al.*; Chapter 6). Finally, there are situations in which a continuous flow of medium into the arterial cannula has advantages; perfusion begins at once and the organ is not totally anoxic for the remainder of the operation (see below).

### DESIGN OF OPERATION

Operative procedures are described in detail for each organ, with alternatives which may be more suited to the individual operator. Some of the more trivial points of technique may be unnecessary, but often decision on such matters requires more control experiments than the outcome warrants and hence the detail survives. Some general points may be made which contribute to successful and reproducible results in the subsequent experiment.

(1) Operations should be as short as possible, consistent with accurate surgery. Evidence on this point is limited, but in the case of liver perfusion it appears that the main cause of failure in otherwise technically successful perfusion may be a prolonged operation, more especially, a prolonged period of anoxia due to reduced or absent blood supply after cannulation of the portal vein (see p. 161). Little is known of the metabolic effects of surgery itself, and it is surmise that any damage is magnified by prolonged operation, or that the effects may take longer to reverse once perfusion begins.

(2) The organ to be perfused should be protected from instruments and not handled directly if this can be avoided. Miller, for example, provides some circumstantial evidence that trauma during the operation alters the metabolic characteristics of the perfused liver (see p. 196). The lung appears to be particularly susceptible to handling, which results in surface exudation (see Chapter 9); the kidney is apparently unaffected (Chapter 4).

(3) To reduce the period of anoxia, the thorax should be opened only if necessary to the operation, and then as late as possible, since the circulation and aeration may remain quite effective so long as the chest is closed.

(4) To reduce the ischaemic period, the efferent vessel should be cannulated before the artery (for example, in kidney perfusion, Chapter 4). If the artery is cannulated early, there may be some advantage in beginning perfusion at once, allowing the medium to go to waste until the circuit is established by venous cannulation. Söling, Willms, Friedrichs, and Kleineke (1968) have modified Schimassek's operation in this way so that the perfused liver is prepared without an anoxic ischaemic interval. However, there is no evidence in this particular case that a great difference is achieved (see p. 195). In the very prolonged operation required for pancreatic perfusion (Vaughan *et al.* 1967; Chapter 6), however, it would appear to be a wise precaution to perfuse from the moment of arterial cannulation.

By contrast, the standard methods of heart perfusion allow for excision of the organ before any cannulae are inserted. Only in the more recent methods of Arnold and Lochner (1965) and of Basar *et al.* (1968) is flow established before excision of the heart, and there is as yet no evidence of metabolic differences between old and new methods. Histologically, there appears to be less interstitial oedema in the latter, and this may be a sign of better oxygenation (see Chapter 5).

(5) *Pre-perfusion.* Once a circuit is established, medium is perfused, and 'standard' conditions should be attained as soon as possible. In incubation experiments, there is sometimes a 'pre-incubation' before substrate is added from a side-arm, to allow for stabilization (and for equilibration of the manometer). In perfusion too there may need to be a pre-perfusion period: this is discussed in detail for the liver, where 30–40 min of pre-perfusion is considered to be essential. In the case of the heart, opinions differ, but 15 min or longer is usually advocated (see Chapter 5). For kidney the pre-perfusion may be unnecessary or even undesirable (see p. 251), and there is no information concerning other organs. It is however, important to establish that the preparation is metabolically stable for the duration of the experiment, and this includes information about the time required to recover from anoxia and surgery.

## SUMMARY

For reasons already discussed (p. 4), organ perfusion techniques have assumed importance in biochemical studies. The number of techniques is increasing rapidly and the problem of comparing results obtained in two different perfusions of the same organ (as, for example, in the Miller and Schimassek methods of liver perfusion) or of perfusion of two different organs (for example, glucose synthesis in liver and kidney) is often impossible. Methods are described here in sufficient detail, it is hoped, for them to be accurately copied. They have been selected, where possible, for direct application to biochemical experiment, and the requirements considered essential for such a purpose are out-lined on p. 11. Wherever the published information is sufficient a comparison, which it is hoped may serve a dual purpose, has been drawn between similar preparations in order to enable an investigator to select the method most suited to his purpose, be it tissue or medium, and brief or prolonged perfusion, and to define the optimum conditions for perfusion for the organ concerned, thereby adding to the accumulating information on the general technique of organ perfusion for biochemical experiment.

For each organ, a list of methods is given at the beginning of the section, with some attempt to classify the methods. This is usually an arbitrary classification, and no attempt has been made to apply this rigidly to different organs. Individual methods are then described under the headings: method, apparatus, perfusion medium, operative procedure, characteristics of perfusion. Under this last heading are included those aspects of function which are thought to give some indications of viability and stability of the preparation, with, in addition, information about the type of investigation to which the preparation is suited. Each section is preceded by a general introduction to the anatomy and physiology of the organ to be described, since it appeared useful to have this information collected into a short paragraph. Only so much anatomy and histology is described as is necessary to the technique of organ perfusion and to the possible interpretation of metabolic data; the medically-based reader may in any event be familiar with these details. Where possible, the descriptions apply to small animals, and in particular to the rat.

It is often tempting to combine methods, taking the best features from the several available, but this is fraught with danger. In so complex a technique each modification must be tested individually and it has been thought more useful, therefore, to describe existing methods in detail, allowing the individual investigator to assess and to test the method closest to his requirement.

### REFERENCES

ABRAHAM, R., DAWSON, W. GRASSO, P., and GOLDBERG, L. (1968) Lysosomal changes associated with hyperoxia in the isolated perfused rat liver. *Exp. & mol. Pathol.* **8**, 370.

AGISHI, T., PEIRCE, E. C., and KENT, B. B. (1969) A comparison of pulsatile and non-pulsatile pumping for *ex-vivo* renal perfusion. *Surg. Res.* **9**, 623.

ALBERTI, K. G. M. M. and BIEBUYCK, J. F. (1971) Personal communication.

ALLISON, P. R., deBURGH DALY, I., and WAALER, B. A. (1961) Bronchial circulation and pulmonary vasomotor nerve response in isolated perfused lungs. *J. Physiol., Lond.* **157**, 462.

ANDERSON, E. and LONG, J. A. (1947) The effect of hyperglycaemia on insulin secretion as determined with the isolated rat pancreas in a perfusion apparatus. *Endocrinology* **40**, 92.

ANDJUS, R. U., SUHARA, K., and SLOVITER, H. A. (1967) An isolated, perfused rat brain preparation, its spontaneous and stimulated activity. *J. appl. Physiol.* **22**, 1033.

ARNOLD, G. and LOCHNER, W. (1965) Die Temperaturabhängigkeit des Sauerstoffverbrauchess tillgestellter, künstlich perfundierter Warmblüterherzen zwischen 34° und 4°C. *Pflügers Arch. ges. Physiol.* **284**, 169.

ATKINS, R. C., ROBINSON, W. A., and EISEMAN, B. (1968) Cytotoxins released from plastic perfusion apparatus. *Lancet* II, 1014.

—— —— —— (1970) The enhancing effect of bone marrow cells on the primary immune response of the isolated perfused spleen. *J. exp. Med.* **131**, 833.

BAGGIOLINI, M. and DEWALD, B. (1968) Stoffwechsel von Pharmaka in der isoliert perfundierten Rattenleber. Untersuchungen über das Methylhydrazinderivat Ibenzmethyzin. *Stoffwechsel der isoliert perfundierten Leber* (eds. W. Staib and R. Scholz), p. 200. Springer-Verlag, Berlin.

BAHLMANN, J., GIEBISCH, G., OCHWADT, B., and SCHOEPPE, W. (1967) Micropuncture study of isolated perfused rat kidney *Am. J. Physiol.* **212**, 77.

BASAR, E., RUEDAS, G., SCHWARZKOPF, H. J., and WEISS, CH. (1968) Untersuchungen des zeitlichen Verhaltens druckabhängiger Änderungen des Strömungswiderstandes in Coronargefäß system des Rattenherzens. *Pflügers Arch. ges. Physiol.* **304**, 189.

BAUMAN, A. W., CLARKSON, T. W., and MILES, F. M. (1963) Functional evaluation of isolated perfused rat kidney. *J. appl. Physiol.* **18**, 1239.

BIEBUYCK, J. F., SAUNDERS, S. J., HARRISON, G. G., and BULL A. B. (1970) Multiple halothane exposure and hepatic bromsulphthalein clearance. *Br. med. J.* **1**, 668.

Biological handbooks (1961) Blood oxygen dissociation: tables 54 and 55. *Blood and other body fluids*, p. 141.

BLAKELEY, G. H. (1965) Neuro-effector mechanisms of the autonomic nervous system. D.Phil. Thesis, University of Oxford.

—— and BROWN, G. L. (1963) A method for perfusion of the isolated spleen. *J. Physiol., Lond.* **169**, 66P.

BLEEHEN N. M. and FISHER, R. B. (1954) The action of insulin on the isolated rat heart. *J. Physiol., Lond.* **123**, 260.

BLOXHAM, D. L. (1967) Effects of halothane, trichloroethylene, pentobarbitone and thiopentone on amino-acid transport in the perfused rat liver. *Biochem. Pharmac.* **16**, 1848.

BONVENTRE, P. E. and OXMAN, E. (1965) Phagocytosis and intracellular disposition of viable bacteria by the isolated perfused rat liver. *J. retic. Soc.* **2**, 313.

BOYD, I. A. and PATHAK, C. L. (1964) The comparative toxicity of silicone rubber and plastic tubing. *Scott. med. J.* **9**, 345.

BOYSE, E. A., OLD, L. J., and THOMAS, G. A report on some observations with a simplified cytotoxic test. (1962) *Plastic reconstr. Surg.* **29**, 435.

BRAUER, R. W., PESSOTTI, R. L., and PIZZOLATO, P. (1951) Isolated rat liver preparation: bile production and other basic properties. *Proc. Soc. exp. Biol. Med.* **18**, 174.

BRODIE, T. G. (1903) The perfusion of surviving organs. *J. Physiol., Lond.* **29**, 266.

BURHOE, S. O. (1940) Methods of securing blood from rats. *J. Hered.* **31**, 445.

CARREL, A. and LINDBERGH, C. A. (1938) The culture of organs. Hoeber, New York.

CHAIN, E. B., MANSFORD, K. R. L., and OPIE, L. H. (1969) Effects of insulin on the pattern of glucose metabolism in the perfused working and Langendorff heart of normal and insulin-deficient rats. *Biochem. J.* **115**, 537.

CHEN, R. F. (1967) Removal of fatty acids from serum albumin by charcoal treatment. *J. biol. Chem.* **242**, 173.

CLARET M. and CORABOEUF E. (1968) Effets de variations du pH sur la polarisation membranaise du foie de rat isolé et perfusé. *C.r. hebd. Séanc. Acad. Sci., Paris* **267**, 642.

COHN, E. J., HUGHES, W. L., and WEARE, J. H. (1947) Preparation and properties of serum and plasma proteins. XIII. Crystallisation of serum albumins from ethanol-water mixtures. *J. Am. chem. Soc.* **69**, 1753.

CURRY, D. L., BENNETT, L. L., and GRODSKY, G. M. (1968) Dynamics of insulin secretion by the perfused rat pancreas. *Endocrinology* **83**, 572.

DAWSON, R. M. C. (ed.), ELLIOTT, D. C., ELLIOTT, W. H., and JONES, K. M. (1969) *Data for biochemical research*, 2nd edn. Clarendon Press, Oxford.

DESPOPOULOS, A. (1966) Consequence of excretory functions in liver and kidney: Hippurates. *Am. J. Physiol.* **210**, 760.

D'SILVA, J. L. and NEIL, M. W. (1954) The potassium, water and glycogen contents of the perfused rat liver. *J. Physiol., Lond.* **124**, 515.

DUBOIS, R. S., VAUGHAN, G. D., and ROY, C. C. (1968) Isolated rat small intestine with intact circulation. *Organ perfusion and preservation* (ed. J. C. Norman), p. 863. Appleton Century Crofts (Meredith Corporation), New York.

EXTON, J. R. and PARK, C. R. (1967) Control of gluconeogenesis in liver. I. General features of gluconeogenesis in the perfused livers of rats. *J. biol. Chem.* **242**, 2622.

FANBURG, B. L. and POSNER, B. I. (1969) Labelling of RNA in the perfused heart: the problem of bacterial contamination. *Biochim. biophys. Acta* **182**, 577.

FELTS, P. A. and MAYES, J. M. (1966) Liver function studied by liver perfusion. *Proc. Eur. Soc. Drug Toxicity* **7**, 16.

FISHER, M. M. and KERLY, M. (1964) Amino acid metabolism in the perfused rat liver. *J. Physiol., Lond.* **174**, 273.

FISHER, R. B. and PARSONS, D. S. (1949) A preparation of surviving rat small-intestine for the study of absorption. *J. Physiol., Lond.* **110**, 36.

FOLKMAN, J., COLE, P., and ZIMMERMAN, S. (1966) Tumour behaviour in isolated perfused organs: *in vitro* growth and metastases of biopsy material in rabbit thyroid and canine intestinal segment. *Ann. Surg.* **164**, 491.

—— WINSEY, S. and MOGLUS, T. (1967) The culture of human leukaemia: a simplified perfusion system for the growth of colonies of human neoplasms. *Surgery, St. Louis* **62**, 110.

—— —— COLE, P. and HODES, R. (1968) Isolated perfusion of thymus. *Exptl cell Res.* **53**, 205.

FORTH, W. (1968) Eisen und Kobalt-Resorption am perfundierten Dunndarmsegment. 3 *Konf der Gesellschaft für Biologische Chemie* (eds. W. Staib and R. Scholz), p. 242. Springer-Verlag, Berlin.

GALLETTI, P. M., SNIDER, M. T., and SILBERT-AIDEN, D. (1966) Gas permeability of plastic membranes for artificial lungs. *Res. Engng* **5**, 30.

GAMBLE, W. J., CONN, P. A., EDALJI-KUMAR, A., PLEUGE, R., and MONROE, R. G. (1970) Myocardial oxygen consumption of blood-perfused, isolated, supported, rat heart. *Am. J. Physiol.* **219**, 604.

GERBER, G. B. (1965) An efficient oxygenator for organ perfusion. *J. appl. Physiol.* **20**, 159.

GOODMAN, D. S. (1957) Preparation of human serum albumin free of long chain fatty acid. *Science, N.Y.* **125**, 1296.

GRODSKY, G. M., BATTS, A., BENNETT, L. L., VCELLA, C., McWILLIAMS, N. B., and SMITH, D. F. (1963) Effects of carbohydrates on secretion of insulin from isolated rat pancreas. *Am. J. Physiol.* **205**, 638.

HAUGE, A., LUNDE, P. K. M., and WAALER, B. A. (1966) Vasoconstriction in isolated blood-perfused rabbit lungs and its inhibition by cresols. *Acta physiol. Scand.* **66**, 226.

HECHTER, O., ZAFFARONI, R., JACOBSEN, R. P., LEVY, H., JEANLOZ, R. W., SCHEAKER, V., and PINCUS, G. (1951) The nature and biogenesis of the adrenal secretory product. *Rec. Prog. Horm. Res.* **6**, 215.

HEMS, R., ROSS, B. D., BERRY, M. N., and KREBS, H. A. (1966) Gluconeogenesis in the perfused rat liver. *Biochem. J.* **101**, 284.

—— STUBBS, M., and KREBS, H. A. (1968) Restricted permeability of the rat liver for glutamate and succinate. *Biochem. J.* **107**, 807.

HEYMANS, L. and NEIL, E. (1958) *Reflexogenic areas of the cardiovascular system.* London.

HILLIER, A. P. (1968) The uptake and release of thyroxine and tri-iodothyronine by the perfused rat heart. *J. Physiol., Lond.* **199**, 151.

HINKE, J. A. M. and WILSON, M. L. (1962) A study of the elastic properties of a 550 $\mu$ artery *in vitro*. *Am. J. Physiol.* **203**, 1153.

HOOKER, D. R. (1915) The perfusion of the mammalian medulla: The effect of calcium and of potassium on the respiratory and cardiac centers. *Am. J. Physiol.* **38**, 201.

HUHMANN, W., NIESEL, W., and GROTE, J. (1967) Untersuchungen über die Bedingungen für die Sauerstoffversorgung des Myokards an perfundierten Rattenherzen. *Pflügers Arch. ges. Physiol.* **294**, 250.

IVERSEN, J. G. and STEEN, J. B. (1966) Oxygen secretion in the isolated, perfused swimbladder. *Hvalräd Skr. Nr.* 48. *Essays in marine physiology*, p. 101. Universitetsforlaget, Oslo.

JOHN, D. W. and MILLER, L. L. (1969) Regulation of net biosynthesis of serum albumin and acute phase plasma proteins (Induction of enhanced net synthesis of fibrinogen, acid glycoprotein, $\alpha2$ (acute phase)–globulin and haptoglobulin by amino acids and hormones during perfusion of the isolated normal rat liver). *J. biol. Chem.* **244**, 6134.

KAVIN, H., LEVIN, N. W., and STANLEY, M. M. Isolated perfused, rat small bowel–technique, studies of viability, glucose absorption. (1967) *J. appl. Physiol.* **22**, 604.

KESSLER, M. and SCHUBOTZ, R. (1968) Die $O_2$-versorgung der hämoglobinfrei perfundierten Rattenleber. *Stoffwechsel der isoliert perfundierten Leber* (eds. W. Staib and R. Scholz), p. 12. Springer-Verlag, Berlin.

KRAMER, K. and DEETJEN, P. (1964) Oxygen consumption and sodium reabsorption in the mammalian kidney. In I.U.B. Symposium Series, Vol. 31. *Oxygen in the animal organism*, p. 411 (eds. F. Dickens and E. Neil). Pergamon Press, London.

KREBS, H. A. (1930) Manometrische Messung des Kohlensauregehaltes von Gasgemisches, modified by W. Gevers and H. A. Krebs, 1966. The effects of adenine nucleotides on carbohydrate metabolism in pigeon-liver homogenates. *Biochem. J.* **98**, 720.

—— and HENSELEIT, K. (1932) Untersuchungen über die Harnstoffbildung im Tierkörper. *Hoppe-Seyler's Z. physiol. Chem.* **210**, 33.

KVETINA, J. and GUAITANI, A. (1969) A versatile method for the *in vitro* perfusion of isolated organs of rats and mice, with particular reference to liver.

LAMBERT, R. (1965) *Surgery of digestive system in the rat.* Thomas.

LANGENDORFF, O. (1895) Untersuchungen am überlebenden Saügertierherzen. *Pflügers Arch. ges. Physiol.* **61**, 291.

LEICHTWEISS H. P. SCHRÖDER H., and WEISS, CH. (1967) Die Beziehung zwischen Perfusionsdruck und Perfusionsstromstärke an der mit Paraffinöl perfundierten isolierten Rattenniere. *Pflügers Arch. ges. Physiol.* **293**, 303

LEVEY, S. and GAST, R. (1966) Isolated perfused rat lung preparation. *J. appl. Physiol.* **21**, 313.

LOCHNER, W., ARNOLD, G., and MULLER-RUCHHOLTZ, E. R. (1968) Metabolism of the artificially arrested heart, and of the gas-perfused heart. *Am. J. Cardiol.* **22**, 299.

LONG, J. A. and LYONS, W. R. (1954) A small perfusion apparatus for the study of surviving isolated organs. *J. Lab. clin. Med.* **44**, 614.

LOUTIT, J. F. and MOLLISON, P. L. (1943). See MOLLISON, P. L. (1967) *Blood transfusion in clinical practice*, 4th edn., p. 68. Blackwell, Oxford and Edinburgh.

LUSCHEI, E. S. and MEHAFFEY, J. J. (1967) Small animal anaesthesia with halothane. *J. appl. Physiol.* **22**, 595.

McCONAGHEY, P. and SLEDGE, C. B. (1970) Production of 'sulphation factor' by the perfused liver. *Nature, Lond.* **225**, 1249.

MAGNUS, R. (1902) Tätigkeit des überlebenden Säugertierherzens bei Dürchströmung mit Gassen. *Arch. exp. Path. Pharmak.* **47**, 200.

MANDELBAUM, I. and BURNS, W. H. (1965) Pulsatile and non-pulsatile blood flows. *J. Am. med. Ass.* **191**, 657.

MANSFORD, K. R. L. (1969) Ph.D. Thesis, University of London.

MILLER, L. L., BLY, C. G., WATSON, M. L., and BALE, W. F. (1951) The dominant role of the liver in plasma protein synthesis. *J. exp. Med.* **94**, 431.

MORGAN, H. E., HENDERSON, M. J., REGEN, D. M., and PARK, C. R. (1961) Regulation of glucose uptake in muscle. I. The effect of insulin and anoxia on glucose transport and phosphorylation in the isolated, perfused heart of normal rats. *J. biol. Chem.* **236**, 253.

MORLAND, A. F. (1965) Collection and withdrawal of body fluids and infusion techniques. *Methods of animal experimentation*, Vol. 1 (ed. W. I. Gay), p. 1.

MORRIS, B. (1960) Some observations on the production of bile by the isolated perfused liver of the rat. *Aust. J. exp. Biol.* **38**, 99.

MORTIMORE, G. B. (1961) Effect of insulin on potassium transfer in isolated rat liver. *Am. J. Physiol.* **200**, 1315

—— and TIETZE, F. (1959) Studies on the mechanism of capture and degradation of insulin [131]I by the cyclically perfused rat liver. *Ann. N.Y. Acad. Sci.* **82**, 329.

MÜLLER, F. (1910) Die künstliche Durchblutung resp. Durchspülung von Organen. *Handb. biol. ArbMeth.* **3**, 327.

NAGASHIMA, R. and LEVY, G. (1968) Effect of perfusion rate and distribution factors on drug elimination characteristics in a perfused organ system. *J. pharmac. Sci.* **57**, 1991.

NAKAYAMA, K. *et al.* (1963) High-amplitude pulsatile pump in extra-corporal circulation with particular reference to haemodynamics. *Surgery, St. Louis* **54**, 798.

NATELSON, S. (1951) Routine use of ultra-micro methods in the clinical laboratory. (A practical micro-gasometer for estimation of carbon dioxide.) *Am. J. clin. Path.* **21**, 1153.

NEELY, J. R., LIEBERMEISTER, H., BATTERSBY, E. J., and MORGAN, H. E. (1967) Effect of pressure development on oxygen consumption by isolated rat heart. *Am. J. Physiol.* **212**, 804.

NISHIITSUTSUJI-UWO, J. M., ROSS, B. D., and KREBS, H. A. (1967) Metabolic activities of the isolated, perfused rat kidney. *Biochem. J.* **103**, 852.

NORMAN, J. C. (ed.) 1968 *Organ perfusion and preservation*. Appleton Century Crofts, New York.

OPIE, L. H. (1965) Coronary flow rate and perfusion pressure as determinants of mechanical function and oxidative metabolism of isolated perfused rat heart. *J. Physiol., Lond.* **180**, 529.

OSKOUI, M. (1968) Pharmacological effects of exposure of various organs to polypropylene tubing. *Archs. int. Pharmacodyn. Thér.* **175**, 223.

PAUL, J. (1970) *Cell and tissue culture*, 4th edn. Livingstone, Edinburgh and London.

PETERS, J. P. and VAN SLYKE, D. D. (1932) *Quantitative clinical chemistry*, 2nd edn. London.

POWIS, G. (1970a) Observations of metabolism and transport in organs perfused *in vitro*. D.Phil. Thesis, University of Oxford.

—— (1970b) Perfusion of rat's liver with blood: transmitter overflows and gluconeogenesis. *Proc. R. Soc.* **B174**, 503.

*Pumping manual* (1968) compiled by the editors of *Pumping* 3rd edn. Morden.

RITTER, E. R. (1952) Pressure/flow relations in the kidney. (Alleged effects of pulse pressure.) *Am. J. Physiol.* **168**, 480.

ROB, C. and SMITH, R. (1968) *Operative surgery*, Vol. 1, 2nd edn. Butterworth, London.

ROBERT, A. and SCOW, R. O. (1963) Perfusion of rat adipose tissue. *Am. J. Physiol.* **205**, 405.

RÖSKENBLECK, H., HUHMANN, W., GLOY, U., and NIESEL, W. (1967) Anwendung eines Dünnschichtdialysators zur Gasäquilibrierung von Perfusionslösungen bei Durchströmungsversuchen an Organen. *Pflügers Arch. ges. Physiol.* **294**, 88.

ROSS, B. D., HEMS, R., and KREBS, H. A. (1967) The rate of gluconeogenesis from various precursors in the perfused rat liver. *Biochem. J.* **102**, 942.

ROWLANDS, S. D. (1969) Oxygen consumption, citrate levels and lactate production of the perfused rat diaphragm. *Comp. Biochem. Physiol.* **29**, 1215.

RYOO, H. and TARVER, H. (1968) Studies on plasma protein synthesis with a new liver perfusion apparatus. *Proc. Soc. exp. Biol. & Med.* **128**, 760.

SCHIMASSEK, H. (1962) Perfusion of isolated rat liver with a semisynthetic medium and control of liver function. *Life Sci.* II, p. 629.

—— (1963) Metabolite des Kohlenhydratstoffwechsels der isoliert perfundierten Rattenleber. *Biochem. Z.* **336**, 460.

—— (1968) Möglichkeiten und Grenzen der Methodik. *Stoffwechsel der isoliert perfundierten Leber* (eds. W. Staib and R. Scholz), p. 1. Springer-Verlag, Berlin.

—— and GEROK, W. (1965) Control of the levels of free amino-acids in plasma by the liver. *Biochem. Z.* **343**, 407.

SCHMIDT, E. (1968) Austritt von Zell-Enzymen aus der isolierten, perfundierten, Rattenleber. *Stoffwechsel der isoliert perfundierten Leber* (eds. W. Staib and R. Scholz), p. 53. Springer-Verlag, Berlin.

SCHNITGER, H., SCHOLZ, R., BÜCHER, TH., and LÜBBERS, D. W. (1965) Comparative fluorometric studies on rat liver *in vivo* and on isolated, perfused, haemoglobin-free liver. *Biochem. Z.* **341**, 334.

SCHOLZ, R. (1968) Untersuchungen zur Redoxkompartmentierung bei der Hämoglobinfrei perfundierten Rattenleber. *Stoffwechsel der isoliert perfundierten Leber* (eds. W. Staib and R. Scholz), p. 25. Springer-Verlag, Berlin.

SEMB, B. K. H., MELVILLE-WILLIAMS, G., and HUME, D. H. (1968) The effect of allogeneic lymphocytes on the isolated, perfused kidney. *Transplantation* **6**, 971.

SLOVITER, H. A. and KAMIMOTO, T. (1967) Erythrocyte substitute for perfusion of brain. *Nature, Lond.* **216**, 458.

SÖLING, H. D., KATTERMAN, R., SCHMIDT, H., and KNEER, P. (1966) The redox state of $NAD^+$–NADH systems in rat liver during ketosis and the so-called 'triosephosphate block'. *Biochim. biophys. Acta* **115**, 1.

—— WILLMS, B., FRIEDRICHS, D., and KLEINEKE, J. (1968) Regulation of gluconeogenesis by fatty acid oxidation in isolated perfused livers of non-starved rats. *Eur. J. Biochem.* **4**, 364.

STAIB, W., STAIB, R., HERRMANN, J., and MEIERS, H. G. (1968) Untersuchungen über die Cortisol-glykoneogenese in der isoliert perfundierten Rattenleber. *Stoffwechsel der isoliert perfundierten Leber*, p. 155. Springer-Verlag, Berlin.

STAIB, R., THIENHAUS, R., AMMEDICK, U., and STAIB, W. (1969) Direkte Einfluss von Insulin und Glucagon auf die Aktivität der Tyrosine-alpha-Ketoglutarate-transaminase. *Eur. J. Biochem.* **8**, 23.

STARLING, E. H. (1895) On the absorption of fluids from the connective tissue spaces. *J. Physiol., Lond.* **19**, 312.

STEIN, O. and STEIN, Y. (1963) Metabolism of fatty acids in the isolated perfused rat heart. *Biochim. biophys. Acta* **70**, 517.

SUSSMAN, K. E., VAUGHAN, G. D., and TIMMER, R. F. (1966) An *in vitro* method for studying insulin secretion in the perfused, isolated rat pancreas. *Metabolism* **15**, 466.

—— —— (1967) Insulin release after ACTH, glucagon and adenosine-3-5-phosphate (cylic AMP) in the perfused isolated rat pancreas. *Diabetes* **16**, 449.

THOMPSON, A. M., ROBERTSON, R. C., and BAUER, T. A. (1968) A rat head-perfusion technique developed for the study of brain uptake of materials. *J. appl. Physiol.* **24**, 407.

UCHIDA, E. and BOHR, D. F. (1969) Myogenic tone in isolated perfused resistance vessels from rats. *Am. J. Physiol.* **216**, 1343.

UMBREIT, W. W., BURRIS, R. H., and STAUFFER, J. F. (1964) *Manometric techniques*. Burgess, Minneapolis.

VALTIS, D. J. and KENNEDY, A. C. (1954) Defective gas transport function of stored red blood-cells. *Lancet* i, 119.

VAN SLYKE, D. D., RHOADS C. P., HILLER, A., and ALVING, A. S. (1934) Relationships between urea excretion, renal blood flow, renal oxygen consumption and diuresis. The mechanism of urea excretion. *Am. J. Physiol.* **109**, 336.

WARBURG, O. (1923) Versuche an überlebendem Carcinomgewebe (Methoden: II. Die Herstellung der Gewebeschnitte). *Biochem. Z.* **142**, 317.

WAYMOUTH, C. (1965) *In* WILLMER, E. N., *Cells and tissues in culture*, Vol. I. Academic Press.

WEIDEMAN, M. J., HEMS, D. A., and KREBS, H. A. (1969) Effects of added adenine nucleotides on renal carbohydrate metabolism. *Biochem. J.* **115**, 1.

WEISS, CH., PASSOW, H., and ROTHSTEIN, A. (1959) Autoregulation of flow in isolated rat kidney, in the absence of red cells. *Am. J. Physiol.* **196**, 1115.

WHITMORE, R. L. (1968) Plasma expanders. *Haemorrheology* (ed. A. L. Copley), *Proceedings of First International Conference Reykjavik* 1966, p. 817.

WILLIAMS, J. A. (1966) Effect of external $K^+$ concentration on transmembrane potentials of rabbit thyroid cells. *Am. J. Physiol.* **211**, 1171.

WILLIAMSON, J. R., KREISBERG, R. A., and FELTS, P. W. (1966) Mechanism for the stimulation of gluconeogenesis by fatty acids in perfused rat liver. *Proc. natn. Acad. Sci. U.S.A.* **56**, 247.

WILLMS, B. and SÖLING, H. D. (1968) Regulation der Gluconeogenese durch Fettsaureoxydation in der isoliert perfundierten Rattenleber. *Stoffwechsel der isoliert perfundierten Leber* (eds. W. Staib and R. Scholz), p. 118, Springer-Verlag, Berlin.

WINDMUELLER, H. G., SPAETH, A. E., and GANOTE, G. E. (1970) Vascular perfusion of the isolated rat gut: norepinephrine and glucocorticoid requirement. *Am. J. Physiol.* **218**, 197.

WINTROBE, M. M. (1967) *Clinical haematology*, p. 429. Kimpson, London.

ZACHARIAH, P. (1961) Contractility and sugar permeability in the perfused rat heart. *J. Physiol., Lond.* **158**, 59.

# 2. Methods of measurement in perfusion

The reintroduction of perfusion techniques, particularly of the liver, reflects their special usefulness in approaching some biochemical problems of current relevance. Many of the possibilities of metabolic study in perfused organs are self-evident; the rates of metabolism of substrates added to the medium, the rate of release of metabolic products into the medium, the effects of hormones on intact tissues, and the influence of the composition of the perfusion medium, by analogy with blood, upon the metabolic function of an organ, may all throw light upon metabolic function at a cellular level. Information that is not available from simpler incubation methods, because the tissue is either damaged or totally disrupted, may be obtained by perfusion. In particular, this allows the inter-relationships of various compartments—tissue, extracellular space, and vascular space—to be studied under conditions *in vitro*.

The choice of perfusion rather than simpler well-tried techniques such as slices, homogenates, or subcellular fractions presupposes that the method of perfusion has some advantages or shows improvements over other methods. This is thought to be the case for the liver, where rates of some synthetic reactions, glucose formation, ketone body formation, and glycogen synthesis (see Chapter 3 for references) are notably higher, and closer to findings *in vivo* than those observed in slices or homogenates. This observation has been disputed more recently (Schimmel and Knobil 1969) on the basis of the higher rates of glucose synthesis from various precursors which are observed in liver slices incubated in smaller numbers per flask than is usual (Hems *et al.* 1966). Nevertheless, the wide range of experimental study already performed with perfused liver, and the special features of the method, have lead to its establishment as a standard method of biochemical research. Comparative data between various experimental techniques is discussed in the sections dealing with individual organs (see Chapter 3 *et seq.*). The present discussion deals with the general method of perfusion and the range of experimental design peculiar to the method. In addition to the experiments devised for use in manometry (see Umbreit, Burris, and Stauffer (1964) for review), most of which may be directly transferred to a perfused system, the methods will be discussed under the following headings.

(1) General methods (applicable to most organs perfused under standard conditions).

   (A)   Physical characteristics—flow and perfusion pressure.
   (B)   Separate treatment of organ and perfusion medium.

       (i)   Medium sampling.
             (a)  Analysis of metabolic products.
             (b)  Loading dose of substrates; calculation of uptake.
             (c)  Continuous infusion of substrates and effectors.
     (ii)  Tissue sampling.
             (a)  Methods.
             (b)  Expression of results.
             (c)  Calculation of gradients.
             (d)  Transport studies.
   (iii)  Gas collection and analysis.

   (C)   Overall 'balance' studies.
   (D)   Studies based upon arterio-venous difference.
   (E)   Measurements at the organ surface.

(2) Special methods which, being related to the specific physiological function of one organ, have no general application. Examples are the use of micro-puncture techniques or clearance studies in the perfused kidney, or the electrocardiogram for the perfused heart.

## METHODS OF INVESTIGATION

The wide variety of methods applied in perfused organ experiments is best summarized by referring to an idealized scheme of organ perfusion, such as has been outlined in Chapter 1 (p. 14). This is summarized further in Fig. 2.1, upon which the general methods or measurement are superimposed. In addition, it is obvious that the experimental system is susceptible of the following variations:

     vary the organ,
     vary the perfusion medium,
     vary the temperature,
     vary the gas-mixture ($O_2$ pressure or $CO_2$ and hence pH).

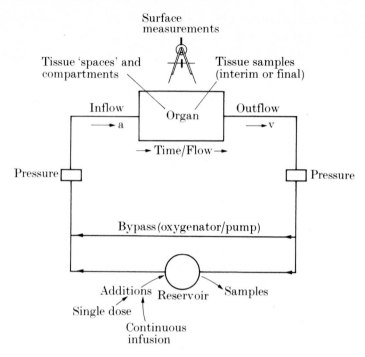

FIG. 2.1.   Idealized perfusion circuit.

## GENERAL METHODS

### Physical characteristics

Using a standard perfusion medium and a suitable organ, the physical characteristics of the system, the pressure, and the flow rate may be defined. This is of interest as a means of assessing the perfusion but, in addition, some metabolic and physiological functions are dependent upon these parameters. Thus, in the heart, perfusion pressure can determine the rate of glucose uptake (see Opie (1965) Chapter 5, p. 294) and in the kidney, flow rate may determine the rate of oxygen uptake (see p. 247).

It is generally accepted that constancy of perfusion pressure and of flow rate during an experiment is a prerequisite of 'normal' function, and it indicates that perfusion is uniform for that period. The converse is not inevitably true; thus a constant flow rate and pressure may be obtained without normal biochemical function (see Chapter 3 *et seq.* for examples). Flow rate and pressure are interdependent when the 'resistance' of the system is constant (Rice and Plaa 1969 and see

p. 105). Finally, flow rate needs to be accurately known for the calculation of rate of substrate uptake when either a once-through perfusion
(q.v.) or arterio–venous difference methods (see below) are used for
this purpose.

*Determination of perfusion pressure*

Inflowing pressure is determined either by gravity in constant head
perfusion systems or by the pump pressure in direct pumping methods.

(i) *Constant head*. When an elevated reservoir of medium is used,
pressure is determined simply by measuring the height of the 'overflow'
above the inflow cannula. The reservoir may be raised or lowered to
alter the pressure for a given purpose (Opie 1965, see Chapter 5);
the multibulb oxygenator with overflow, used by Miller, Schimassek,
Hems, and others may also be elevated, and the distance above the
organ measured. In the apparatus of Hems *et al.* a vertical rod is
provided upon which the 'lung' is mounted by an adjustable clamp.

(ii) *Pump pressure*. A manometer is required to determine the pressure between the pump and the organ, and this may be situated either
directly at the cannula (see Weiss *et al.* (1959) for special design) or on
the pump outflow, usually as a side-arm (for example, that of Ruderman *et al.*). Manometers may be the simple mercury type (see Starling
and Lovatt Evans (1968) for review) and a trap should be provided
between the side-arm conveying medium and the main circuit. Alternatively, the aneroid manometer is suitable to most purposes (see
previous references). In special situations, in which rapid response is
required, transducers have been used; their use in determining the
out-flow pressure in working heart perfusion is discussed in Chapter 5.
A transducer must be calibrated by ordinary physical methods if
work is to be assessed on the basis of pressure developed.

In general perfusion studies no great accuracy is required for the
determination of pressure and there is little to be gained from using
sophisticated apparatus for this purpose, outside the use of transducers in heart perfusion already mentioned.

*Determination of flow rate*

Flow rate should be known for two reasons; to assess its constancy
throughout an experiment and to allow calculation of rates of substrate uptake (see above) in special situations. Differing degrees of

accuracy are required for these two purposes, the latter requiring considerably greater precision than the former. Several methods are in common use.

(i) *Collection.* Medium is collected as it emerges from the perfused organ into a graduated receiver or test-tube. This is simple when applied to once-through systems, but it introduces error when used in a re-circulation system. Especially if the recirculating volume is small, the

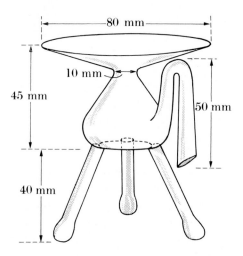

Fɪɢ. 2.2.   Syphon flow-meter. A glass funnel with a syphon side-arm contain-ing a calibrated volume of medium (10 ml). The chamber stands on 3 'legs' within the reservoir (Fig. 2.3) with space for free movement of the magnetic stirrer.

volume required for an accurate timed collection—5–10 ml is a mini-mum—may seriously deplete the circulating volume. The medium must then be rapidly returned to the apparatus, which requires that the apparatus be opened, and there will always be some losses during transfer. Nevertheless, this method is sufficiently accurate for the purpose of most metabolic studies in the perfused heart. An alternative, which 'collects' medium but does not require that it be removed from the apparatus, is a graduated vessel with a tap, which may be filled and timed during a perfusion. Reopening the tap returns the medium to the circuit and a simple extension of this idea is the syphon used by Hems *et al.* (1966) and illustrated in Fig. 2.2. This collects medium flowing from the liver (or kidney) and overflows only at the calibrated volume. A minimum of medium is temporarily excluded from the circuit and the method is semi-continuous.

(ii) *Bubble flow-meters.* One of the simplest methods of determining flow is to inject a bubble of air into the tubing at a chosen site and to time its progress through a suitable length of tube; usually a graduated pipette is suitable (Hems *et al.* (1966); see Fig. 3.9). This method is applied to the outflowing medium, when the bubble of gas presents no problem (it must be of sufficient size to occupy the entire lumen of the pipette, so that it faithfully reflects the rate of passage of medium).

'Proximal' bubble flow-meters have been used (for example, by Bleehen and Fisher 1954; Chapter 5), in which case a fine polythene tube allows a bubble of gas to be introduced into the inflow and, after it has traversed the calibrated tube, the bubble escapes into a bubble-trap.

(iii) *Other flow-meters.* More elaborate flow-meters are, of course, widely available (see Bain and Harper (1968) for review of this field) but are not usually indicated for biochemical studies. The use of perfused kidney to study pressure–flow relationships has, however, biochemical implications (see Chapter 4), and for this purpose, Leichtweiss, Weiss, Basar, and others have used electromagnetic flow-meters.

## Relationship between flow rate and perfusion pressure

The simple linear relationship between perfusion pressure and flow through an isolated organ depends upon the resistance of the system remaining constant. This has been tested in perfused liver (Rice and Plaa 1969), heart (Opie 1965 and Basar *et al.* 1968 (for pulse wave)), and kidney (Kinter and Pappenheimer 1956, and Weiss *et al.* 1959). With the exception of the kidney, in which a special relationship exists (flow remains constant over a considerable range of perfusion pressures (Van Slyke *et al.* 1934)), the expected linearity may be assumed.

An extension of the concept of flow rate is the time taken for a given sample of medium to traverse the entire vascular bed of the organ perfused, 'the perfusion time'. This gives some idea of the length of time for which a drug or a substrate may be in contact with the tissue in a flowing system. (As will be discussed later, this time appears to be so long as to play no significant part in 'rates of uptake'.)

Perfusion 'time' is determined as follows (Hillier 1969). Blue dye (0·5 ml) is injected into the inflow tubing. The outflow is collected, and the time that elapses until the maximum colour is recovered is recorded. On this basis, perfusion time is found to be 10 s, when the flow rate is 0·6 ml/min per g of liver. As expected, a linear relationship exists for the perfusion time versus flow rate, and a single determination of the former allows this characteristic to be calculated for any flow rate. As yet, this concept has not been applied to metabolic studies. However, it should be possible also to derive the volume of the vascular bed from such data, viz. flow rate,

$$\frac{V}{T} = \frac{v}{t},$$

where $V$ = volume in ml, $T$ = time in s, $t$ = perfusion time, and $v$ represents the volume occupied by medium for this time, i.e. volume of the vascular bed of the perfused organ. Therefore,

$$v = \text{flow rate} \times \text{perfusion time}.$$

### Flow rate and substrate uptake

Whether or not flow rate affects the rate of elimination or extraction of a drug (or substrate) is discussed by Nagashima and Levy (1968a). Using a model system of the removal of a drug from the perfusate in a liver perfusion system and making the following assumptions: (i) that the flow rate is high, relative to the volume of medium within the organ (see p. 109 for this calculation) and (ii) that drug (substrate) transfer from the medium to the liver is not rate-limiting, then the 'apparent rate constant of removal' is inversely proportional to the perfusion volume and virtually independent of flow rate.

Analysis of the distribution of a drug into the perfused organ yields mathematical expressions that may be useful in ascertaining whether its distribution is limited by flow, by the diffusion of the drug from the medium into the organ, or by a function of both processes (Nagashima and Levy 1968b). The extreme case, in which flow rate limits substrate uptake by limiting supply, or, more significantly, by inducing anoxia in the preparation is illustrated by the observations of Morris (1963) on the uptake of chylomicrons. At the lowest flow rates the rate of uptake appears to fall, while at higher flow rates the rate of extraction is inversely proportional to flow, i.e. constant uptake per unit time.

**Methods which depend upon a separate treatment of organ and perfusion medium**

An essential difference between the method of organ perfusion and other incubation techniques used *in vitro* is the ease with which, in perfusion, tissue and bathing medium are separately treated and analysed. Many of the special methods to be discussed depend upon this feature.

*Perfusion medium*

*Formation of 'metabolites'.* Products of metabolism of intracellular enzymes, 'secretions' that normally appear in the blood-stream, will appear in progressively increased concentrations in a recirculated perfusion medium. The medium may be sampled, either from the out-flow (v in Fig. 2.1) (for example by Morgan, Opie, and others in heart perfusion) or from a reservoir, which is inserted into the circuit to contain the larger portion of the medium perfused in any one experiment (for example, as by Hems *et al.*; and see Fig. 2.3). Sampling from the once-through method presents no special problems.

*Methods of sampling the medium.* Suitable designs of a reservoir from which samples may be taken by pipette without disturbing the perfusion system are included in the apparatus of Miller and of others. That introduced by Hems *et al.* (1966) and used in many other perfusion techniques is shown in Fig. 2.3.

Mixing is important if a truly representative sample is to be obtained (and, in addition, to prevent 'settling' of erythrocytes in the medium). A magnetic stirrer with a glass-covered magnet in the reservoir is the method commonly in use; other apparatuses combine reservoir and oxygenator, and mixing is achieved by rotation about a longitudinal axis (see p. 53). Mixing assumes special importance if the recirculating volume is small: Morgan *et al.* suggest 15–20 ml recirculating volume; Rabitzsch (1968) has designed an apparatus which permits the perfusion volume to be reduced to that of the blood-volume of a rat, about 10 ml (see p. 278).

Medium usually contains protein, whether added, such as bovine albumin and red cells with haemoglobin or, as occurs in most perfusion systems, protein or enzymes released from the organ during the course of an experiment. Unless contra-indicated, therefore, rapid deprotein-ization is a part of the technique of sampling of the perfusion medium.

For example, 0·5 ml medium is pipetted directly into 4·0 ml 4% $HClO_4$, and mixed and centrifuged after standing for a few minutes. A measured quantity of the supernatant is then neutralized with KOH (not NaOH), which removes perchlorate ions from solution.

**Type A**

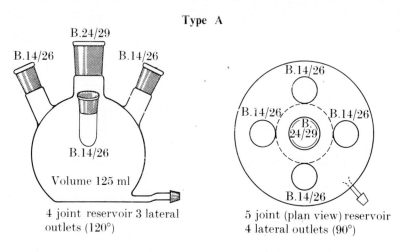

4 joint reservoir 3 lateral outlets (120°)

5 joint (plan view) reservoir 4 lateral outlets (90°)

**Type B**

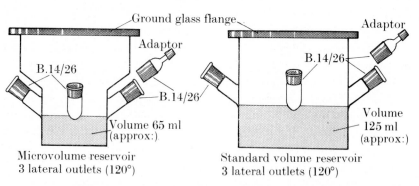

Microvolume reservoir 3 lateral outlets (120°)

Standard volume reservoir 3 lateral outlets (120°)

FIG. 2.3.   Glass reservoirs for perfusion apparatus. A variety of designs suited to the Hems perfusion apparatus. Not to scale (see Chapter 3, p. 167).

The sample thus neutralized may then be stored at 4°C, deep-frozen at −20°C, or analysed immediately. The dilution corrections are calculated in the usual way.

Sample-volume corrections are made as follows. Regular sampling of medium is a feature of the method of perfusion, and unless the volume removed at each sample is very small in relation to the circulating volume, a correction must be applied. In its simplest form, 'total content' of a substance in the perfusion medium $(T)$ is obtained by multiplying perfusion volume $\times$ concentration in an aliquot. Perfusion volume $(V_0)$ is diminished by the aggregated sample volume $(v)$. Thus initial content

$$T_0 = V_0 \times \text{conc}_0;$$

for later samples

$$T_1 = (V_0 - v) \times \text{conc}_1.$$

On the basis of experiments with dicoumarin removal by the perfused liver, Nagashima *et al.* (1968) advocate a more elaborate treatment; the equation that they suggest is similar in form to that discussed above:

$$C_2{}^c = C_1{}^c - V/V_0 \ (C_1 - C_2),$$

where $C_2{}^c$ = corrected plasma concentration,

$C_2$ = actual plasma concentration in that sample,

$C_1$ = actual plasma concentration in sample preceding,

$C_1{}^c$ = corrected plasma concentration in sample $C_1$ (or the actual concentration if this is initial sample),

$V$ = volume of perfusion medium before withdrawal of $C_2$, and

$V_0$ = initial volume.

The authors point out, however, that this correction may still not give a true rate of removal. The corrected line appears to deviate from the true after about two half-lives of the substrate. It does not deviate if the rate of elimination of the substrate (or drug) is constant. No further suggestions are made to correct for this additional error and it is suggested that calculations of rate of removal, particularly in such cases as exhibit an exponential fall in medium content, should be based only upon the early part of curves of removal versus time.

A further treatment of the case, for the clearance of iodoinsulin in a recirculation perfusion, is offered by Mortimore *et al.* (1959). It is assumed that (a) iodoinsulin is destroyed by the liver at a rate proportional to its instantaneous concentration in the blood presented to the liver, and that (b) the recycling rate is sufficient to bring about essentially instantaneous mixing between reservoir blood and the blood in the liver. Consequently the rate of decline in concentration of iodoinsulin in the blood is inversely proportional to the volume of the recirculating medium.

It is an advantage of modern enzymic techniques of analysis that serial analyses may be carried out on very small samples, so that errors arising from samples subtracted from the total medium are likely to be small. Clearly, very large samples, even if replaced with an equal volume of standard perfusion medium may introduce unknown variations into the conditions of an experiment, and when such large samples are required they should best be 'terminal' samples. This may in turn necessitate a change in experimental design, a group of perfusions of different duration being required to construct a time-curve of reaction, something that is normally achieved with a single perfusion. A similar limitation applies to large tissue samples.

Substrates added to the perfusion medium may be taken up and metabolized by the organ perfused in the same way as by tissue slices. The course of such reactions may be simply followed by regular sampling of the medium. The corrections that may be necessary are discussed above. There are various means of adding substrates to the perfusion medium which may lead to differences in results. The problem is usually associated with recirculation perfusion since in the once-through method, substrate added to the reservoir reaches the organ at a constant concentration. In recirculation perfusion, substrate may be added as a single dose (by analogy with most closed incubation methods), in which case the concentration of substrate reaching the organ is progressively decreased in proportion to the rate of its metabolism. When only the products of such reactions are followed, a fall in the rate of their appearance may reflect either 'failure' of perfusion, or exhaustion of the substrate. To determine this it may be necessary to follow substrate removal or to conduct a series of experiments to measure the $K_m$ of the substrate for this particular system. Experiments should then be carried out at a concentration above this figure. Empirically, it is sometimes simpler to test the response of the perfused organ to a second 'dose' of substrate (see, for example, Hems *et al.* (1966) and Schölz and Bücher (1965)), which may convincingly demonstrate the viability of the organ. Consideration of the other methods of adding substrate to maintain constant concentration in the in-flowing medium will be postponed until the methods of calculation of the loading-dose method have been discussed.

*Calculation of rate of metabolism of loading-dose*

Although very simple, the minor pitfalls of this method will be discussed in the light of an example (Hems *et al.* 1966). The required initial (loading) dose, for example 10 mM lactate, is calculated from the concentration determined in the initial sample and the known initial volume of the perfusion medium. Allowance must be made for any samples removed during the pre-perfusion (see above); more important is the timing of the initial sample if a true figure for the loading dose is to be obtained. Ideally, it should be taken early enough for there to have been no significant uptake, but late enough to allow for complete mixing within the circuit. This time depends upon the volume of medium, the rate of mixing, and the rate of circulation through the apparatus. A useful technique in an apparatus that permits this (for example, that of Hems *et al.*; see the Appendix) is to increase both the rate of stirring of the reservoir and the rate of circulation of medium by increasing the speed of the pump. This does not involve altering the rate of perfusion through the organ.

Despite these precautions, it is unusual to achieve a precise value for the initial substrate concentration in a recirculation perfusion, and for calculation of balance (see below) the dose added should be known. Subsequent samples, taken at the required intervals during a perfusion do not suffer from this problem, and may be corrected as discussed above. Plotting concentration against time allows the straight portion of the curve to be used to calculate a rate of reaction, a reliable time-course being obtained in each experiment. The problem of the doubtful first sample after substrate addition is thereby avoided. A fall in the rate of the reaction within a single perfusion is indicative of shortage of substrate (see above) and it may be necessary to determine the $K_m$ for a given substrate in the system (thus Exton and Park calculate a $K_m$ 1·2 mM for L-lactate in the perfused starved rat liver). In addition, since the rate of selected reactions may be used as tests of function (see p. 189), an unexpectedly low rate may reflect some abnormality in the conditions of perfusion.

*Continuous infusion of substrate or effector*

Under special circumstances, the concentration of a substrate or effector which is normally employed in incubation experiments *in vitro* is unsuitable for use in perfusion. The usual reason is that it has vasomotor effects which reduce the rate of perfusion below the permissible minimum (for example, palmitic acid in the liver). Some substrates may be 'toxic', resulting in failure of perfusion even when in incubation experiments there may be no indication of such toxicity. To circumvent this problem, substrates may be added at a low but constant rate by means of an infusion pump. This may be used in combination with a small loading-dose. In either case, the method results in a different form of experiment, in which the substrate concentration is kept nearly constant throughout the experiment. Where this, rather than the avoidance of toxicity, is the main reason for infusion of substrates, the rate of uptake of substrate, or removal of effector from the medium must be known from preliminary experiments. Even with accurate information on this point, it is almost impossible to achieve uniform concentration during an experiment and the method introduces some variation that is not found in the loading-dose method. Many workers are prepared to accept this, as it is a more physiological method of presenting substrates to the perfused organ (Schimassek 1968).

In one other situation, the continuous, or at least the repeated addition of substrate is necessary; this is the case where the substrate has very short half-life, either as a metabolite or due to chemical degradation, for example, oxaloacetate→pyruvate.

For accuracy, the rate of uptake in such a system should be calculated from determinations of substrate concentration in the medium at intervals during perfusion, together with the concentration × rate of addition of the stock solution; initial and final volumes of substrate solution should be determined as well as the rate of infusion, which may vary during an experiment. When the rate of formation of product(s) is the main concern of the experiment, no such problems arise. Provided that 'substrate inhibition' does not occur, the results of loading-dose and continuous infusion experiments are similar (compare Exton and Park with Teufel *et al.* (1967); refs. Chapter 3).

*Tissue*

Studies on the tissue perfused, separated from the medium, are possible and crucial to many of the special methods used in perfusion. Thus, balance studies, the determination of gradients between extra- and intra-cellular compartments, as well as studies on enzymes and overall function within the surviving organ are possible. Although most of these methods depend upon tissue sampling techniques, a number of methods have been introduced for studies at the organ surface and these will be discussed separately p. 121.

(i) *Tissue-sampling techniques.* Different methods of sampling may be applied, depending upon the subsequent fate of the tissue obtained.

Standard treatments are applied to samples required for enzyme studies (Colowick and Kaplan 1955), subcellular fractionation (Schneider 1948), mitochondrial preparations (Schimassek 1968), and slices or homogenates may be prepared in the usual way from tissue already studied in perfusion.

Electron microscopy, after rapid fixation in perfusion, has been discussed as an application of perfusion studies (see Torack 1969, Orth and Morgan 1962, Hamilton *et al.* 1967), and special techniques are established for this purpose.

In addition, perfusion has shown itself to be particularly well suited to the rapid tissue-sampling technique introduced by Wollenberger *et al.* (1960). This method, which is widely used in the study of tissue contents of unstable metabolic intermediates, involves the immediate cooling of tissue samples between two blocks of aluminium, pre-cooled in liquid nitrogen. The blocks are mounted in a large pair of tongs so that tissue is compressed very rapidly to a thickness of 1–2 mm, a dimension that allows cooling of the deepest cells within seconds. The method is described in detail by Wollenberger *et al.* (1960) and by Newsholme and Randle (1964) who applied it to the perfused rat heart.

As an adjunct to the cooling of tissue, a practical means of obtaining a working powder from which extracts may be made without the tissue warming sufficiently to permit metabolic changes within it, is the percussion mortar described by Neely *et al.* (1967). The details of this apparatus, which has found wide application in perfusion studies for metabolic purposes are shown in Fig. 2.4.

Freeze-clamping in perfusion has been applied by Morgan, Randle, and others to the perfused heart; by Schimassek, Hems, Exton and Park, and others to the liver; and by Nishiitsutsuji–Uwo *et al.* to the kidney. Crudely applied, the method must yield a terminal tissue sample, since it is impossible to continue perfusion. In the special case of the liver (and by extension, possibly also the lung, pancreas, or intestine) the lobular structure, with a discrete vascular supply to each lobe, allows an interim sample of tissue to be taken, and perfusion continues to the remaining tissue. Serial samples of liver have been taken in this way (Menahan *et al.* 1968), illustrating the advantage of perfusion in establishing constant experimental conditions. Some authors suggest that differences occur between lobes (Morris 1963, D. A. Hems, personal communication) and it is important to establish this fact for any tissue component of interest, since the design of experiments may have to be altered accordingly. However, there appears to

9

be no significant difference between the acetyl CoA content of different lobes of liver sampled during perfusion (Menahan *et al.* 1968).

In the case of the heart, with medium still flowing, the bung that supports the cannula(e) is lifted from the heart chamber, and the whole heart is quickly clamped with pre-cooled Wollenberger clamps. The clamps are immediately returned to liquid nitrogen to keep the

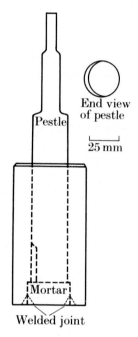

Pestle

End view of pestle

$\llcorner$_____$\lrcorner$
25 mm

Mortar

Welded joint

FIG. 2.4. Percussion mortar. Diagram of apparatus originally constructed by Neely, Liebermeister, and Morgan (1967). The apparatus is made of stainless steel. The top of the pestle fitted into an adaptor of the Skil Rotohammer. One or two grammes of frozen tissue can be homogenized at a time.

specimen cold. The subsequent handling of the heart sample is discussed in Chapter 5.

For the perfused kidney a similar procedure was adopted by Nishiitsutsuji–Uwo *et al.* (1967) with satisfactory results, comparable with those obtained from the kidney 'freeze-clamped' *in vivo*. The glass arterial cannula may not survive the clamping.

In the isolated liver preparation (see Preps. I and II, Chapter 3) the whole organ may be 'clamped' in the way described above. More elegantly, clamped immediately, single lobes may be excised either as a terminal or an interim sample. The results from the two methods are not significantly different (compare Hems *et al.* (1966) and Schimassek (1962); see Chapter 3) suggesting that the minimal delay involved in the latter and more generally useful method is acceptable. In the isolated preparation (see Chapter 3), there is no need to tie the base of the lobe, although this is desirable, as the medium flow rate then truly reflects the rate of perfusion. For liver *in situ*, however, it is essential to prevent loss of medium after sampling, if the perfusion is to continue. A loose loop of 3/0 silk is placed around the pedicle of the lobe chosen. Fig. 3.1 illustrates the position of the papilliform lobes, those most conveniently sampled by this method. With the help of an assistant to manipulate the clamps, the tip of the lobe is lifted in forceps, the lobe is cut across about 2 mm distal to the ligature, and transferred as rapidly as possible to the open clamps. The operator now tightens the prepared ligature around the base of the lobe, sufficiently to prevent further blood loss. The tension required is critical, since the friable liver is easily lacerated with a renewal of bleeding. The papilliform lobe provides some 100–200 mg wet weight of liver. If more is required, the outer half of the right main lobe presents itself as an easily accessible site and blood loss may readily be prevented by a ligature.

(ii) *Methods of expressing results.* It is frequently necessary to refer results to the amount of tissue perfused in order to compare a series of experiments and it is usual to weigh the organ at the end of a perfusion. For consistent results the organ should be handled in the same way each time, that is, drained free of medium under gravity and gently blotted, or freeze-clamped while still perfusing.

Results may now be expressed in a number of ways, referring for example to the 'fresh' weight of tissue perfused, the dry weight, the total protein, DNA or RNA. These are all standard procedures that depend upon simple analysis of the perfused tissue, a sample of which may be taken after weighing the whole.

In the special case of the freeze-clamped sample, perfusion medium is retained within the organ and results in a different wet/dry weight ratio from that obtained after blotting the organ. In these instances, as well as in those organs that gain weight through oedema during perfusion (for example, kidney and pancreas), results of metabolic reaction rates or tissue contents are better expressed on the basis of dry weight. Results may then be expected to be comparable with figures obtained *in vivo* and expressed in the same way.

There is some discussion about other means of expressing metabolic results. The conclusions are more relevant to the perfused organ than to tissue slices, homogenates, or sub-cellular fractions, where the relationship to the entire organ *in vivo* is lost. Morgan, Henderson, Regan, and Park (1961) comment that results are better correlated if hearts of equal weight are perfused, rather than hearts from rats of equal body weight. In the case of liver, Miller's group of workers suggests using the surface area of the donor rat (calculated from a formula of Lee and Clarke (1929) based upon body weight). This gave more reproducible results for the endogenous rate of urea synthesis of different livers than did liver weight, body weight, or liver protein. Most other workers with the perfused liver have preferred to express results on the basis of wet or dry weight (see Chapter 3; Schimassek, Hems *et al.*, Menahan *et al.*, and Exton and Park) and obtain remarkably consistent results for the rates of reactions from saturating doses of substrate. Since the wet/dry weight ratio for liver does not significantly differ from the liver *in vivo* when standard conditions of perfusion are used, there is no significant difference between results expressed per wet, rather than per dry weight. Some critical definitions—tissue content, concentration, or 'level'—are made by Hohorst, Kreutz, and Bücher (1959), while a method of correcting such results for the blood content of the liver is proposed by Holzer *et al.* (1956). The errors involved in this case appear to be very small—4–6 per cent for most substrates and enzymes—and no such correction has been applied in most published results of liver perfusion.

If no definite decision upon the ideal method of expressing results can be made, the need for consistency is self-evident; where possible, the animal's weight, as well as the wet and dry weights of the tissue perfused should be given. This allows comparison with other results in perfusion and with experiments *in vivo*. Additional information may include the protein content (or DNA), if comparison with other systems used *in vitro* is to be made.

The problem is illustrated by the results of Murad and Freedland (1967) *in vivo* and of Freedland and Krebs (1967) in the perfused liver. If normal and thyroidectomized rats are compared, glycogen content is higher in the former when expressed per 100 g body weight, but lower when expressed per gramme of wet liver. Freedland and Krebs (1967) comment that 'Since the weight of liver in relation to total body weight is 15–20% greater in thyroxine treated rats, the effect of thyroxine (on gluconeogenesis), is even greater when expressed per unit body weight than per unit liver weight.'

(iii) *Calculation of gradients.* If the tissue content of a substance is determined by such sampling methods as those outlined above, simultaneous sampling of the perfusion medium allows the existence of a gradient of concentration across the cell to be determined. Such information is of obvious importance in the interpretation of data that are based upon determinations on the medium alone. Experiments of this sort have been conducted in the perfused heart by Morgan *et al.* (1961), using glucose as a substrate. These authors discuss the extra information required for the true determination of tissue content. Calculation of intra- and extra-cellular space is undertaken, using sorbitol or chloride as markers, as is done *in vivo* (Davson 1970, Johnson 1966). Such work has not been carried out in other perfused organs but this would be required for the accurate interpretation of data on metabolic gradients.

Gradients of concentration between the different intra-cellular compartments are of increasing interest, but perfusion at present offers no special methods for their detection.

(iv) *Transport studies.* Perfusion offers special possibilities for the study of transport and metabolism, since analysis of various compartments allows the two processes to be distinguished. The heart has been studied in particular (Morgan *et al.* 1961). In the special sense applied to intestinal absorption, perfused intestinal preparations have been similarly used, but are not at present superior to earlier methods (for example, sacs or luminal irrigation *in vivo*). Temperature variation has been used to distinguish between uptake and metabolism of drugs in the perfused liver (Kalser *et al.* 1968) and heart (Lochner *et al.* 1968). Thus, disappearance of a drug from the perfusion medium at 4°C is taken to indicate transport (or binding), while the uptake at more physiological temperatures represents both transport and metabolism.

*Gas collection and analysis*

Finally, for balance studies in perfused organs, analysis is required of the gas which has been used to oxygenate the organ and which contains all the $CO_2$ released during perfusion. Problems in the total collection of gas during perfusion experiments are discussed in Chapter 1 (oxygenators). In a well-designed apparatus (for example, those of Morris and French 1958, Schimassek, and Miller), it should be possible to collect gases with great accuracy, so that losses of $CO_2$, in particular, are minimal. Using $^{14}CO_2$, up to 95 per cent of $^{14}C$ label in a perfusion system may be recovered (Miller *et al.* 1951, Exton and Park 1967). Unlike those used in manometry, perfusion apparatuses are usually 'open', and the volume of gas introduced is very large. This poses special problems in collection and analysis (see Chapter 1, p. 59).

## Balance studies

By combining the methods of sampling and analysis outlined on p. 101 it is possible to conduct total balance studies analogous to those available in closed manometric systems (see Umbreit *et al.* (1964)). The added advantage of separate analysis of medium and tissue has already been discussed. A good example of total analysis of this sort is provided by Chain, Mansford and Opie or Morgan and others (see Chapter 5), for the metabolism of $^{14}C$ glucose by the perfused heart. Teufel *et al.* (1967) undertook a similar analysis for $^{14}C$ pyruvate in the perfused liver. Provided that $CO_2$ analysis is not essential to the balance, there is no necessity to use isotopic methods, and the experiments of Hems *et al.* (1966) in the liver illustrate such balance studies.

## Studies based upon arterio–venous differences

A useful method applicable to the isolated perfused organ is the synchronous determination of arterial and venous concentration. When the flow rate of medium is known, arterio–venous difference allows calculation of the rate of uptake or release by the organ at that moment in time.

uptake (units/min) = flow rate (ml/min)
$$\times (\text{arterial conc./ml} - \text{venous conc./ml.})$$

Sequential determinations of arterial and venous concentration in a recirculation system, or of venous concentrations alone in a once-through system allows a time course of uptake or release to be constructed. The most useful metabolic parameter calculated on this basis is the oxygen uptake. With suitable apparatus for perfusion (see Oxygenators and Materials, Chapter 1), accurately reproducible values may be obtained, equivalent to those in a closed manometric technique.

### Methods of determining oxygen content and oxygen uptake

Continuous monitoring methods of oxygen analysis have been widely used. Alternatively, a number of standard discontinuous methods are available which have been applied to perfusion studies.

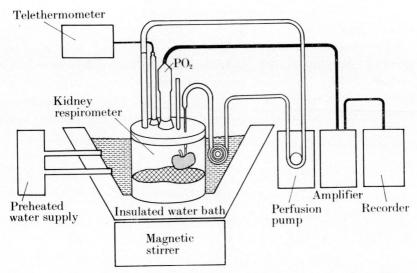

FIG. 2.6.   A polarographic respirometer. The isolated kidney is perfused in a sealed chamber containing a known amount of oxygen. Oxygen tension is recorded by the electrode mounted in the lid of the chamber. For details see Turner *et al.* (1968 and Chapter 4).

(Hems *et al.* 1966). Repeated venous oxygen determinations are then referred to this single value, and uptake calculated from the flow rate.

*Other measurements by arterio–venous difference*

All metabolic measurements upon the medium in a once-through system fall into this group. In addition, uptake studies have been conducted in recirculation perfusion by determining arterio–venous differences. In each case, an accurate estimate of flow rate is crucial to the calculation of rate of uptake or production.

**Measurements at the organ surface**

An interesting use of perfused organs has been the introduction of surface recording by physical methods which allow metabolic or physical parameters of organ function to be assessed without direct sampling and without interruption of perfusion. One in particular, the surface fluorimeter devised by Chance, Schoener, Krejci, Russmann, Wessmann, Schnitger, and Bücher (1965) has contributed metabolic information on the respiratory chain in the perfused liver, and in the perfused heart (Chance, Williamson, Jamieson, and Schoener 1965), which could not be obtained readily by any sampling technique. Most of the methods to be described briefly in this section have found application in the perfused liver but could be readily transferred to other organs.

(i) *Surface oxygen electrode.* Kessler and Schubotz (1968) describe a multiple platinum oxygen electrode which, when applied to an organ surface, is able to record oxygen tension simultaneously in eight independent, adjacent fields (see Fig. 2.7). The total area occupied by the electrode is a circle of 1500 $\mu$m diameter, so that disturbance to the

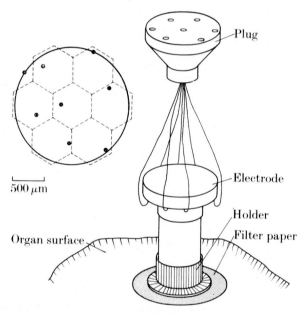

500 $\mu$m

Plug

Electrode

Holder

Filter paper

Organ surface

F IG . 2.7.    Surface oxygen electrode (Kessler and Schubotz 1968). An electrode with multiple recording points (inset shows the surface area covered) may be applied to the surface of an organ to record oxygen tension locally. The holder is applied to moist filter paper.

perfused organ is minimal. Till now this apparatus has been used for the technical purpose of establishing when perfusion is uniform, but in an already established technique its use might be foreseen as a monitor of oxygen tension under varying experimental and metabolic conditions.

The one firm conclusion resulting from studies with this electrode is discussed in Chapter 3, namely, that in haemoglobin-free perfusion, uniform perfusion and adequacy of oxygen supply to the tissue cannot be implied from measurement of an excess of oxygen in the outflowing medium. As a result, shunts have been postulated to be present in the liver (Kessler and Schubotz 1968) and kidney (Leichtweiss *et al.* 1969) but not in the heart (Huhmann, Niesel, and Grote 1967).

(ii) *Surface* pH *electrode and surface 'redox' electrode.* These may both have application in studies with perfused organs.

(iii) *Surface spectroscopy.* Grote, Huhmann, and Niesel (1967) have demonstrated the use of surface spectroscopy in the analysis of myo-globin in the perfused heart.

(iv) *Intra-cellular micro-electrode.* The technique of recording mem-brane potential across the cell wall in intact tissues is well established (Gesteland, Howland, Lettvin, and Pitts 1959). Micro-electrodes may be inserted with some certainty into individual cells, and the technique has been applied to the perfused liver (Claret and Coraboeuf 1968, 1970) and to the perfused rabbit thyroid (Williams 1966, see Chapter 6). Under some condition the results may be used as a test of viability of the perfused organ, but it is also envisaged that intra-cellular measure-ments and determination of membrane potential may have metabolic applications in these preparations.

(v) *Surface fluorimetry.* In haemoglobin-free perfusions it has been possible to follow the reduction and re-oxygenation of pyridine nucleo-tides by means of a fluorimeter poised above the organ surface (Chance

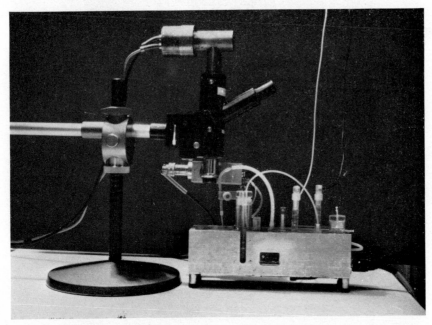

FIG. 2.8.   Fluorescence microscope (Scholz 1968). A binocular microscope ('Ultropak' Leitz, Wetzlar) is fitted with a mercury lamp (ST 40 Haraeus, Honau); primary filter at 366 $\mu$m, secondary filter, Wratten 2C with narrow band at 400 $\mu$m, and a photomultiplier (RCA 1 P21).

*et al.* 1965*a*, *b*). While this is a highly specialized method, and depends for its use upon haemoglobin-free perfusion techniques which may not be ideal (see Chapter 1 p. 32), the insight that such experiments have given into metabolic changes in anoxia and after addition of chosen substrates has been considerable. It must be included as one of the most useful special techniques in use with perfused organs.

Figs. 2.8 and 2.9, taken from Scholz (1968), illustrate the instrument in use with a liver perfusion system and technical details are given by Schnitger *et al.* (1965). Some of the findings are discussed in Chapter 3,

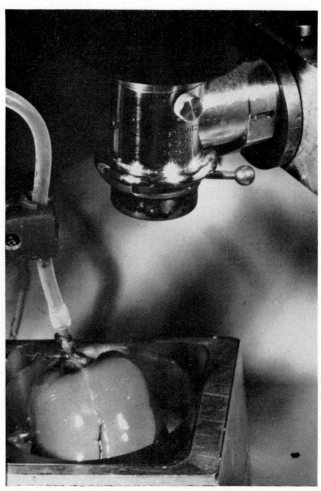

FIG. 2.9.    Fluorescence microscope in use with the perfused rat liver (Scholz 1968). The haemoglobin-free perfused rat liver is shown suspended from the portal vein cannula and resting in a temperature-controlled chamber.

where its use in assessing organ viability is discussed in relation to the liver. Provided the organ can be perfused with a haemoglobin-free medium and kept immobile during the experiment, there seems to be no limitation to its use in other systems.

(vi) *Other methods.* Isotope counting at the organ surface (Owen and Hazelrig 1966) has been used in studies on uptake of labelled substances from the perfusion medium (the continuous monitoring of isotopes in the perfusion medium has been used by Hillier (1968) in the perfused heart, and technical details are included in this reference).

Electron-spin-resonance and paramagnetic resonance techniques in use in analytical chemistry have been applied to the perfused liver by Tenconi *et al.* (1969).

## Methods applicable to specific organs

The range of special techniques applied to each organ perfused is discussed in the relevant chapters; many are related to the obvious physiological tests of function. In the heart, such methods as measurement of tension and force of contraction, ejection pressure, cardiac output, and electro-cardiography have been applied to perfusion (Chapter 5). In the kidney, a number of especially useful techniques based upon the micro-analysis of tubular contents includes micropuncture (Richards 1938) and a newer technique of micro-perfusion of tubules (Sonnenberg and Deetjen 1964). These techniques, which have been so productive in experiments *in vivo*, are equally applicable to the perfused organ (Bahlmann *et al.* 1967) and may contribute information on metabolic aspects of tubular function. In addition, clearance studies are a special feature of renal function which may be profitably studied in perfusion (Smith 1951, Nishiitsutsuji–Uwo *et al.* 1967). In the liver, the special function of the biliary tree may be studied in perfusion; bromsulphthalein clearance, bile formation, and secretion of drugs, and the oxygen tension in bile have all been studied.

### Altering the conditions of perfusion

Referring back to Fig. 2.1 the further range of experiment offered by varying the organ, the state of the donor animal, the composition of the perfusion medium, the temperature or the gas tensions (and possibly also the pH) will be briefly discussed.

(i) *Varying the organ.* Considerable evidence is available to illustrate that the expected variation in metabolism induced by diet (Ross *et al.* 1967; Nishiitsutsuji–Uwo *et al.* 1967) by hormone treatment (Freedland

and Krebs 1967), by the sex of the animal (Schriefers and Korus 1968), or by other manoeuvres conducted *in vivo* before removal of the organ for a perfusion experiment, is then observed under *in vitro* conditions of perfusion. The examples quoted all refer to the liver or kidney, but there is no reason to doubt that other organs will reflect the same pattern of variation, and experiments may therefore be based upon pre-treatment of animals in a variety of ways. Subsequently perfusion may be conducted, adding another set of variables *in vitro*.

(ii) *Varying the perfusion medium.* Enough has been said on this subject in Chapter 1 to make its importance obvious. To a large extent, the metabolic response of the perfused organ is dependent upon the composition of the medium (Schimassek 1968). An important example, in which this influence is used to illustrate the metabolic dependence of the organ upon its medium is the experiment by Khayambashi and Lyman (1969) with the perfused pancreas. By perfusing with serum from an animal fed on soya-bean, the pancreas can be induced to increase its output of amylase. Similarly, Gordon (1968) and other workers have been able to show that perfusion with blood from partially hepatectomized rats results in differences in protein synthesis in the normal liver under study. At a more defined level, the rate of a metabolic pathway may be altered by altering the composition of a semi-synthetic perfusion medium in minor, but specific ways (as, for example, by Hems *et al.* 1966, see Chapter 3, p. 175).

(iii) *Varying the temperature.* While in physiological or pharmacological experiments it may be acceptable to choose a convenient working temperature, the well-known temperature-dependence of biochemical reactions makes this a critical feature of perfusion conditions. It is usual to work at 37°C, but for specific purposes workers in the metabolic field have introduced other temperatures. As a result of the work of Kessler, quoted above, it is generally concluded that haemoglobin-free perfusion should be conducted at lower temperatures, when the oxygen consumption is significantly lower and can be met by the oxygen delivered in solution. The kidney perfusion technique of Turner *et al.* (1968) (see Chapter 4) appears to be limited to work at temperatures that allow the required oxygen for the period of an experiment to be enclosed in the chamber. The $Q_{10}$ of kidney is such that these conditions are met by working at 24°C. Nevertheless, 37°C remains the temperature of choice if metabolic data is to be correctly interpreted. In another related field, experiments at lower temperatures (4°C) have been used to distinguish between uptake and metabolism of a drug. As discussed above, the uptake or 'binding' is considered to be non-energy dependent and hence independent of temperature. 'Metabolism' is considerably reduced at these low temperatures (see Kalser, Kelvington, and Randolph 1968, for discussion). A special application of hypothermia in organ preservation for transplantation (Perry 1970) makes this an interesting field for metabolic study and special techniques of cooling for preservation; re-warming for use or for study are discussed by Carruthers *et al.* (1969). Most of this work is at present carried out with larger animals, but the rat may be more suitable (see Chapter 1, Introduction) for pilot and preliminary studies.

(iv) *Varying the oxygen or gas pressure.* Oxygen pressure may have metabolic effects, within the range 0–1 atm (see Abraham *et al.* (1968) and Chapter 3), and the possibility of conducting experiments under anoxic conditions is discussed for the liver (p. 205) and for the heart (p. 295). However, perfusion, unlike other incubations may be acutely sensitive to lack of oxygen and this limits experiments of this type (Chapter 3). By varying the composition of the gas entering the apparatus, oxygen tension and pH may be independently altered, allowing a wide range of experimental conditions (see Chapter 1, Table 1.4 for examples). By modifying the perfusion apparatus, and by including the whole within a pressure chamber, it is possible to conduct experiments at more than atmospheric pressure, and thereby to test the effects of hyperbaric oxygen. Studies of this type have been conducted in the heart (Dundas and White 1968) and the liver (Brauer *et al.* 1956) and are applicable to most perfusion systems.

Variation of gas composition in the perfused lung may be applied to the gas supplying the medium or to the gas used to ventilate (see Chapter 9).

## Tests of function

Metabolic methods have proved of special value in testing the viability of perfused organs, and in each section such tests are discussed. There remain a number of newer systems which have never been tested in this way, and highly specialized studies have been undertaken without confirming the viability of the organ perfused. Any of the methods of metabolic methods of measurement in perfused organs may be applied as a test of function, and the discussion of such tests is to be found in Chapter 3, p. 189, since it is in the perfused liver that most comparative work has been done.

SUMMARY

This chapter is intended to point out the range and ease of metabolic measurement available in perfused organs, and a stylized model of an organ perfusion has been used as a basis. Most of the work already published in this field fits into this pattern, but inevitably, new investigative techniques are being applied as the method comes into more general use. Specific discussion of experiments is out of place in this review of the method. However, some small but important features of design of apparatus and of experiments become apparent. Thus, access to the medium, possibly at more than one point in the circuit, and to the organ either throughout the experiment or at least at the moments chosen for measurement or sampling, and a satisfactory gas collection are essential if an apparatus is to be applied to metabolic experiment.

When considering a choice of method, either a choice between two methods of perfusion or between perfusion and some other method, it is important to consider the types of measurement to be made. From the preceding discussion it is clear that some types of study, especially those that require 'terminal' sampling and therefore a separate experiment for each point on a time curve, may be particularly time-consuming in perfusion, and may be more easily accomplished in another system used *in vitro*. Schimmel and Knobil (1969) are at pains to point out that by incubating liver slices very many more determinations are available in a given time than those obtained using the more laborious perfusion method of Hems *et al.*

However, it is the aim of this review to point out the special advantages of perfusion as a metabolic tool and these are considered to warrant the extra apparatus and surgical techniques involved.

REFERENCES

ABRAHAM, R., DAWSON, W., GRASSO, P., and GOLDBERG, L. (1968) Lysosomal changes associated with hyperoxia in the isolated perfused rat liver. *Exp. & Mol. Pathol.* 8, 370.

BAHLMANN, J., GIEBISCH, G., OCHWADT, B., and SCHOEPPE, W. (1967) Micropuncture study of isolated perfused rat kidney. *Am. J. Physiol.* 212, 77.

BAIN, W. H. and HARPER, A. M. (ed.) (1968) *Blood flow through organs and tissues.* Livingstone, Edinburgh.

BASAR, E., RUEDAS, G., SCHWARZKOPF, H. J., and WEISS, C. (1968) Untersuchungen des zeitlichen Verhaltens druckabhängiger Änderungen des Stromwiderstandes im Coronargefassystem des Rattenherzens. *Pflügers Arch. ges. Physiol.* 304, 189.

BLEEHEN, N. M. and FISHER, R. B. (1954) The action of insulin on the isolated rat heart. *J. Physiol., Lond.* **123**, 260.

BRAUER, R. W., LEONG, G. F., McELROY, R. F., and HOLLOWAY, R. J. (1956) Haemodynamics of the vascular tree of the isolated rat liver preparation. *Am. J. Physiol.* **186**, 537.

CARRUTHERS, R. K., CLARK, P. B., ANDERSON, C. K., and PARSONS, F. M. (1969) Tubular function demonstrated in rat kidneys after storage at −79 degrees C. *Br. J. Urol.* **41**, 186.

CHANCE, B., SCHOENER, B., KREJCI, K., RUSSMANN, W., WESSMANN, W., SCHNITGER, H., and BÜCHER, TH. (1965) Kinetics of fluorescence and metabolite changes in rat liver during a cycle of ischaemia. *Biochem. J.* **341**, 325.

—— WILLIAMSON, J. R., JAMIESON, D., and SCHOENER, B. (1965) Properties and kinetics of reduced pyridine nucleotide fluorescence of the isolated and *in vivo* rat heart. *Biochem. Z.* **341**, 357.

CLARET, M. and CORABOEUF, E. (1968) Effects of the variation of pH on the membrane polarisation of isolated and perfused rat liver. *C. R. hebd. Séanc. Acad. Sci., Paris* **267**, 642.

—— —— (1970) Membrane potential of perfused and isolated rat liver. *J. Physiol., Lond.* **210**, 137P.

COLOWICK, S. P. and KAPLAN, N. O. (eds.) (1955) General preparative procedures. In *Methods in enzymology*, Vol. 1, Chap. 1. Academic Press, New York.

DAVSON, H. (1970) *A textbook of general physiology*, 4th edn., Chap. 9, p. 536.

DUNDAS, C. R. and WHITE, D. C. (1968) The oxygen consumption of isolated, perfused rat heart at 1, 2 and 3 atmospheres. *Br. J. Surg.* **55**, 862.

EXTON, J. R. and PARK, C. R. (1967) Control of gluconeogenesis in liver. I. General features of gluconeogenesis in the perfused livers of rats. *J. Biol. Chem.* **242**, 2622.

FREEDLAND, R. A. and KREBS, H. A. (1967) The effect of thyroxine treatment of the rate of gluconeogenesis in the perfused rat liver. *Biochem. J.* **104**, 45P.

GESTELAND, R. C., HOWLAND, B., LETTVIN, J. Y., and PITTS, W. H. (1959) Comments on microelectrodes. *Proc. Inst. Radio Engrs* **47**, 1856.

GORDON, A. H. (1968) The limitations and special advantages of the perfused liver in relation to the synthesis and catabolism of the plasma proteins. In *Stoffwechsel der isoliert perfundierten Leber* (eds. W. Staib and R. Scholz), p. 90. Springer-Verlag, Berlin.

GROTE, J., HUHMANN, J., and NIESEL, W. (1967) Untersuchungen über die Bedingungen fur die Sauerstoffversorgung des Myokards an perfundierten Rattenherzen. II. Zur Funktion des Myoglobins. *Pflügers Arch. ges. Physiol.* **294**, 256.

HAMILTON, R. L., REGEN, D. M., GRAY, M. E., and LEQUIRE, V. S. (1967) Lipid transport in liver: I. Electron microscopic identification of very low density lipo-proteins in perfused rat liver. *Lab. Invest.* **16**, 305.

HEMS, R., ROSS, B. D., BERRY, M. N., and KREBS, H. A. (1966) Gluconeogenesis in the perfused rat liver. *Biochem. J.* **101**, 284.

HILLIER, A. P. (1968) The uptake and release of thyroxine and tri-iodothyronine by the perfused rat heart. *J. Physiol., Lond.* **199**, 151.

—— (1969) The release of thyroxine from serum protein in the vessels of the liver. *J. Physiol., Lond.* **203**, 419.

HOHORST, H-J, KREUTZ, F. H., and BÜCHER, TH. (1959) Über Metabolitgehalte und Metabolit-Konzentrationen in der Leber der Ratte. *Biochem. Z.* **332**, 18.

HOLZER, H., SEDLMEYER, G., and KIESE, M. (1956) Bestimmung des Blutgehaltes von Leberproben zur Korrektur biochemischer Analysen. *Biochem. Z.* **328**, 176.

HUHMANN, W., NIESEL, W., and GROTE, J. (1967) Untersuchungen über die Bedingungen fur die Sauerstoffversorgung des Myokards an perfundierten Rattenherzen. I. Zur Frage nach dem Vorliegen einer Gegenströmversorgung. *Pflügers Arch. ges. Physiol.* **294**, 250.

JOHNSON, J. A. (1966) Capillary permeability, extra cellular space estimation, and lymph flow. *Am. J. Physiol.* **211**, 1261.

KALSER, S. C., KELVINGTON, E. J., and RANDOLPH, M. M. (1968) Drug metabolism in hypothermia. Uptake, metabolism and excretion of $S_{35}$-. *J. Pharmac. exp. Ther.* **159**, 389.

KESSLER, M. and SCHUBOTZ, R. (1968) Die $O_2$-versorgung der hämoglobinfrei perfundierten Rattenleber. *Stoffwechsel der isoliert perfundierten Leber* (eds. W. Staib and R. Scholz), p. 12. Springer-Verlag, Berlin.

KHAYAMBASHI, M. and LYMAN, R. L. (1969) Secretion of rat pancreas perfused with plasma from rats fed soyabean trypsin inhibitor. *Am. J. Physiol.* **217**, 646.

KINTER, W. B. and PAPPENHEIMER, J. R. (1956) Role of red blood corpuscles in regulation of renal blood flow and glomerular filtration rate. *Am. J. Physiol.* **185**, 399.

LEE, M. O. and CLARKE, E. (1929) Determination of the surface area of the white rat with its application to the expression of metabolic results. *Am. J. Physiol.* **89**, 24.

LEICHTWEISS, H. P., LÜBBERS, D. W., WEISS, CH., BAUMGARTL, H., and RESCHKE, W. (1969) The oxygen supply of the rat kidney; measurements of intrarenal $pO_2$. *Pflügers Arch. ges. Physiol.* **309**, 328.

LOCHNER, W., ARNOLD, G., and MÜLLER-RUCHHOLTZ, E. R. (1968) Metabolism of the artificially arrested heart and of the gas-perfused heart. *Am. J. Cardiol.* **22**, 299.

MENAHAN, L. A., ROSS, B. D., and WIELAND, O. (1968) Acetyl CoA level in perfused rat liver during gluconeogenesis and ketogenesis. *Biochem. biophys. Res. Commun.* **30**, 38.

MILLER, L. L., BLY, C. G., WATSON, M. L., and BALE, W. F. (1951) The dominant role of the liver in plasma protein synthesis. *J. exp. Med.* **94**, 431.

MORGAN, H. E., HENDERSON, M. J., REGAN, D. M., and PARK, C. R. (1961) Regulation of glucose uptake in muscle. I. The effect of insulin and anoxia on glucose transport and phosphorylation in the isolated perfused heart of normal rats. *J. biol. Chem.* **236**, 253.

MORRIS, B. (1963) The metabolism of free fatty acids and chylomicron tryglycerides by the isolated perfused liver of the rat. *J. Physiol., Lond.* **168**, 564.

—— and FRENCH, J. E. (1958) The uptake and metabolism of [14]C labelled chylomicron fat by the isolated perfused liver of the rat. *Q. Jl. exp. Physiol.* **43**, 180.

MORTIMORE, G. E., TIETZE, F., and STETTEN, D. (1959) Metabolism of insulin $I^{131}$: studies in isolated, perfused rat liver and hind-limb preparations. *Diabetes* **8**, 307.

MURAD, S. and FREEDLAND, R. A. (1967) Effect of thyroxine administration on enzymes associated with glucose metabolism in the liver. *Proc. Soc. exp. Biol. Med.* **124**, 1176.

NAGASHIMA, R. and LEVY, G. (1968a) Effect of perfusion rate and distribution factors on drug elimination characteristics in a perfused organ system. *J. pharmac. Sci.* **57**, 1991.

—— —— (1968b) Effect of flow rate on the distribution kinetics from perfusate to a perfused organ. *J. pharmac. Sci.* **57**, 2000.

—— —— and SARCIONE, E. J. (1968) Comparative pharmacokinetics of coumarin anticoagulants. III. Factors affecting the distribution and elimination of bishydroxycoumarin (BHC) in isolated liver perfusion studies. *J. pharmac. Sci.* **57**, 1881.

NEELY, J. R., LIEBERMEISTER, H., and MORGAN, H. E. (1967) Effects of pressure development on membrane transport of glucose in isolated rat heart. *Am. J. Physiol.* **212**, 815.

NEWSHOLME, E. A. and RANDLE, P. J. (1964) Regulations of glucose uptake by muscle. *Biochem. J.* **93**, 641.

NISHIITSUTSUJI-UWO, J. M., ROSS, B. D., and KREBS, H. A. (1967) Metabolic activities of the isolated, perfused rat kidney. *Biochem. J.* **103**, 852.

OPIE, L. H. (1965) Coronary flow rate and perfusion pressure as determinants of mechanical function and oxidative metabolism of isolated perfused rat heart. *J. Physiol., Lond.* **180**, 529.

ORTH, D. N. and MORGAN, H. G. (1962) The effect of insulin, alloxan diabetes, and anoxia on the ultrastructure of the rat heart. *J. Cell Biol.* **15**, 509.

OWEN, C. A. and HAZELRIG, J. B. (1966) Metabolism of $^{64}$Cu-labelled copper by the isolated rat liver. *Am. J. Physiol.* **210**, 1059.

PERRY, V. P. (1970) Review of method of organ perfusion and organ culture. In *Microcirculation, perfusion and transplantation of organs* (eds. R. I. Malinin, B. S. Linn, A. B. Callahan, and W. D. Warren), p. 243. Academic Press, New York.

RABITZSCH, G. (1968) Koronarreperfusion isolierten Warmbluterherzen mit geringen Umlaufsvolumina und Kontinuierlicher Kontraktions und Koronarflussregistrierung. *Acta biol. med. germ.* **20**, 33.

RICE, A. J. and PLAA, G. L. (1969) The role of triglyceride accumulation and of necrosis in the haemodynamic responses of the isolated perfused rat liver after administration of carbon tetrachloride. *Toxic. appl. Pharmac.* **14**, 151.

RICHARDS, A. N. (1938) Croonian lecture. Processes of urine formation. *Proc. R. Soc.* **B126**, 398.

ROSS, B. D., HEMS, R., FREEDLAND, R. A., and KREBS, H. A. (1967) Carbohydrate metabolism of the perfused rat liver. *Biochem. J.* **105**, 869.

RUDERMAN, N. R., HOUGHTON, C. R. S., and HEMS, R. (1971) Evaluation of the isolated perfused rat hind-quarter for the study of muscle metabolism. *Biochem. J.* **124**, 639.

SCHIMASSEK, H. (1962) Perfusion of isolated rat liver with a semi-synthetic medium and control of liver function. *Life Sci.* **11**, 629.

—— (1968) Möglichkeiten und Grenzen der Methodik (erläutert am Beispiel verschiedener Perfusionsmedien). In *Stoffwechsel der isoliert perfundierten Leber* (eds. W. Staib and R. Scholz), p. 1. Springer-Verlag, Berlin.

SCHIMMEL, R. J. and KNOBIL, E. (1969) Role of free fatty acid in stimulation of gluconeogenesis during fasting. *Am. J. Physiol.* **217**, 1803.

SCHNEIDER, W. C. (1948) Intracellular distribution of enzymes. III. The oxidation of octanoic acid by rat-liver fractions. *J. biol. Chem.* **176**, 259.

SCHOLZ, R. (1960) Untersuchungen zur Redoxkompartmentierung bei der Hämoglobinfrei perfundierten Rattenleber. *Stoffwechsel der isoliert perfundierten Leber* (eds. W. Staib and R. Scholz), p. 25. Springer-Verlag, Berlin.

—— and BÜCHER, TH. (1965) In *Control of energy metabolism* (ed. B. Chance), p. 393. Academic Press.

SCHRIEFERS, H. and KORUS, R. (1968) Pattern of metabolites of the glucuronide fraction following perfusion of the liver of male rats with testosterone. *Hoppe-Seyler's Z. physiol. Chem.* **349**, 1391.

SMITH, HOMER W. (1951) second printing 1955. *The kidney: structure and function in health and disease.* Oxford University Press, New York.

SONNENBERG, H. and DEETJEN, P. (1964) Methode zur Durchströming einzelner Nephronabschnitte. *Pflügers Arch. ges. Physiol.* **278**, 669.

SPROULE, B. J., MILLER, W. F., CUSHING, I. E., and CHAPMAN, C. B. (1957) An improved polarographic method for measuring oxygen tension in whole blood. *J. appl. Physiol.* **11**, 365.

STARLING, E. H. and LOVATT EVANS, C. (1968) Principles of human physiology (eds. H. Davson and M. G. Eggleton), pp. 236. Churchill, London.

TENCONI, L., PROCACCIA, S., CIVARDI, F., and MASSARI, N. (1969) Perfusion of the isolated liver: changes in time of the electronic paramagnetic resonance (or ESR, electrospin resonance). *Minerva med., Roma* **60**, 2244.

TEUFEL, H., MENAHAN, L. A., SHIPP, J. C., BÖNING, S., and WIELAND, O. (1967) Effect of oleic acid on the oxidation and gluconeogenesis from [1-$^{14}$C] pyruvate in the perfused rat liver. *Eur. J. Biochem.* **2**, 182.

TORACK, R. M. (1969) Sodium demonstration in rat cerebrum following perfusion with hydroxy-adipaldehyde-antimonate. *Acta Neuropathol.* **12**, 173.

TURNER, M. D., NEELY, W. A., and PITTS, H. R. (1968) Polarographic respirometer for the isolated perfused rat kidney. *J. appl. physiol.* **24**, 102.

UMBREIT, W. W., BURRIS, R. H., and STAUFFER, J. F. (1964) *Manometric techniques*, 4th edn. Burgess Life Science Series, Minneapolis.

VAN SLYKE, D. D., RHOADS, C. P., HILLER, A., and ALVING, A. S. (1934) Relationships between urea excretion, renal blood flow, renal oxygen consumption and diuresis. The mechanisms of urea excretion. *Am. J. Physiol.* **109**, 336.

WEISS, CH., PASSOW, H., and ROTHSTEIN, A. (1959) Autoregulation of flow in isolated rat kidney, in the absence of red cells. *Am. J. Physiol.* **196**, 1115.

WILLIAMS, J. A. (1966) Effect of external K$^+$ concentration on transmembrane potentials of rabbit thyroid cells. *Am. J. Physiol.* **211**, 1171.

WOLLENBERGER, A., RISTAU, O., and SCHOFFA, G. (1960) Eine einfache Technik der extrem Schnellen Abkühlung grosserer Gewebestücke. *Pflügers Arch. ges. Physiol.* **270**, 399.

# Part 2: Perfusion of Individual Organs

# 3. Liver

THE liver was the first organ to be used as a biochemical tool when perfused through the vascular system. The report by Claude Bernard appeared in 1855:

> Pour cela, je pris un tube de gutta percha long de 1 mètre environs et portant à ces deux extremités des ajutage en cuivre. Le tube étant préalablement rempli d'eau, une de ces extrémités fut solidement fixée sur le tronc de la veine porte à son entrée dans le foie, et l'autre fut ajustée au robinet de la laboratoire de médicine du collège de France. En ouvrant le robinet, l'eau traversa le foie avec une grande rapidité . . .

This preparation was used to demonstrate that the liver, when washed free of blood, and apparently of glucose too, then released a further quantity of glucose if perfused after lying on the bench at room temperature overnight. The study of glycogen synthesis and release, and of glucose formation, still occupies a central position in modern uses of the isolated perfused liver (see review: Kestens 1964, Staib and Scholz 1968). More recently, the isolated perfused rat liver has come to be established as an important research tool in other biochemical investigations and may in some fields supplant the ubiquitous liver slice as the preparation of choice.

The introduction by Krebs (1968), at a meeting to discuss the biochemical uses of the isolated perfused liver (Oestrich am Rhein 1967) might be taken as the 'text' for the present chapter, . . .

> I would like to begin by explaining why I turned to the perfused whole organ after having used tissue slices, tissue homogenates, mitochondria and isolated enzyme systems for over 30 years. These various tissue preparations have proved satisfactory for the study of many aspects of metabolism, but they are of limited use in the study of the regulation of metabolic processes, especially the integrated control mechanisms of the functionally active organ. As is now known, key factors in these control mechanisms are the concentrations of intermediary metabolites and these depend on the intactness of the tissue structure. Metabolites are diluted on homogenization, and are washed out of slices and isolated mitochondria. This can cause major losses of metabolic activities and what remains may represent only a small fraction of the original capacity.

Miller's view (1961) of perfused liver is 'that functional performance of liver cells is best studied in the isolated liver, perfused with whole, homologous, continuously oxygenated blood under physical conditions closely approximating those in the intact animal'. He attributes earlier failure to (a) failure to carry out perfusion for more than $1\frac{1}{2}$–2 h; (b) the use of aqueous perfusion media such as Ringer's solution, in place of blood; (c) the lack of mechanical filters to remove tiny fibrin clots which quickly plug the hepatic circulatory tree, and (d), the relative unavailability of inexpensive, non-toxic anticoagulants, such as heparin. In addition, the use of reverse perfusion and of single doses of hormones in place of continuous infusion may explain the failure to observe effects of such hormones of the liver (Miller 1961).

Bücher's (1968) introductory remarks at the symposium takes a more extreme line: 'Without prejudice to reproducibility, the perfused organ may be subjected to conditions unacceptable to the physiologist, provided that they contribute to a clearer enunciation of the problem and simplification of the answers.'

In essence this is also Hechter's view, i.e. that perfusion must be adapted to the experiment rather than (slavishly) to reproduce physiological conditions (Hechter *et al.* 1951). Much of the remaining controversy concerning the technique of liver perfusion in the rat centres around the two points of view expressed by Bücher and Hechter on the one hand, and by Miller on the other. It is perhaps not surprising that the former authors suggest perfusion with red-cell free media, while the latter suggests a medium that includes rat blood. Schimassek (see review, 1968) and Krebs (see Hems *et al.* 1966) have done much to resolve the two extremes and to indicate the ways in which perfused liver may be of biochemical use. In the course of the past 20 years, since Miller reopened the subject of standardized technique in organ perfusion (Miller *et al.* 1951), many of the theoretical and practical problems encountered in such isolated organ preparations have been discussed. The reaction of the organ to perfusion, and comparison of the liver *in vivo* with preparations *in vitro* have been studied, often in great detail, and a large body of information is now available upon which to base the choice of method suitable for biochemical experiment. This information may be of great help in establishing perfusion techniques for other organs and for other species.

The analogy should not be taken too far, however, as liver has special features which dictate the form of perfusion. Its blood supply is divided between the portal vein and the hepatic artery. Such a high proportion of blood is derived from the portal vein (about 80 per cent in the dog (Fischer 1963), and 75–80 per cent in the rat (Abraham, Dawson, Grasso, and Goldberg 1968; Powis 1970)) that it is now generally agreed that perfusion via the portal vein alone suffices in the rat (see p. 212). As a consequence, the standard liver perfusion is a 'low-pressure' system; medium flows at the required rate when a pressure of only 20 cm $H_2O$ is applied to the inflow. This, in turn, has important consequences for the choice of apparatus and, more especially, in the choice of a perfusion medium. Consideration of Starling's law (see Chapter 1) suggests that with a low perfusion pressure, the colloid osmotic pressure of the perfusion medium which is required to prevent the formation of oedema is likely to be lower than that required in arterial perfusion system, such as kidney or heart.

The fact remains, however, that much that will be said of liver perfusion in the rat may be applied to other organ perfusion systems which are as yet not as firmly established in biochemical method.

## Anatomy

The liver, 5–10 g in weight in rats of from 150–400 g body weight lies in the upper right quadrant of the abdomen. Its relative weight varies with the age of the animal, and the liver is particularly large in new-born rats. Guinea pigs (10–15 g liver) and mouse (1–2 g liver) are sufficiently similar in size to be suitable for perfusion by the techniques to be discussed here. Pigeon liver, although anatomically quite different from that of the mammal, has also been used (Söling, personal communication).

The liver of the rat is divided into major left and right lobes by a groove, visible on its lower surface. Each major lobe is divided into an accessory and a main lobe. While the left main lobe is larger than the right, the left is the smaller of the two accessory lobes (Rubarth 1958). The right main lobe gives origin to two minor lobes—the caudate and the papilliform (or papillary) lobe. The papillary lobe is divided into two lobes of rather characteristic shape, so that they are readily recognizable in the perfusing liver. Fig. 3.1 illustrates this; a feature that is of importance when sampling of liver lobes is contemplated. Note that Green (1963) gives a rather simplified description of the anatomy of the rat liver, which is, however, less accurate than that given by Trowell (1942) or by Rubarth (1958), upon which this description is based.

A noticeable feature of the gross anatomy of the liver is the porta hepatis—an area of the inferior aspect of the liver, which receives the portal vein and the two main branches of the hepatic artery and is the area from which the two bile ducts emerge. No dissection of this area is required during the preparation of a liver perfusion, and none is desirable, since all the structures named are delicate and liable to be damaged.

*Ligaments.* The liver is held in position by ligaments, essentially condensations of the two layers of peritoneum. The falciform ligament suspends the organ from the lower surface of the diaphragm; the left coronary ligament from the upper border of the left lobe to the diaphragm (a right coronary ligament, from the right main lobe to the

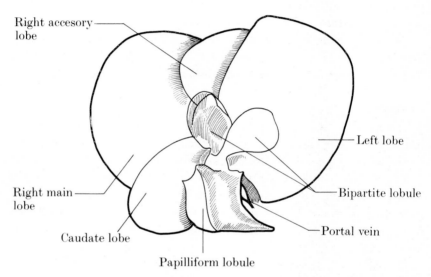

FIG. 3.1.   Rat liver viewed from below. The minor liver lobes are illustrated as suitable for biopsy with continuing perfusion.

diaphragm is absent in the rat, but present in the guinea pig); the gastro-hepatic ligament from the porta hepatis to the lesser curve of the stomach, and hepato-colic and hepato-renal ligaments passing from the liver to the colon and right kidney respectively. In the mouse only the falciform and a cauda-inferior vena caval ligaments are present.

The inferior vena cava lies between the caudate and papillary lobes (part of which lies behind the vein and can only be freed by cutting the overlying connective tissue condensation) and runs in a furrow against the posterior surface of the liver.

*Biliary system.* The liver has an excretory function, (see p. 146) and the bile pigments, salts, and other metabolic products appear in the fine system of bile canaliculi within the liver. The canaliculi coalesce to form at first fine bile ducts, and later a main hepatic duct which emerges from each main lobe at the portal hepatis. The common bile duct is formed by the union of these smaller ducts and this is now seen coursing across the ventral surface of the pancreas to reach the duodenum, where it opens into the lumen. The rat has no gall bladder and the flow of bile is uninterrupted. In other species, for example guinea-pig and mouse, a gall bladder is present as a sac into which bile passes before entering the duodenum. The composition of bile, especially its water content, is altered during its passage through the gall bladder, and this must be considered in liver perfusion in animals possessing one. The importance of cannulating the bile duct in perfusion studies depends upon the operative techniques adopted. In the isolated liver preparation to be described (Miller, Schimassek, and others) bile flows into the perfusion medium if the duct is not cannulated. This should be taken into consideration when interpreting results. It may indeed be preferable to recommend cannulation of the bile duct in all such studies, so that this variable is excluded. In the technique used *in situ* by Mortimore, Hems, and others, bile continues to drain into the duodenum if the duct is not cannulated, and it therefore presents no problem. Nevertheless, in this technique too, most investigators cannulate the bile duct for the sake of the extra information that may thereby be obtained.

## Blood supply

There are two sources of blood supply to the liver, the portal vein (80 per cent) and the hepatic artery (*c.* 20 per cent). The relationship between these two circulations within the liver is discussed in the following section. The portal vein arises from the main veins draining the spleen, pancreas, and gastro-intestinal tract (splenic, pancreatico-duodenal, superior and inferior mesenteric veins), most of which are described in Chapters 6 and 7. It carries deoxygenated blood at low pressure and in a steady non-pulsatile stream to the liver. The normal portal vein pressure in the rat is 12–14 cm blood (see Brauer, Leong, and Pessotti 1953). The hepatic artery arises from the coeliac axis, a branch of the aorta that arises just beneath the diaphragm. Fig. 6.2 illustrates its origins. Its course from there to the liver is of some importance. At its origin from the gastroduodenal artery it lies deep behind the portal vein, but it approaches the posterior aspects of the vein some 5 mm below the liver, and in the abdomen opened and prepared for portal vein cannulation (see Fig. 3.2) it lies parallel to and behind the portal vein. In this situation the hepatic artery can easily be included in ligatures which are passed around the portal vein.

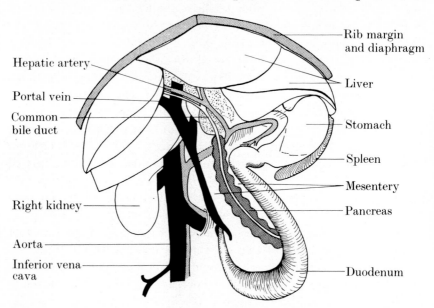

Fig. 3.2.  Abdominal contents exposed for liver perfusion. The liver and its related vessels are exposed through a mid-ventral incision. The intestine is diverted to the animal's left. Note especially the close relationship between portal vein and hepatic artery. The ligatures required are shown in Fig. 3.10.

This is of consequence if the hepatic artery circuit is to be perfused (Abraham *et al.* 1968, Powis 1970).

*Venous drainage of the liver*

Venous blood emerges from the sinusoids (see Histology) to enter the hepatic veins. These vessels have only a very short course outside the liver before entering the inferior vena cava just below the diaphragm. There is no second venous outflow analogous to the double inflow, and blood from these two sources is mixed by the time it enters the liver sinusoids (Hase and Brim 1966). The hepatic veins are multiple and are both too short and too friable for direct cannulation. The inferior vena cava has no tributaries of note between the renal veins below and its entry to the right atrium above (but see Trowell (1942) and Fig. 5.16 for exception) and can be cannualted either below or above the diaphragm to collect all of hepatic outflow.

*Lymphatic drainage of liver*

There is a copious lymphatic drainage to the liver, in the spaces of Disse and beneath the liver capsule, finally draining into a hepatic lymph trunk (Bollman, Cain, and Grindlay 1948). However, no studies of the lymphatics have been undertaken in perfused rat liver preparations and fluid losses by this route are probably too small to be of significance in short-term perfusions.

*Nerve supply*

Sympathetic and parasympathetic nerves supply the liver, largely as vasomotor nerves. However, in a recent study (Powis 1970) a metabolic function of nerve stimulation in an isolated perfused liver preparation was demonstrated and, in consequence, the hepatic nerve may assume somewhat greater significance in future studies. The nerve, which contains fibres from the coeliac ganglion (situated around the origin of the coeliac artery) runs with the hepatic artery. It may be stimulated by applying two electrodes directly to the artery (Powis 1970). Alternatively, the nerve can be isolated and separated from the artery, by displacing the spleen to the animal's left.

## Histology

Detailed histology of the liver, including the recent information obtained by electron microscopy, is discussed in monographs on liver (for example, by Schiff 1969, Elias and Sherrick 1969). There are minor species differences and Fig. 3.3 is from the dog liver. This illustration shows the hepatic arterial and portal venous blood supplies mixing before entering the sinusoids. Histologically, the liver is a continuous mass of parenchymal cells, tunnelled by vessels and Fig. 3.4 shows the current view of the arrangement of cells and sinusoids. Bile canaliculi occur at the interfaces between adjacent cells. At lower magnification there is a lobular structure based, according to the current view, either upon the portal triads (vein, artery, and bile duct) or upon the central vein (tributaries of the hepatic vein), with sinusoids radiating outwards, between plates of liver cells (Fig. 3.5).

Specific studies in rat liver (Gershbein and Elias 1954, Hase and Brim 1966) underline the differences in vascular anatomy between the rat and other species. Centrilobular radicles of the hepatic veins receive sinusoids and appear to be the skeleton about which the liver is constructed. Portal venous branches and hepatic artery branches travel together in the portal canals and have anastomoses by means of capillaries. The termination of such capillaries has not been closely examined. The studies of Hase and Brim (1966) indicated that medium (silicone rubber) injected into the portal vein does not enter the capillaries of the hepatic arterial system (a point of some importance in

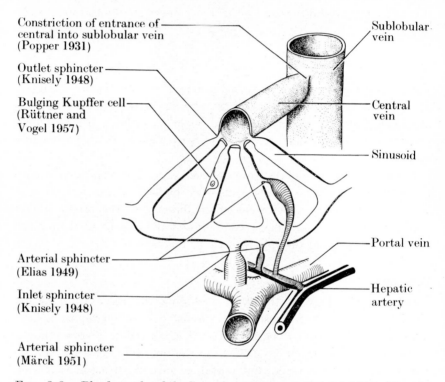

Constriction of entrance of central into sublobular vein (Popper 1931)

Outlet sphincter (Knisely 1948)

Bulging Kupffer cell (Rüttner and Vogel 1957)

Arterial sphincter (Elias 1949)

Inlet sphincter (Knisely 1948)

Arterial sphincter (Märck 1951)

Sublobular vein

Central vein

Sinusoid

Portal vein

Hepatic artery

Fig. 3.3.  Blood supply of the liver (from Elias and Sherrick 1969). The relationship between portal venous and hepatic arterial supply is shown; both systems supply the sinusoids which separate sheets of liver parenchyma cells.

portal vein perfusion of rat liver, as discussed below), while injection into the hepatic artery does fill part of this portal venous system and, in addition, easily fills the more central sinusoids.

*Cellular heterogeneity.* Apart from the liver parenchyma cells, there are flat 'litoral' or Kupffer cells, which line the sinusoids and have a phagocytic function. In addition, from time to time, there is a suggestion that the liver lobes differ from one another, so that different enzymes are thought to predominate in one lobe or another. There

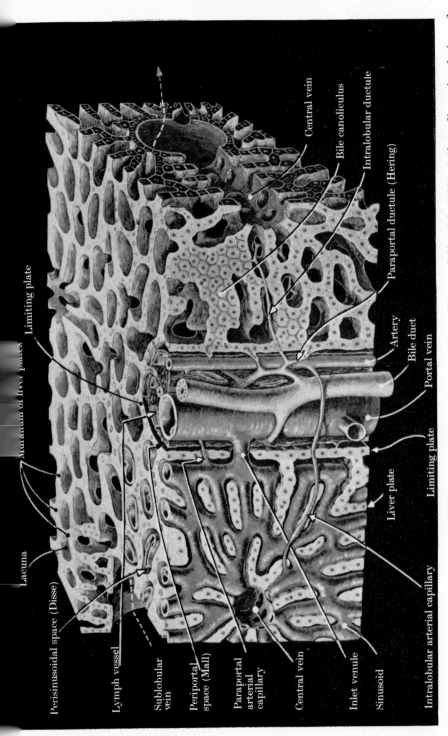

Limiting plate

Murarum of liver plates

Lacuna

Perisinusoidal space (Disse)

Lymph vessel

Sublobular vein

Periportal space (Mall)

Paraportal arterial capillary

Central vein

Inlet venule

Sinusoid

Intralobular arterial capillary

Liver plate

Limiting plate

Portal vein

Bile duct

Artery

Paraportal ductule (Hering)

Intralobular ductule

Bile canoliculus

Central vein

FIG. 3.4. Relationship between the biliary and vascular systems of liver (Elias and Sherrick 1969). A three-dimensional view of liver structure showing the origins of the biliary ducts.

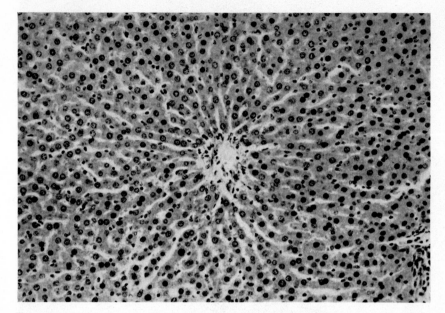

FIG. 3.5.    Rat liver on light microscopy. The lobular structure and radially arranged sinusoids are seen in this section of liver after 130 min perfusion by the method of Hems *et al.* (1966) (p. 173) (magnification ×50).

is little objective evidence for this, except from Morris (1963) (see p. 208), and sampling of liver lobes during or after perfusion does not have to be specially matched (see, however, p. 113).

## Physiology

The main aspects of liver function to be covered in this section are the blood supply of the liver, the pressure and flow in the two afferent systems, and the excretory function in the form of bile flow. In addition, the use of bromsulphthalein excretion as a test of hepatic function will be discussed.

### Blood flow

A detailed account of recent views on the physiology of hepatic blood flow is given by Fischer (1963) in Rouiller's treatise on liver; this is a useful source of information on hepatic microscopy and physiology and although much of the discussion centres on studies in man or the dog, it is generally applicable to the rat.

The hepatic artery provides only a minor part of the total blood which is received by the liver. The remainder, variously estimated at

66–86 per cent (Fischer 1963), is derived from the splanchnic bed, via the hepatic portal vein, and is relatively desaturated of oxygen. However, an oxygen content of only 3·3 volumes per cent below that in the arterial circuit is observed in the portal venous blood of the dog.

The liver therefore receives oxygen in considerable excess of its requirements from this source. This argument is developed further on p. 179 where the question of perfusion of the hepatic artery is discussed; it is concluded that, from the point of view of providing oxygen, perfusion of the portal vein alone is adequate. Nor is there any evidence, in the rat at least, that the hepatic artery is necessary for normal biochemical function.

Factors that influence the hepatic blood flow *in vivo*, and may therefore be of interest to investigators using the isolated perfused liver, include anaesthesia, posture, arterial blood pressure, nervous control, and various hormones (adrenaline, histamine) and are discussed at length by Fischer (q.v.). More relevant to the present discussion, is the work of Brauer, Pessotti, and Pizzolato (1951) who established the normal haemodynamic parameters of liver for the rat and compared their findings in an isolated perfused rat liver with those *in vivo*. Table 3.1 summarizes their findings (conditions of perfusion are given by Brauer *et al.* (1953)).

### TABLE 3.1

*Normal haemodynamic parameters of rat liver*

(from Brauer *et al.* 1953)

(1) Time course. Three major transitions are described:

animal→ perfusion

peak bile flow → steady state

steady state → terminal phase

(2) Flow-pressure relations

|  | Flow (cm³/min/g liver) | Pressure (cm blood) |
|---|---|---|
| *In vivo* | 1·0 | 12–14 |
| Perfused (with 'hepatic vasoconstrictor substance') | 3·0 | 12–14 |
| Perfused with whole blood | 1·0 | 5 |
|  | 3·0 | 8 |
|  | 6·0 | 12 |
|  | 7·0 | 14 |
|  | 12·0 | 18 |

II

*Bile flow*

Bile is a complex, isotonic solution containing pigments (derived from the breakdown of porphyrins), bile salts, cholesterol, and lecithins, in the form of a micelle and inorganic electrolytes. It is produced in the canaliculi and modified as it passes through the duct system. Discrete transport mechanisms are present which control the secretion of a wide range of substances, both physiological and foreign, into the bile, and there may be parallels between most of the function of the bile passages and that of the renal tubules (Sperber 1959). A variety of factors control and modify bile production *in vivo*. Some of these have been studied in the perfused rat liver (Clodi and Schnack 1966, Entemnan, Holloway, Albright, and Leong 1968), and give comparable results. Particularly, the rate of bile production and its salt and solid content appear to be normal and may serve as a test of liver function in perfusion (Brauer *et al.* 1951).

*Bromsulphthalein secretion*

Devised as a liver function test for use in man, the clearance of this dye from the blood and its appearance in the bile is a function which is present in the isolated perfused liver of the rat (Brauer *et al.* 1951). The rate of clearance, expressed as the half-life of BSP in the perfusion medium is similar in the perfused liver and in the liver *in vivo* (Table 3.2). It may be used as a test of liver function for perfusion studies, but no systematic investigation has been made using this parameter. Most workers (Bel 1969, McGraw 1968) have been content merely to report that clearance occurs in their preparation and more sensitive tests of liver function have been devised (see Chapter 3 p. 190). Other compounds which are secreted in the bile include phenolphthalein derivatives, hippuric acid derivatives, sulphonamides, and penicillin; some of these compounds have been studied in the perfused liver and their secretion may form the basis of functional tests.

*Phagocytosis*

The Kupffer cells lining the liver sinusoids have the capacity to engulf bacteria, inorganic matter, and other cells. This aspect of liver function too may be quantified (see Filkins and Smith 1965) and the half-life of carbon particles added to the medium has been used in the perfused organ as a test of function (Filkins, Chase, and Smith 1965).

### TECHNIQUE OF LIVER PERFUSION

No single method covers all the possible metabolic applications of liver perfusion, and four main methods will be described. They represent what most experimenters may require from an isolated perfused rat liver. One is a technique used *in situ*, which has now been well documented in biochemical terms. The second is the fully isolated liver technique, which is rather more widely used than the former and has therefore a copious literature. As should be convincing from the discussion, there are no metabolic differences between these two approaches to perfusion of the liver and their individual importance rests upon purely technical considerations, such as the ease of preparation, sampling techniques, and the particular study that is to be undertaken. There remains the problem of presenting and discussing, often to reject, the innumerable modifications to which these two basic methods have been subject. In the following section the methods of the most commonly cited authors are described briefly, with reference to the two basic methods described above. It should be mentioned that this does not imply chronological precedence, but is merely a convenient method of classifying and ordering the numerous methods described.

### Classification of methods of perfusion

On an arbitrary basis, the numerous methods offered in the literature may be reduced to those introducing differences in principle. Apart from the physical characteristics—isolated liver, liver *in situ*, or 'reverse perfusion' through the inferior vena cava—the introduction of defined, semi-synthetic perfusion media by Schimassek (1962) appears to have opened the method to a breadth of biochemical investigation not hitherto considered. This method is therefore considered in addition to the basic, isolated preparation of Miller (see Method IA).

### Methods of perfusion

I. Isolated liver
   A. General technique. Method—Miller *et al.* (1951).
   B. Modification of Schimassek (1962–3); semisynthetic, defined perfusion medium.
II. *In situ* liver (Mortimore 1959). Method—modification of Hems *et al.* (1966).
III. Reverse perfusion. Method—Trowell (1942).
IV. Hepatic artery perfusion technique.

**Method 1A. Isolated liver perfusion technique** (Miller *et al.* 1951)

In the liver perfusion technique of Miller *et al.* the organ is perfused at a constant hydrostatic pressure with a medium of diluted rat blood which has passed through a multibulb oxygenator. It enters the portal vein and drains, either from the cut inferior vena cava, or from a cannula inserted into it, and thence into a reservoir. From the reservoir it is pumped through filters to the top of the oxygenator. In this group of methods, the liver is removed from the animal after the cannulation procedure and perfusion is conducted with the liver in an enclosed organ chamber. It therefore differs from the methods described in Method II in which the proposal of Mortimore and Tietze (1959) is followed and the organ is perfused in isolation, but *in situ* (p. 165 *et seq*).

*Apparatus*

The apparatus designed by Miller has been almost universally adopted by workers using this technique, and it is also available commercially (see Fig. 1.10). Fig. 3.6 illustrates Miller's apparatus. Few workers now use the piston type (Dale–Schuster) pump illustrated in this figure, and in the commercial apparatus a standard roller pump has been substituted (compare Fig. 1.10). Later modifications by Miller's group incorporated the rotary pump of Crisp and DeBroske (1952). There is provision, illustrated here (Fig. 1.10) to pass gas first into the lower reservoir, whence it passes into the multibulb oxygenator. Incoming gas thereby flushes accumulated $CO_2$ from the lower reservoir and should ensure more efficient total gas collections (Miller, Burke, and Haft 1955). Despite this precaution, Miller *et al.* report up to 20 per cent of the $^{14}CO_2$ unaccounted for in experiments using substrate quantities of $^{14}C$ lysine. Their suggestion (based upon an observation of Rostorfer, Edwards, and Murlin 1943) that $CO_2$ is lost through the liver capsule cannot be accepted without reservation since Exton and Park report 95 per cent $CO_2$ recovery (see Chapter 2).

*Oxygenator.* The multibulb glass oxygenator used in this apparatus is described in Chapter 1. Miller gives no dimensions and the description is based upon diagrams, and a photograph reproduced more recently by Gans and Lowman (1967). The column is long and slim, with almost eliptical bulbs; workers using this method have sometimes experienced difficulties in producing a suitable oxygenator with the properties that

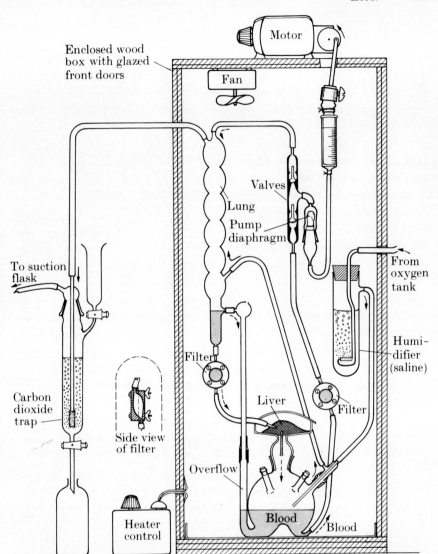

FIG. 3.6.   Liver perfusion apparatus of Miller *et al.* (1951). The apparatus, enclosed in a temperature-regulated cabinet, is described in detail in the text.

Miller described (see, for example, Fisher and Kerly 1964). In Chapter 1 the dimensions are given of an alternative and probably more efficient multibulb oxygenator used by Hems *et al.* (1966) (p. 50).

*Organ chamber*. The liver is contained in a glass organ chamber as illustrated, and covered with a watch-glass to minimize evaporation from the organ surface. The chamber is positioned directly over a reservoir into which outflowing medium drops.

The requirements of the organ chamber are that the liver should be supported with the minimum obstruction to blood flow, and that it should be enclosed or covered by some means to prevent its drying in the warm atmosphere of the perfusion cabinet. Other precautions may be taken, such as the injection of steam into the perfusion cabinet to maintain a high humidity as, for example, Kavin, Levin, and Stanley (1967). Miller covers the chamber with a simple watch-glass; others cover the chamber with damp gauze or with Parafilm. More important is the means by which the liver is supported within the chamber. Ideally the portal vein cannula is held in a movable clamp (for example, a ball joint) at a height a few centimetres above the chamber, so that the liver lobes hang down freely. The inferior vena cava cannula then hangs free for most of its length and there is no danger of 'kinking'.

In discussion of the relative merits of isolated techniques and those used *in situ* (see p. 194) the extra problems introduced by having the liver hanging free in this way are of more importance. Since perfusion experiments last for several hours and it is a great advantage to be able to leave the organ unattended and to know that changes in perfusion rate, unevenness of perfusion, and other mechanical features will not bring the experiment to a premature end. As in other perfusion experiments (e.g. kidney), 'tricks' whereby this is achieved develop within research groups and do not usually appear in the published method; a useful idea in this context is the magnetic cannula support (see Blakeley, Brown, and Ferry (1963) and Powis (1970)) in which the cannula support, mounted on a small magnet, can be slid over the surface of a vertically placed steel sheet to give the required tension or angulation to the cannula and vessel.

*Cannulae.* For the portal vein a polythene cannula of o.d. 2 mm is suitable. (The metal Frankis–Evans cannula used in the *in situ* method is too unwieldy for use here.) For the inferior vena cava the polythene cannula of o.d. 3 mm recommended by Hems *et al.* (see Method II) may be used here. The bile duct was not cannulated in the original method of Miller *et al.* (1951), but can be cannulated with PP10 polyethylene tubing (see Method IB).

*Filter.* Efficient filtration of the blood-perfusion medium is considered to be crucial to successful perfusion in Miller's hands (see Miller

(1961)). Two discs of Lucite are compressed together to hold a disc of white silk, $100 \times 150$ mesh per inch, and two such filters are incorporated, one below the oxygenator and the other beyond the pump.

*Temperature control.* Miller (1951) recommends the use of a controlled-temperature cabinet, in this case a Perspex cabinet with a filament heater applied to one wall and a fan that ensures the even distribution of the air thus warmed. The temperature is maintained at 38–40°C. The design details of this cabinet are included in Fig. 1.10 (Metalloglass Ltd.), while those of an alternative cabinet described by Hems *et al.* (1966) are also given in some detail in the Appendix (see, in addition, Chapter 1, p. 76). Temperature control by water-jacketing of the individual components of the apparatus (for example, Staib, Staib, Herrmann, and Meiers 1968), or by using the standard apparatus in a climatically controlled room (for example, Gans and Lowman (1967)) have been described.

## Perfusion medium

Miller used a medium of fresh, heparinized rat blood, usually diluted to an haematocrit of 25–40 per cent with Ringer's solution. Minor variations in this medium have appeared in the course of the publications from Miller's laboratory and these are included here as alternative media. The more significant variations introduced by other workers with this method are discussed on p. 197, while the semi-synthetic perfusion medium introduced by Schimassek is treated as a special case in Method IB.

*Medium* I (Miller *et al.* 1951, Miller, Burke, and Haft 1955). Fresh rat blood, obtained by cardiac puncture (see p. 33) in ether-anaesthetized animals after 18 h starvation is used. To 130 ml blood is added 12 ml of a heparin solution containing 5 mg heparin/ml of Ringer's solution; this is then diluted by one-third with Ringer's solution, the resultant haematocrit varying between 25 and 40 per cent. When the medium is prepared in this way the haemoglobin content may vary equally widely.

Medium I is gassed with 100% $O_2$, while 95% $O_2$ plus 5% $CO_2$ is suggested by the authors only if large amounts of lactate, pyruvate, or α-oxoglutarate are added as substrate in order 'to keep the pH down' (Miller *et al.* 1955).

*Medium* II (Green and Miller 1960). For use in 'fed rats', this medium is basically as Medium I, two parts heparinized rat blood plus one part Ringer's solution, to a total volume of 91 ml. To this is added 500 mg/D-glucose (approximately 35 mM glucose final concentration).

*Medium* III (John and Miller 1969; Goldstein, Stella, and Knox 1962). In later experiments in which protein synthesis or enzyme induction is to be studied, authors have used an amino-acid supplement:

Amino acids (in mg) added as constant infusion to a volume of 100 ml perfusion medium, during 12 hours perfusion. L. asparagine* 34·2, threonine 16·4, serine 19·7, glutamine* 42·2, L. proline 10·6, alanine 9·4, glycine 19·2, valine 12·6, L. methionine 3·6, iso-leucine 16·2, leucine 21·7, tyrosine 11·3, L. phenylalanine 12·4, lysine 17·0, histidine-HCl 10·9, L-arginine HCl 35·4, tryptophan 8·0, prepared as a mixture = 312·3 mg plus cysteine-HCl 7·7 mg = 320 mg total.

* L. asparagine and L. glutamine replace aspartic and glutamic acids.

An alternative semi-synthetic medium described as 'full-supplementation' conditions, is described by John and Miller:

*Medium* IV. *Full supplementation medium* (John and Miller 1969)

For 100 ml medium: washed bovine red cells 38 ml, Krebs–Ringer bicarbonate 50 ml, bovine serum albumin (Armour Fraction V) approx. 3 g, glucose 100 mg, heparin 10 000 units, penicillin 3 000 units, streptomycin 0·5 g.

Hormones: insulin 5·1 units, bovine growth hormone 0·5 mg, cortisol 5·0 mg.

Further, a continuous infusion of 19 ml Ringer solution, at 1·5 ml/h, with the following ingredients completes the additions: glucose 500 mg, penicillin 3000 units, streptomycin 3·0 mg (L-lysine-1-$^{14}$C HCl, 15 $\mu$Ci (0·62 mg)), insulin 6·8 units, bovine growth hormone 1·0 mg, cortisol 5·0 mg, plus 320 mg amino-acid mixture as above (Medium III).

*Operative procedure*

Wistar rats (250–350 g), starved for 18 h before use are anaesthetized with ether (or, in some experiments, with 50% $CO_2$ in oxygen). With the animal lying on its back, the limbs are fixed in extension, the anterior abdomen and chest are wetted with spirit or shaved free of hair and a mid-line abdominal incision through the skin is made. The incision extends from the pubis to the upper chest and is followed by a similar incision through the linea alba to expose the abdomen (see p. 171 for detailed procedure). This incision is extended laterally between clamps and the intestine thus exposed is diverted to the animal's left. Swabs moistened with saline are applied to the exposed viscera and used to define the operative field. Ligatures are placed around the gastrohepatic and the gastroduodenal ligaments, two around the latter to include the pancreatico-duodenal vessels, and the ligaments are divided. In addition, other connections of the liver, for example, a fine ligament to the inferior vena cava, are divided; these preliminaries allow the liver to be removed with the minimum of delay when cannulation is complete.

The remaining dissection is described in more detail on p. 172 and does not differ from the operation required for perfusion *in situ*, except in the last stages. The following ligatures are positioned.

(i) Around the common bile duct, seen as it lies on the anterior surface of the pancreas (note that Miller's original description does not include cannulation of the common bile duct, and described in addition the isolation and ligation of the portal lymphatics. These vessels are not seen in the normal course of events and do not require special attention). The bile duct is cut beyond this ligature.

(ii) Around the portal vein; three loose ligatures are placed as shown for Method II (Ligs. 3, 4, and 5 in Fig. 3.10).

(iii) Around the inferior vena cava, below the liver and above the right renal vein (Lig. (1) in Fig. 3.10).

Lig. (1) is tied and the inferior vena cava cut below. The distal ligature on the portal vein is tied and the vein is cannulated (see p. 171 for closer detail). The remaining two ligatures are tied to hold the cannula in place and the distal ligature, tied around the cannula, helps to prevent the vessel from kinking.

The rat is turned to bring the head to the operator's right, the thorax is widely opened (see p. 171), and the heart, including the right atrium and the thoracic vena cava are exposed. A loose ligature is placed around the inferior vena cava and the outflow cannula is inserted through the right atrium (Lig. 6, Fig. 3.10). (Some accounts describe the direct cannulation of the thoracic vena cava, but this procedure is more difficult, time-consuming, and more likely to fail than the alternative suggested.) The Lig. 6 is tied to hold the cannula in place.

*To isolate the liver*. The inferior vena cava is cut between the heart and the cannula. In Miller's description the liver is dissected out, together with the diaphragm; other authors suggest an incision through the anterior diaphragm leaving only a collar of the diaphragm attached to the inferior vena cava. The liver, with or without the diaphragm is then lifted free of the abdomen, together with its cannula and transferred to the organ chamber, where, once the portal cannula is attached to the apparatus, perfusion begins at once. Removal of the liver requires skill and should be accomplished with the minimum of handling of the organ.

In later work (Haft and Miller 1958) it is suggested that the degree of trauma to which the organ is subjected at this stage may influence the results of metabolic experiments. Be this as it may, it is certainly preferable that the liver is not handled, and some authors employ an assistant at this stage to lift the liver lobes gently with spatulae of plastic or glass during the dissection. Adjustments of the cannulae requires further attention if optimal flow rate and uniform perfusion of the liver are to be achieved. The portal cannula is best held a few (3–4) cm above the organ tray so that the lobes of the liver hang freely and the outflow cannula is not occluded. In apparatus without a flow-meter, the drop rate from the inferior vena cava cannula is used as a measure of perfusion rate (Miller *et al.* 1951) and the cannula should in consequence be held just a little higher so that drops form freely and fall far enough to be counted by eye. In Miller's hands, the operative time was 25–35 min, with an anoxic period of from 6–8 min. To avoid the lengthy period without flow, other authors have instituted a flow during the operation (Söling, Willms, Friedrichs, and Kleineke 1968), but somewhat improved operative technique and attention to speed can so reduce the period of anoxia that this precaution is of limited value (see Schimassek, p. 161).

*Characteristics of perfusion*

At the hydrostatic pressure chosen for these experiments 20–25 cm blood, the flow rate is 1·0–3·0 ml/min per g of liver, i.e. the average 250 g rat, with 8 g liver, gives a flow between 8 and 24 ml/min. With this pressure and flow rate, Miller reports successful perfusion for from 6 to 9 h. Macroscopic and microscopic appearances are taken to confirm the success of the perfusion.

Macroscopically, the liver is reported to appear quite like the organ before perfusion, except for (a) small peripheral areas in which wrinkling of the surface suggests drying, (b) scattered and occasional petechiae, 1–3 mm in diameter, (c) rarely, an apparently collapsed, small proximal lobe. Oedema is not a noticeable feature of this preparation. Microscopically, no difference is detected on haematoxylin and eosin, Sudan IV, or Bauer–Feulgen stained sections between the perfused and non-perfused liver, although the petechiae ((b) above) can be located.

Electron microscopic appearances are not recorded by Miller, but are discussed on p. 200. Much of the work of Miller's group has been directed to establishing the function of the perfused liver preparation in a qualitative and a semi-quantitative way. The limitations of this approach are discussed on pp. 164 and 189; but the evidence is overwhelming that the liver perfused in this way is capable of all the physiological and metabolic functions known of the organ *in vivo*. Some of the tests applied by Miller and other workers who have accurately followed the method here described are as follows.

(a) $^{14}C$ lysine is incorporated into protein which appears in the medium; the process is not linear with time and is therefore difficult to quantitate and to apply as a test of function.

(b) $^{14}CO_2$ appears when $^{14}C$ lysine is added to the perfusion medium.

(c) L-lysine is removed from the medium, exponentially with time over the course of 6–7 h perfusion.
There is a lag period in each of the events (a)–(c) which is a further disadvantage in applying these as tests of liver function or viability.

(d) The rate of protein synthesis (measured as protein which appears in the perfusion medium) is increased when a mixture of amino acids is added to the perfusion medium (see Medium III). This is a non-linear effect and cannot be easily quantified as a test of function. As a specific test of liver viability after a period of perfusion, Miller suggests the response of the organ to a second addition of amino acids. After the first addition protein synthesis increases but reaches a plateau; the second addition, at 4 h, results in an almost linear increase in the rate of protein synthesis which continues for 3 h. This response is not preceded by a lag period and the rate is easily determined.

*Bile production.* The bile duct was only cannulated in a small proportion of the experiments reported by Miller, and no time course of bile formation is given. The rate, 0·25–0·5 $\mu l$/min per g of liver appears to be lower than that reported by other workers, and this section will therefore be based upon the more detailed observations of Brauer *et al.* (1951) in a preparation which is probably interchangeable (see

p. 184 for method). Table 3.2 records the principal features of biliary function in the rat liver, perfused with homologous blood.

TABLE 3.2

*Physiological function in the perfused rat liver*

(from Brauer *et al.* 1951)

| | Isolated liver | *in vivo* |
|---|---|---|
| Body weight (g) | $275 \pm 8\cdot9$ | 250–310 |
| Plasma volume (cm³) | $72\cdot1 \pm 1\cdot27$ | 9·6–11·8 |
| Haematocrit (%) | $11\cdot3 \pm 0\cdot34$ | 47 |
| Plasma protein (g/100 ml) | $4\cdot8 \pm 0\cdot19$ | 5·6 |
| Perfusate through liver (cm³/min) | $9\cdot3 \pm 5\cdot8$ | 8–12 |
| Glucose in perfusate (mg %) | $210 \pm 13\cdot7$ | 115–140 |
| Bile flow (µl/min)    Male | $6\cdot4 \pm 0\cdot60$ | 17·3 |
|                     Female | $5\cdot1 \pm 0\cdot54$ | 12·0 |
| Bile solids (g %)    Male | $2\cdot7 \pm 0\cdot2$ | $3\cdot2 \pm 0\cdot3$ |
|                     Female | $3\cdot6 \pm 0\cdot3$ | $5\cdot2 \pm 0\cdot29$ |
| Secretion pressure of bile (cm bile) | $14\cdot6 \pm 0\cdot82$ | $20\cdot3 \pm 1\cdot1$ |
| Time to remove 50% of BSP from circulation (min) | $17\cdot0 \pm 2\cdot5$ | $5\cdot6 \pm 0\cdot5$ |
| BSP in plasma; (mg/litre) with continuous infusion of 0·2 mg/kg/min | $24\cdot7 \pm 4\cdot7$ | $17\cdot4 \pm 3\cdot1$ |
| BSP in bile (g/litre); conditions q.v. | $5\cdot7 \pm 0\cdot59$ | $4\cdot1 \pm 0\cdot42$ |

### Synthetic reactions

*Urea synthesis.* (a) *Endogenous.* The rate of urea synthesis without supplementation of the perfusion medium is initially rapid, but falls after 1 h, presumably when the endogenous substrates and those substrates present initially in rat blood are exhausted. The results illustrated by Miller are consistent but show a wide scatter in the rate. Only in perfusion of diabetic rat liver, or perfusion of normal livers when amino acid supplements are added to the medium is there a linear rate of urea synthesis with time upon which a quantitative test of liver function might be based. (b) *Maximum rate from added substrates.* Staib *et al.* (1968) find the *maximum* rate of urea synthesis to be the same, whether the semi-synthetic medium of Schimassek, or a rat-blood medium as suggested by Miller (Medium I) is used. Rates are linear and thus provide a good test of liver function.

*Glycogen synthesis* (Sokal, Miller, and Sarcione 1958). While conclusively demonstrating a net synthesis of glycogen in serial samples of liver taken during the course of a perfusion, an observation which has been difficult to reproduce in other perfused liver preparations, these workers fail to provide a quantitative basis upon which a test of liver function might be based. The results were achieved by using a continuous infusion of glucose, and glycogen synthesis was linear in only one of the experiments illustrated. Hence rate of synthesis and stoichometry of the reaction remain difficult to assess. For later comparison, the rate of glycogen synthesis in the one 'linear' experiment mentioned gives a rate of approximately 0·5 $\mu$mol/g per min, in the liver of rats starved for 24 h, perfused with a medium containing insulin and 22–23 mM glucose.

*Effect of hormones*. Qualitative effects of glucagon, adrenaline, and insulin in the perfused liver are discussed by Miller, with reference to the rate of endogenous urea synthesis. These effects might serve as a test of liver viability if carried out during the course of perfusion. The effect of glucagon in producing a regular and dramatic release of glucose by glycogen breakdown in the fed rat liver (Sokal *et al.* 1958) might also be a test of function.

*Clearance studies*. Two methods of 'clearance' of the perfusion medium form the basis of tests of liver viability in perfusion. Phagocytosis is responsible for the removal of [198]Au (Jeunet, Shoemaker, and Good 1967) and of thorotrast (Agarwal, Hoffman, and Rosen 1969) and of carbon particles (Filkins *et al.* 1965) from the medium and the rate of disappearance from the medium has been used as a test of function. Clearance of curare (Meyer and Scaf 1968) is inhibited by strophanthin and probably represents a different and energy-dependent transport into bile. As such, the rate may be taken as an index of function. Penicillin clearance (Kind, Beaty, Fenster, and Kirby 1968) and insulin removal (Solomon, Fenster, Ensinck, and Williams 1967; Burgi, Kopetz, Schwartz, and Froesch 1963) are more complex, since both substances are metabolized as well as being transported into bile. The simple half-life of these substances in the perfusion medium could be used as a test of function.

*Enzyme induction*. The demonstration that specific enzyme synthesis occurs in response to a number of chemical or hormonal stimuli in the perfused organ is a useful addition to the information concerning the organ's viability. Tryptophan pyrrolase, the enzyme most consistently studied, is synthesized at a rate linear with time, after an initial lag phase (Barnabei and Sereni 1960, Goldstein, Stella, and Knox 1962) and this rate may be interpreted as a test of a number of linked liver functions responsible for *de novo* protein synthesis. If serial samples of liver are taken during a perfusion, the concentration of another enzyme, known not to be induced by the stimulus provided, may serve as a control of the basal liver activity (Hager and Kenney 1968).

*Conclusions*

The aim of this section has been to present the evidence of the physiological and biochemical normality of the blood-perfused liver of the rat. No significant differences between this preparation and the liver *in vivo* have been found by the numerous investigators using it. In addition to its obvious application the information thereby obtained provides a most useful base-line to methods that seek to modify, simplify, or standardize the technique of liver perfusion. However, it is the complexity of the existing perfusion medium, in short, its similarity to the situation *in vivo*, which makes quantitative tests of biochemical function difficult to apply. This has been done more rigorously in the later preparations to be discussed in this section.

**Method IB. Isolated liver perfusion with semi-synthetic medium** (Schimassek 1962, 1963)

As already discussed in the introduction, the different experimental approach introduced with this method has opened the technique of liver perfusion (and indirectly, therefore, of other organ perfusions) to controlled biochemical investigation.

In principle, the method is as described for Method IA. The liver is perfused through the portal vein under a constant hydrostatic pressure, in a recirculating system maintained by a roller pump. A multibulb oxygenator similar to that of Miller is used, and the whole apparatus is contained within a temperature-controlled cabinet. Using a perfusion medium which contains washed heterologous red cells, bovine serum albumin, and a balanced salt solution (Tyrode solution), perfusion of 3–4 h duration have been performed and a variety of biochemical parameters by which the preparation may be assessed have been described (Schimassek 1963).

*Apparatus*

Fig. 3.7 illustrates the apparatus, which is simpler than that described in Method 1A, but essentially similar. A roller-pump replaces
the piston pump used by Miller; the reservoir is closed, and stirred by
means of a small magnet, with the stirrer motor in the floor of the
cabinet (cf. Hems *et al.* p. 166). The oxygenator is of glass, and of a
multibulb design, similar to that of Miller *et al.* (1951). While no detailed

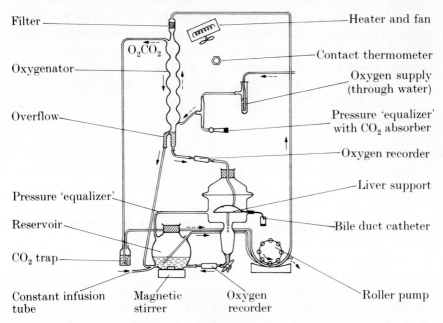

FIG. 3.7. Perfusion apparatus of Schimassek (1963). The apparatus is contained in a temperature-controlled cabinet (not shown).

description is given, and dimensions appear not to have been critically
assessed (cf. Chapter 1), this oxygenator is shorter by half than that of
Miller and is, nevertheless, adequate for the purpose (see Characteristics
of perfusion). Oxygen tension of the medium leaving the oxygenator
is determined continuously by means of a haemoreflectometer. Improvements in the design of the organ chamber allow continuous
sampling of the outflowing medium, continuous collection of bile,
and prevent drying of the surface of the liver. Sampling for venous
oxygen tension must, however, be suspect in this system without
modification, since effluent medium is in contact with the air or with a
gas mixture of unknown composition before reaching the recording

electrode. Access to the liver, for adjustments or for sampling of a lobe during perfusion, is possible by removing the lid which is sealed to the chamber by its ground-glass edges, lightly greased.

*Gas collection.* The apparatus illustrated is sealed, and is designed for total gas collection. This necessitates a number of pressure-equalizing connections (illustrated). Moistened gas flows into the bottom of the oxygenator at a rate of *c.* 200 ml/min (Hems *et al.* 1966) to emerge at the top. It is drawn through the $CO_2$ trap by passing a further polythene tube through the roller pump; this ensures a constant flow of gas. No figures for the recovery of $CO_2$ or oxygen by this apparatus are available; the losses through polythene tubing may be significant (see p. 82). In addition, the effluent medium passes through a chamber in which the gas mixture is not controlled (q.v.). Cannulae are as in Method IA.

*Perfusion medium* (see Table 1.1)

*Medium* I. Bovine blood is collected in the standard acid-glucose mixture with tetracycline (Terramycin) added to a concentration of 40 mg/litre. Blood may be used up to 24 h after collection. Cells are washed twice in 0·9 NaCl and then suspended in Tyrode solution containing tetracycline (10 mg/litre). Bovine cells are said to be particularly suitable as they are of smaller diameter than rat cells and do not readily form rouleaux (they have no special advantage over human red-cells, however, if these are readily available (Hems *et al.*, see Method II)). Finally, bovine albumin (2·6 per cent) is added and the medium gassed with 100 per cent oxygen.

A detailed discussion of the composition of this medium is included (in Chapter 1, Media), but it may be pointed out here that Tyrode solution ($HCO_3 = 11·9$ mM) gassed with oxygen does not provide suitable buffering in the system. Concerning the sterility of medium, the author refers to bacterial counting tests performed upon medium after 3 h of perfusion. Rarely was there any sign of bacterial growth. A further discussion of the use of antibiotics in organ perfusion is contained in Chapter 1.

*Medium* II. Menahan, Ross, and Wieland (1968) report experiments in which the medium described by Hems *et al.* (see p. 169) was used in a perfused liver prepared in every other respect according to Schimassek's method.

*Medium* III (Williamson, Kreisberg, and Felts 1966). A haemo-globin-free medium in this perfusion system is used; the composition is as follows: Krebs–Henseleit medium, plus 4 g % #V bovine serum albumin (Sigma); $5 \times 10^{-5}$ M EDTA, and 10 mM L (+) alanine. The 100 ml recirculating volume is filtered through a Millipore filter (0·45 $\mu$m) before use.

*Operative procedure*

Schimassek reports an operative procedure 'after Miller (1951)', but certainly among the many followers of the present method (for example, Söling *et al.* 1968 and Struck, Ashmore, and Wieland 1965), there are minor differences in the operation which, although of no special importance in themselves, may be helpful to workers using this preparation.

The alternative operative procedure is as follows. Male Wistar rats, 220–280 g are used, anaesthetized with Evipan intra-peritoneally (15 mg/100 g rat). The operation differs from that described in Method IA in that:

(i) After opening the abdomen two ligatures are tied around the stomach, one around the oesophageal end to include the adjacent blood-vessels (L. gastric a.) and the other around the pylorus. The stomach is removed between these ligatures, a procedure that facilitates the remaining dissection and especially the subsequent removal of the liver from the animal.

(ii) A ligature is placed around the bile duct, at some distance from the liver, and the slightly distended duct is cannulated with PP10 tubing (see p. 171 for details).

(iii) The portal vein is tied, incised, and cannulated as already described and, in the modification of Söling *et al.* (1968), medium is flowing through the cannula from the moment of its introduction into the vein.

(iv) Opening of the thorax and cannulation of the inferior vena cava via the right atrium proceeds in the usual way (p. 171). The lower in-ferior vena cava is tied and with the help of an assistant the liver is removed from the rat. The assistant supports the lobes of the liver with plastic spatulae while the operator locates and cuts all the remain-ing attachments. The period of anoxia is reduced in this way.

In Schimassek's original account the operation time is said to be 15–20 min, with anoxia lasting for 7–9 min.

*Characteristics of perfusion*

Flow begins at once, and is allowed to proceed at the maximum rate possible—2–3 ml/min per g for the first 20 min of perfusion. Thereafter, the rate is reduced by partly occluding the portal inflow (tap) and is allowed to remain as 1–1·5 ml/min per g of liver for the remainder of the experiment, usually 180 min in all. Standard conditions of perfusion are defined by Schimassek as follows.

(a)  The standard perfusion medium (above) is used.

(b)  Flow rate is 2–3 ml/min per g for the first 20 min of perfusion, and

(c)  flow rate is 1–1·5 ml/min per g for the rest of the perfusion.

(d)  Perfusion is for less than 3 h.

(e)  Test substances are added only after 60 min of perfusion.

Under these conditions, the following parameters are obtained.

The external appearance of the organ remains normal for 3 h; light microscopy after this time shows some distension of the inter-cellular spaces, but the liver is otherwise normal.

Bile flow is 1–1·3 g/3 h (=1–1·5 $\mu$l/min per g although no details of linearity with time are given), a figure close to that found *in vivo* (see p. 156).

Oxygen consumption, initially higher, reaches the *in vivo* level of 2·2 $\mu$mol/min per g of liver after 30 min of perfusion and is maintained at this rate thereafter. It is therefore a useful parameter of liver survival in this preparation (Schimassek 1962).

The author lays particular stress upon the use of biochemical parameters of liver function, and two aspects are dealt with in the original publications.

(i) *Enzyme profile of the organ.* Since the maintenance of an enzyme pattern similar to that *in vivo* may be taken to indicate that normal metabolic pathways and energy-producing mechanisms are available, this is a certain test that the perfused liver has the capabilities of the organ *in vivo*.

(ii) *Tissue content of substrates and cofactors.* Shortly before the results of Schimassek's work were published, there was elaborated by Hohorst, Kreutz, and Bücher (1959) the suggestion that oxygenation of a tissue *in vivo* might be assessed by determining the tissue concentrations of various components of redox couples. Using the assay methods developed in this work, Schimassek has shown conclusively that the isolated perfused liver (under standard conditions) has concentrations of glycolytic intermediates, quotients of redox couples, and adenine nucleotide contents very close to those found *in vivo*.

In subsequent work by Schimassek and by others, this general approach has enabled 'normal' and 'abnormal' conditions of perfusion to be assessed.

Table 3.3 shows the principle findings discussed by Schimassek for the 'standard conditions' of this preparation. The limitations of 'tissue concentration' and the use instead of rates of reactions is discussed below (p. 190).

TABLE 3.3

*Substrate content of isolated, perfused rat liver, after various times of perfusion; according to Schimassek (1963)*

| Metabolite | Condition | | | |
|---|---|---|---|---|
| | *in vivo* | before perfusion | after 30 min perfusion | after 120 min perfusion |
| Lactate | 1450 | 12000 | 3570 | 3400 |
| Pyruvate | 145 | 50 | 277 | 345 |
| α-glycero-P | 253 | 1560 | 304 | 450 |
| DAP | 38 | 21 | 45 | 67 |
| Malate | 443 | 750 | 281 | 280 |
| Oxaloacetate | 7 | 1 | 4·2 | 4 |
| FDP | 22 | 17 | 20 | 41 |
| F-6-P | 75 | — | — | 41 |
| G-6-P | 370 | 986 | 141 | 206 |
| Glucose | 8600 | 25900 | 11600 | 11500 |
| Glycogen | 340000 | 320000 | 214000 | 185000 |
| AMP | 300 | — | 209 | 280 |
| ADP | 900 | 1853 | 684 | 620 |
| ATP | 2900 | 710 | 2270 | 2060 |
| L/P | 10 | 239 | 13 | 10 |
| G/D | 6·7 | 72 | 6·7 | 6·5 |
| M/O | 64 | — | 70 | 70 |
| ATP/ADP | 3·3 | 0·4 | 3·3 | 3·3 |

Figures are given in $m\mu mol/g$ wet weight of liver.

## Further tests of liver function

*State of mitochondria.* Refined and quantitative tests of mitochondrial function are available (Chance and Williams 1956) by which their functional state may be assessed. Studies of the sort suggested by Bücher and Klingenberg (1958) have been carried out on mitochondria of normal livers and livers after 3 h perfusion (Schimassek 1963). The mitochondria show respiratory control, viz. addition of ADP stimulates respiration (cf. Lardy, and others), but are not normal in that the degree of stimulation achieved thereby is less than that observed in fresh liver mitochondria prepared in the same way. The respiratory control ratio (=rate of oxygen uptake+ADP/rate − ADP) is 4, against the expected figure of from 6 to 8. Why this abnormality occurs is not known: no time course studies are available, but this may be a very critical parameter of liver function. Its use as a test, although specialized apparatus is required, should be considered. Alternatively P/O ratio might be used: in the present study the P/O ratio remained normal, although the abnormality discussed above was present, and it is therefore a less sensitive indicator of function. In addition, its value as a test of 'efficiency' of phosphorylation is in some doubt (Chance and Williams 1956).

These studies by Schimassek have not been repeated in this or in other perfusion systems, and should therefore be considered as preliminary.

*Tests on the perfusion medium*

(i) *Oxygen content and oxygen consumption.* See above.

(ii) *Lactate/pyruvate (L/P) ratio.* Schimassek has found that the L/P ratio of the medium reflects accurately the L/P ratio of the tissue and suggests that this is a useful test of the functional integrity of the liver in perfusion. The normal L/P ratio is taken as 10 and rises in anoxia. It is also a characteristic of the perfused liver that the content of these two intermediates in the medium can be restored to a ratio of 10 very rapidly after either one has been added to the perfusion medium to alter the ratio. However this is clearly difficult to quantify as a 'test'.

(iia) *Maintenance of a constant level of a metabolite in the medium.* This presupposes a regulating mechanism and may therefore be used as a test of function. Schimassek demonstrates the liver's capacity to 'control' medium lactate concentrations in this way. It is difficult to apply quantitatively.

(iii) *Appearance of enzymes in the medium.* This rather crude test of tissue integrity is used by Schimassek (quoting Herfarth 1966) to assess the usefulness of alternative osmotic substances to bovine albumin. More elaborate studies have since been undertaken by Schmidt, Schmidt, and Herfarth (1966) and Schmidt (1968), but are again difficult to quantify. Schimassek takes as a sign of failure the appearance of enzyme in the medium after 4 h perfusion.

(iv) *Release of amino acids from the liver.* Schimassek and Gerok (1965) report that the liver perfused under their 'standard' conditions releases amino acids into the perfusion medium at a reproducible rate, and to an extent which represents a constant ratio of amino acid inside and outside the liver. This again is a difficult observation to employ as a quantitative 'test' of liver function. Nevertheless, these workers report differences between normal, hypothyroid, and hyper-thyroid animals, particularly in the low rate of release of alanine in hyperthyroid, and of glutamine in hypothyroid animals.

*Conclusions*

The uniformity of conditions of perfusion established by Schimassek is confirmed by the tests discussed. With the possible exception of the respiratory control ratio of the mitochondria, the tests discussed are not as stringent as those proposed in the following section, based upon the high energy required for multi-step biosynthetic reactions, but acceptable rates of glucose synthesis from lactate, and of urea synthesis, have been convincingly demonstrated in the same system (Söling *et al.* 1968, Menahan *et al.* 1968). Schimassek's work satisfactorily introduced the use of semi-synthetic perfusion media into studies of liver biochemistry.

**Method II. Liver** *in situ* (method of Hems *et al.* (1966))

This method combines the techniques introduced by Miller and Schimassek (q.v.) with the modification (proposed by Mortimore and Tietze (1959)) that the liver, although isolated from the rest of the animal, remains *in situ*. The liver is perfused through the portal vein, with a semi-synthetic medium, similar to that of Schimassek (1963), under a hydrostatic pressure of about 20 cm of water, maintained by a reservoir of adjustable height. The medium leaves the liver through the vena cava and drops into a collecting vessel. From here it is pumped to the top of a multibulb oxygenator and then returned to the reservoir.

*Apparatus*

The apparatus, based on the designs of Miller *et al.* (1951) and of Schimassek (1963), is housed in a cabinet provided with sash Perspex windows at the front and rear and heated by a thermostatically controlled fan heater (see Fig. 3.8). Discussion concerning the design of such a cabinet is included in Appendix 1.

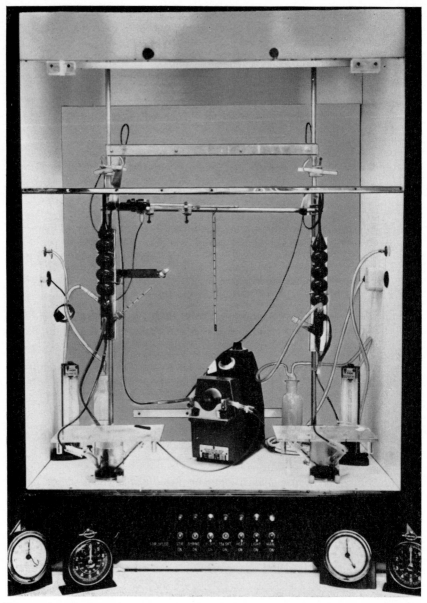

FIG. 3.8.    Perfusion cabinet with apparatus for liver perfusion *in situ* (Hems *et al.* (1966)). The temperature-controlled cabinet contains apparatus for two simultaneous and separate liver perfusions, maintained by a single pump. For details of cabinet see the Appendix and for details of apparatus see text and Fig. 3.9.

The multibulb glass oxygenator is of critical design to ensure an even flow of medium without frothing (see Chapter 1, Oxygenator). It is attached to a stainless steel plate which is mounted on a ball-and-socket joint clamped to a vertical rod. The joint makes it possible to adjust the position of the oxygenator accurately to the vertical, which

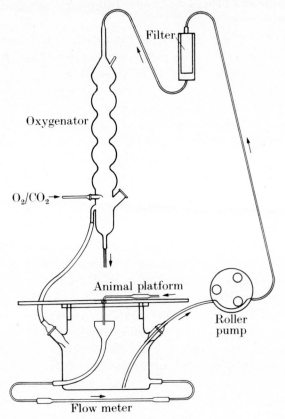

FIG. 3.9. Circuit diagram of apparatus for isolated liver perfusion *in situ* (Hems *et al.* 1966).

ensures complete spread of the perfused fluid over the oxygenator surface. At the bottom of the oxygenator is a reservoir which contains oxygenated medium. The height of the reservoir is adjusted to give a sufficient hydrostatic pressure for optimal flow rates without swelling of the liver. A height between 20 and 30 cm is generally accepted. The circuit is illustrated diagrammatically in Fig. 3.9.

This design has been abandoned in more recent experiments, since the special glassware required (particularly the funnel fused to the inner surface of the reservoir) is difficult to make and is particularly fragile. Alternative flow meters, for example, the syphon (see Chapter 2) permit the use of a reservoir of similar dimensions without either the inflow or outflow ports or the special funnel (illustrated on p. 108). Nevertheless, the arrangement illustrated here has special advantages for experiments in which rapid venous sampling is required.

*Pump.* A roller pump MHRE Watson (Marlow) Ltd. (see Chapter 1) is used in this preparation. Its pulsatile pressure wave is obviously not transmitted to the liver in the circuit illustrated.

*Filter.* A plastic mesh filter, taken from a disposable blood transfusion set (e.g. Capon Heaton & Co. Ltd ) is used, and is changed every few days. Although adequate for the task, this filter with its wide mesh is less efficient than that employed by Miller (see Method IA) and may in part account for the shorter duration of perfusion

*Gas supply.* The gas used is usually 95% $O_2$ 5% $CO_2$, the concentration required to provide a pH of 7·4 with 25 mM bicarbonate buffer used in this system (Krebs–Henseleit) It is saturated with water before entering the oxygenator, by bubbling through a wash-bottle fitted with a sintered-glass distributor. It enters the oxygenator at the bottom and leaves through an outlet at the top. At the bottom of the oxygenator is a reservoir in which medium collects, and an overflow returns excess medium to the collecting vessel. A B.14 standard joint at the bottom of the oxygenator allows a thermometer, thermostat, or a combination pH electrode to be introduced into the medium. The collecting vessel is provided with a flange top on which the Perspex operation platform rests, and a Perspex ring, fixed beneath the platform, fits closely inside the collecting vessel so that this vessel is effectively airtight. A centrally placed hole in the platform takes the outlet tube by which outflowing perfusion medium from the liver returns to the collecting vessel. As described (see Fig. 3.9) medium is fed into a funnel, the outlet of which is fused into the wall of the collecting vessel itself. The medium then passes through a flow meter, in which flow rate is measured by injecting an air bubble into the rubber tubing at the beginning of the flow meter and timing it to travel along the 5-ml calibrated glass tube (a pipette).

Transparent Vinyl tubing No. BT/6 (Portland Plastics Ltd.) is used throughout except for the connection between the overflow on the reservoir and the collecting vessel where a wider gauge (NT/18) tubing is used to prevent the formation of bubbles. A shorter length of rubber tubing is required at the beginning of the flow meter. In most experiments the total volume of perfusion medium of 150 ml was found suitable and allowed detection of small changes in metabolite concentration while providing sufficient buffer capacity for even the most rapid metabolic changes observed in the experiments.

*Perfusion medium*

Details will be given here for convenience. Further information is contained in Chapter 1, Media. Following the method of Schimassek (1963) who used a semi-synthetic medium containing bovine albumin, ox erythrocytes and physiological saline (q.v. p. 160) in place of rat blood, the medium adopted had the following constitutents: (1) physiological saline of Krebs and Henseleit (1932); (2) bovine serum albumin powder Fraction V (Armour or Pentex); (3) washed human red cells. Aged human red cells, stored 4 to 5 weeks at 4°C and no longer suitable for blood transfusion were used. They still possess full oxygen-carrying capacity but do not glycolyse (the rate of glycolysis in each batch of cells may be measured before use). Originally it was thought undesirable to have glycolysing red cells in the perfusion system since gluconeogenesis was to be studied (Hems *et al.* 1966). Since, by the criteria discussed below perfusion is satisfactory, it would seem preferable to adopt this metabolically inert addition to the perfusion medium; in fresh red cells of whatever species glycolysis proceeds at a rate which cannot be ignored in relation to the mass of liver perfused (Exton and Park 1967; Exton, Corbin, and Park 1969). Red cells from 100–200 ml of blood are washed (see Media) and made up to 100 ml with Krebs–Henseleit saline, to give a haemoglobin concentration between 10 and 15 g per 100 ml. This stock suspension may be used on the day of preparation if kept at 4°C, but storage beyond 24 h results in a more fragile population of red cells, and greater haemolysis during perfusion.

For one perfusion 150 ml medium is required (but see modifications below). This is prepared from the above three stock reagents by dissolving 3·9 g bovine albumin in about 120 ml Krebs–Henseleit saline. Since albumin is acid, the pH must be adjusted; about 0·5 ml N NaOH is added to bring the pH to 7·4, as measured by a glass electrode. Then enough of the red cell suspension is added to give a final haemoglobin concentrating of 2·5 g/100 ml. Further alkali, in the form of

M NaHCO3, usually 2·0 ml, is required to bring the pH to 7·3–7·4 again, and the bicarbonate concentration to 25 mM. It is crucial to the method that the total bicarbonate concentration be checked in the laboratory in a Van Slyke or a Natelson apparatus after equilibration of the medium with $O_2$ and $CO_2$ (95% : 5%). With each new charge of albumin, since this is the more variable component of the medium, bicarbonate concentration should be checked. Without this precaution, the medium pH is unlikely to be comparable between successive experiments. The final concentration in this medium are albumin 2·6 g % (weight/volume), haemoglobin 2·5% (w/v), bicarbonate 25 mM.

*Operative procedure* (see Fig. 3.10)

The rat is anaesthetized for the operation by intraperitoneal injection of Nembutal (0·1 ml 6% solution per 100 g body weight). Then heparin

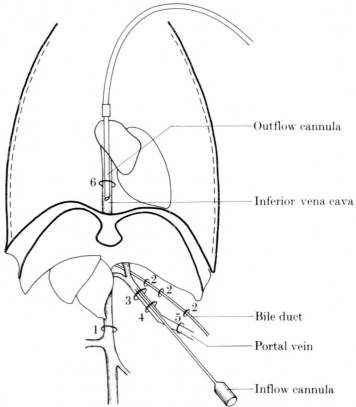

FIG. 3.10.  Operative preparation for liver perfusion *in situ*. The abdomen and thorax are exposed as described in the text. Ligatures 1–6 are placed and the portal vein and inferior vena cava cannulated.

(0·1 ml, 100 units per rat) is injected into the saphenous vein. The abdomen is opened through a mid-line incision, and mid-transverse incisions to right and left of the mid-line are made avoiding the larger vessels. Bleeding is minimized by clamping the major vessels of the abdominal wall with four artery forceps. The intestines are then placed to the animal's left, between layers of tissue wetted with saline so that the liver, portal vein, right kidney, inferior vena cava and the bile duct become exposed. The thin strands of connective tissue between the right lobe of the liver and the vena cava are cut and a loose liga-ture of silk (Lig. 1), size 3/0 (Ethicon Ltd.) is placed around the cava above the right renal vein.

Next, the bile duct is cannulated by a length of Portex tubing size PP10 cut at an angle to provide a sharp point. The bile duct is cut across half its diameter with fine scissors and the cannula is inserted and pushed to the point where the duct arises from its branches. Here it is secured with a ligature (Lig. 2). Two loose ligatures (Ligs. 3 and 4), are passed around the portal vein at intervals of 3–4 mm below the point where the vein divides to enter the separate lobes of the liver and a third ligature (Lig. 5) is placed around the vein at a point distal to the liver. The portal vein is then cannulated with a No. 16 Frankis–Evans needle; a double cannula, from which the sharpened central cannula can be withdrawn after insertion into the vein. With care the blood loss is negligible. The two loose ligatures are tied securing the cannula in place and the third ligature is tied, shutting off the blood supply from the viscera to the portal vein. At this stage the inside needle is removed leaving the cannula in position. The thorax is opened by a transverse incision just above and along the line of the insertion of the diaphragm and by two longitudinal incisions towards the head from the two ends of the transverse incision. The chest wall is flapped back towards the head and a large (15 cm) pair of artery forceps is placed along the base of the flap and locked in position. The flap is then cut off. The vagus and phrenic nerves and the oesophagus are cut about 1 cm above the diaphragm in order to paralyse the diaphragm and to eliminate possible vasomotor effects of the vagus. A loose ligature is placed around the inferior vena cava close to the heart (Lig. 6). The cannula which is placed in the inferior vena cava consists of a 5-cm length of Portex tubing (PP 270) which has been heated in a gas flame, drawn out to an outside diameter of 2 mm and cut off at an angle to form a sharp tip. This is sharp enough to penetrate the right atrium; it is pushed down the vein towards the diaphragm and tied in position

(Fig. 3.11). At this stage the loose ligature around the abdominal vena cava is tied. The cannula in the portal vein usually fills with blood, especially if the hepatic artery remains open to supply the liver with blood throughout the operation, but if this is not the case perfusion medium in injected with a syringe through the inner needle of the cannula. The medium is injected continuously while withdrawing the

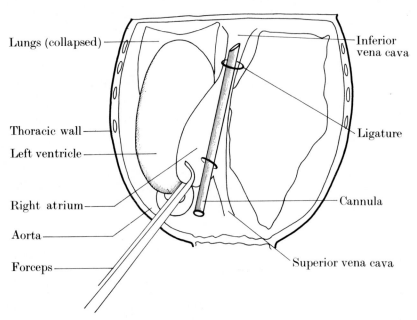

Lungs (collapsed)

Thoracic wall

Left ventricle

Right atrium

Aorta

Forceps

Inferior vena cava

Ligature

Cannula

Superior vena cava

FIG. 3.11.   Cannulation of the right atrium and inferior vena cava. With the head of the animal towards the operator the chest is opened and a loose ligature placed round the inferior vena cava. The right atrium is grasped in curved forceps and drawn to the animal's left. The sharpened end of the cannula is forced through the atrial wall and passed into the lumen of the inferior vena cava.

needle; care is taken to avoid air entering the portal vein. The preparation is then connected to the perfusion apparatus and the circulation started. The first 5 ml of venous blood (mainly rat blood) are discarded.

The whole operation takes about 10 min. There is an interval of about 2·5 min from the introduction of the cannula into the portal vein during which the circulation is maintained by the hepatic artery alone, and a period of about 1 min between the cessation of breathing (caused by the opening of the thorax) and the start of the perfusion.

The admission of an air bubble to the portal vein or an extended interruption of the liver circulation may cause an uneven perfusion, indicated by uneven colouring, from which the liver may not completely recover. Such perfusions are best rejected. An indication of the success of the operation is a uniform red colour of the perfused liver, similar to the colour *in vivo*.

*Characteristics of perfusion*

Perfusion pressure (20–30 cm $H_2O$), and flow (15–25 ml/min) remain constant for the normal experimental period which is suggested for this preparation, namely 130 min. Bile flow too is within the expected range (67 mg/h per g wet weight of liver), and remains constant for the 130-min period tested. Additional information is included in Table 3.6 (p. 195). Fig. 3.5 illustrates the normal histology of the liver perfused in this way.

Oxygen consumption is 2·2 $\mu$mol/min per g of wet liver in the starved rat and remains steady for the experimental period. The expected increases in oxygen consumption occur in response to added substrate (see Table 3.5). The test suggested by Schimassek (q.v.) to determine the content of adenine nucleotides in the liver during the course of a perfusion gives substantially the same results. Perhaps as a result of the shorter period of anoxia during operation, effectively normal concentrations are observed within 5 min of starting perfusion. However, Schimassek too observes such figures under his standard conditions of perfusion and the preparations should not be considered to be different on this basis.

Table 3.4 records the results of experiments in the preparation *in situ* of Hems *et al.* and records in addition the effect of adding a gluconeogenic substrate such as L-lactate to the medium. However, these authors have concentrated particularly upon the biosynthetic functions of the perfused liver. Glucose formation from a loading dose of lactate occurs in a linear fashion with time, without a lag period, and falls only when substrate concentrations have fallen below critical levels. Similarly, the rate of urea synthesis from $NH_4Cl$ appears to be linear with time. A second 'dose' of $NH_4Cl$ added to the perfusion medium after 85 min results in a rate of urea synthesis identical to the first; and, in addition, ornithine has the dramatic stimulatory effect on urea synthesis which is expected from its catalytic role demonstrated in liver slices (Krebs and Henseleit 1932).

Ross *et al.* (1967) report the rates of glucose synthesis found with a

## TABLE 3.4

### Adenine nucleotides in perfused rat liver

| Substrate added to medium | Duration of perfusion (min.) | No. of observations | ATP ($\mu$mol/g) | ADP ($\mu$mol/g) | AMP ($\mu$mol/g) | Ratio $\dfrac{ATP}{ADP}$ | Total adenine nucleotides ($\mu$mol/g) |
|---|---|---|---|---|---|---|---|
| Nil | 5 | 3 | 2·18±0·10 | 0·89±0·06 | 0·37±0·03 | 2·46±0·20 | 3·44±0·12 |
| Nil | 15 | 3 | 2·17±0·15 | 0·86±0·03 | 0·36±0·02 | 2·54±0·10 | 3·39±0·16 |
| Nil | 85 | 3 | 2·05±0·15 | 0·66±0·01 | 0·44±0·03 | 3·11±0·18 | 3·15±0·12 |
| Lactate | 85 | 5 | 1·70±0·25 | 1·00±0·10 | 0·30±0·04 | 1·76±0·18 | 3·00±0·20 |
| Lactate | 130 | 4 | 1·70±0·08 | 0·86±0·02 | 0·26±0·04 | 0·99±0·07 | 2·82±0·12 |
| Initial values | 0 | 5 | 2·53±0·16 | 0·94±0·09 | 0·21±0·06 | 2·70±0·26 | 3·68±0·31 |

Livers of rats starved for 48 h were perfused with the standard medium. Lactate (1500 $\mu$mol) was added at 38 min. For the analytical procedures see Chapter 2. The initial values refer to livers of rats starved for 48 h and anaesthetized with ether. Results are given as means±S.E.M.

range of gluconeogenic precursors, confirming that the perfused liver metabolizes this range of substrates. Only those substrates for which permeability barriers in liver are known to occur (Hems *et al.* 1968) failed to be taken up by this preparation. Glutamate, aspartate, and acids of the tricarboxylic acid cycle are included in this group, and failure to metabolize these substrates should not therefore be taken as an indication of failure of a perfused liver preparation.

Table 3.5 shows the results of experiments devised to test the maximum biosynthetic capacity of the liver in perfusion. The rates observed equal or exceed those observed in the rat *in vivo* of the rate-limiting enzyme(s) of each pathway. These features are discussed by Hems and others, and it is sufficient to record here the apparently normal activity of the perfused liver in this context. The importance of the perfusion medium (see Chapter 1, for discussion) is illustrated with this preparation by experiments from which one ion is lacking from the medium, or in which the pH is far from 7·4. In either case, the rate of glucose synthesis from lactate is depressed.

Further discussion of the features of this preparation are included in a later section, in which several preparations are compared (see p. 193).

### Method III. Reverse perfusion (Method of Trowell (1942))

It is of both theoretical and practical interest that the direction of blood flow through the liver should be reversible. There are no valves in the larger veins between the inferior vena cava and the portal vein; flow into the capillaries of the hepatic artery itself is unlikely, in view of the muscular 'sphincters' described by Popper (see Fig. 3.4).

Initially (Trowell 1942) this was a method of circumventing the apparently insoluble problem of 'inflow' or 'outflow' block which thwarted the earlier attempts at isolated liver perfusion in the rat (Bauer, Dale, Poulson, and Richards 1932 for review). It will be abundantly clear from Methods I and II that this is no longer an objection to perfusions performed in the physiological direction. Nevertheless, there have been some recent revivals of the technique of reverse perfusion (Schreifers 1968; Mayes and Felts 1967; Thomas, Ganz, and Buschemann 1968) and the technical details of the method are therefore of relevance.

Medium is allowed to flow under gravity (4–8 cm $H_2O$) into the thoracic inferior vena cava, the abdominal end of the vessel being ligated below the liver. The outflow is via the portal vein, which is not cannulated, and the liver is contained in an organ bath which allows

TABLE 3.5

*Performance of the perfused rat liver as tested by gluconeogenesis from lactate and urea synthesis from ammonia*

| Substrates added | Glucose formed | Lactate used | Urea formed | O$_2$ used | Extra O$_2$ used addition of substrates | Extra O$_2$ uptake calc. for ATP needs |
|---|---|---|---|---|---|---|
| None | 0·14±0·026 | — | 0·09±0·01 | 2·20±0·22 | — | — |
| L-lactate | 1·06±0·09 | 1·95±0·23 | — | 3·50±0·20 | 1·30±0·22 | 0·92 |
| NH$_4$Cl | — | — | 0·80±0·14 | 3·64±0·18 | 1·66±0·34 | 0·48 |
| L-Lactate; NH$_4$Cl | 1·05±0·06 | 2·24±0·11 | 0·45±0·04 | 6·02±0·28 | 3·44±0·18 | 1·25 |
| NH$_4$Cl; L-ornithine | 0·185±0·05 | — | 1·87±0·27 | 4·60±0·40 | 2·60±0·14 | 1·19 |
| NH$_4$Cl; L-ornithine; L-lactate | 1·05±0·05 | 2·19±0·17 | 1·95±0·26 | 8·00±0·80 | 5·60±0·93 | 2·15 |

Livers of rats starved for 48 h were perfused as described (Hems *et al.*). The quantities of lactate and ammonia added in 150 ml of perfusion fluid were 1500 $\mu$mol and that of ornithine was 400 $\mu$mol. The results are $\mu$mol/g wet wt/min (mean ± S.E.M.). The extra O$_2$ used is calculated in each experiment from the O$_2$ uptake expected for meeting the extra ATP needs, calculated on the assumption that the synthesis of a glucose from lactate requires 1 O$_2$ (6 ATP) and the synthesis of 1 urea from NH$_3$ requires 4 ATP ($\frac{2}{3}$ O$_2$).

it to be bathed in effluent medium. A recirculation system is provided by means of a pump and the liver may be maintained for 2–4 h.

*Apparatus*

The original apparatus of Trowell contains an organ chamber, a reservoir which, shaken in the presence of gas, serves as an oxygenator and a rotary pump to return outflowing medium to the reservoir, and to maintain a constant level of medium in the organ chamber. Temperature is controlled by immersing all the components, except the pump, in a water bath at 38°C.

*Cannula.* A glass cannula, i.d. 1 mm at the tip, with a terminal flange of o.d. 2 mm, is used for the inferior vena cava.

More recent apparatus used with this preparation has been the conventional apparatus of Felts and Mayes (1966) (see p. 189) and the very simple gravity-feed apparatus used by Schreifers and Korus (1960). The latter differs only in the liver being free, in air rather than immersed in medium, and the use of a more formal oxygenator, based on the design of Hooker (1915) (see Oxygenators, Chapter 1).

*Perfusion medium*

*Medium* I (Trowell 1942). Krebs–Henseleit medium, gassed with 95% $O_2$:5% $CO_2$, was used without additions.

*Medium* II (Hechter *et al.* 1953). After an initial perfusion with physiological saline, these workers used citrated bovine blood, which had first been filtered through muslin and china silk.

*Medium* III (Schreifers and Korus 1960). A haemoglobin-free medium used in this preparation is composed: (in g/litre): glucose 2·0; NaCl 8·2; KCl 0·25; $CaCl_2$, 0·25; $MgSO_4$ 0·13; $NaHCO_3$ 0·60; $NaH_2 PO_4$ 0·05, plus penicillin 20 000 units/litre; streptomycin 0·1 g/litre, and gassed with 95% $O_2$:5% $CO_2$.

*Medium* IV. Thomas *et al.* (1968) used the medium of Schnitger (see Chapter 1, p. 22) in the preparation of Schreifers and Korus.

*Operative procedure* (see Fig. 3.12)

Under ether anaesthesia, the abdomen and thorax are opened (combining the procedures described on p. 171) to expose the liver, diaphragm, and heart. The anterior part of the diaphragm is incised, almost down to the inferior vena cava, and a loose ligature (Lig. 1) is positioned around the vessel, above the liver but below the diaphragm. Using a curved forceps it is possible to include in addition the left phrenic branch of the inferior vena cava in this ligature, a point of which Trowell makes great play.

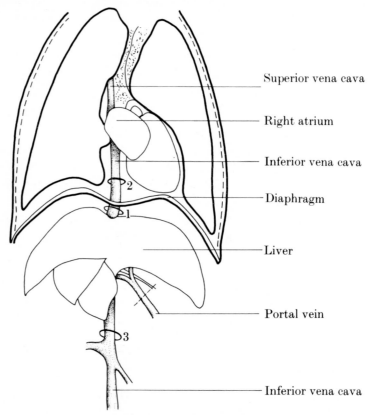

Superior vena cava

Right atrium

Inferior vena cava

Diaphragm

Liver

Portal vein

Inferior vena cava

FIG. 3.12.    Operative procedure for retrograde perfusion of liver. The cannula introduced through the right atrium (cf. Fig. 3.11) is held in place by Ligatures 1 and 2, below and above the diaphragm. The lower inferior vena cava is closed with Ligature 3 and the portal vein and bile duct are cut. For details see p. 179.

Further ligatures are placed around the inferior vena cava, below the liver, and above the left renal vein (Lig. 2), and another round the same vein, above the diaphragm (Lig. 3). The phrenic nerves are cut as they lie either side of the inferior vena cava, to 'immobilize the diaphragm', and an internal iliac vein is cut to relieve venous congestion. The cannula is inserted into the upper inferior vena cava through the right atrium (see p. 171 and Fig. 3.11), and advanced just beyond the diaphragm to be tied in place with Lig. 1. This is further than is normally done for the forward perfusion of the liver (see Methods I and II). Flow begins at once, and the inferior vena cava is incised below Lig. 2 to allow blood to be washed out. Then Lig. 2 is tied, and an incision across the portal vein quite close to the liver is made. Medium now flows through the liver under a pressure head of 8 cm $H_2O$. When blood has been washed out the pressure is reduced to 4 cm $H_2O$ and the liver may now be removed from the animal. It is transferred to the organ chamber where it is arranged to lie with its portal vein uppermost, i.e. dorsal surface underneath, and medium is allowed to accumulate in the chamber until the liver is just submerged.

*Characteristics of perfusion*

Prepared in this way, liver perfusions of 2–4 h duration are possible, with constant perfusion pressure and flow rates. Histology apparently shows uniform perfusion, but the cells are oedematous and rounded. This is not perhaps surprising with the perfusion medium employed. Hechter and Schreifers do not report on the histology of their preparations. Urea synthesis from $NH_4Cl$ and ornithine, with the expected catalytic effect of the latter amino acid, has been demonstrated, and appears to be linear with time. It is interesting to compare the rate of urea synthesis with that observed in a 'forward' perfusion: Trowell's figures are (approximately) 0·5 $\mu$mol/min per g with $NH_4Cl$ alone, increasing to 1·5 $\mu$mol/min per g in the presence of ornithine. Hems *et al.* (1966) report figures of 0·8 and 1·9 $\mu$mol respectively; such differences might be attributable to the difference in perfusion medium rather than to the different direction of perfusion, but no strictly comparable investigations have been undertaken.

No other points of comparison are available; Schreifers'ss tudies demonstrate the linear uptake of cortisone from the perfusate (with a not precisely linear formation of products), but no comparable data is available for orthodox perfusions. Similarly, the demonstration of the biosynthesis of homovanylloylglycine in a reverse perfusion system by Thomas, Ganz, and Buschemann (1968) has no parallel in the field of orthodox liver perfusion. See p. 197 for further discussion.

**Method IV. Perfusion of the hepatic artery** (Methods of Ross *et al.* (1967) and Powis (1970))

As discussed in the Introduction, most workers have ignored the hepatic arterial supply in perfusing the rat liver. It is estimated to carry no more than 20 per cent of the blood to the liver, and oxygenation is adequate via the portal vein supply. Although metabolic experiments confirm that the liver functions normally without perfusion of the hepatic artery, techniques are available which allow this very small vessel to be perfused at arterial pressure, either under hydrostatic conditions (Ross *et al.* 1967) or by direct pumping (Powis 1970) and these methods will be briefly described. The metabolic characteristics of the preparation are compared with those of the liver perfused through the portal vein alone on p. 183.

(a) *Hydrostatic perfusion* (Ross *et al.* 1967)

This method is adapted to apply in the method described by Hems *et al.* (Method II above).

*Method.* The apparatus is modified as shown in Fig. 3.13 by raising the oxygenator to a height of 120 cm above the liver to provide the

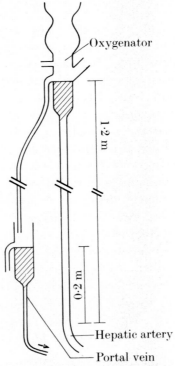

FIG. 3.13.    Apparatus for combined perfusion of hepatic artery and portal vein. This figure should be compared with Fig. 3.9. Only the arrangement of oxygenator and the reservoirs is shown. The oxygenator is raised to a height of 120 cm above the liver, and supplies the hepatic artery. The overflow of the reservoir falls to the second reservoir shown, which is only 20 cm above the liver. This reservoir supplies the portal vein. Note the second overflow, which enters the collecting vessel.

necessary pressure to perfuse the hepatic artery. The portal vein is supplied from the same oxygenator but from a second reservoir which is only 15–20 cm above the liver, and the overflow from the second reservoir is returned to the collecting vessel. The preparation of the animal is the same as for a portal vein perfusion, placing loose ligatures around the inferior vena cava and portal vein. Care is taken to exclude the hepatic artery from these ligatures. To cannulate the hepatic artery the abdominal contents are now deflected to the animal's right and the lower oesophagus is cut between ligatures. The stomach is deflected to the right, exposing the aorta and coeliac artery to its division into gastroduodenal and hepatic branches. The two major branches of the coeliac artery (left gastric and gastroduodenal) and one small branch are tied leaving only the hepatic artery. At this stage heparin is injected into the lower inferior vena cava. The aorta is tied above and below the coeliac artery and a fine glass cannula is introduced, through a cut in the aorta, into the coeliac artery where it is tied in place. The portal vein and inferior vena cava are cannulated as usual and the liver can then be perfused *in situ* via the hepatic artery, or the portal vein, or both together.

## (b) *Direct pumping method* (Powis 1970)

The method advocated by Powis involves pulsatile perfusion of the portal and hepatic arterial circuits with dialysed whole rat blood. The apparatus, based upon that used for perfusion of the cat spleen, and discussed in the context of intestinal perfusion (Powis 1970, Chapter 7), is illustrated in Fig. 3.14.

*Perfusion medium.* 80–100 ml heparinized rat blood, obtained from the carotid artery, is dialysed for 24 h at 4°C against Krebs–Henseleit buffer containing 1 per cent glucose ($2 \times 1$ litre). Prostaglandin $E_I$ (60 $\mu$g) is added after collecting blood, and again after dialysis.

*Operative procedure.* To allow for a long operation time (45 min) the animals are anaesthetized, ventilated through a tracheotomy (see Chapter 9) and then pithed (i.e. the brain stem and spinal cord are destroyed by inserting a probe through the foramen magnum). The operative procedure is more elaborate than others discussed above, and will be described. However, there appears to be no need for the elaborate method, since Abraham *et al.* (1968) describe an isolated liver perfused through both routes and prepared in a rapid operation.

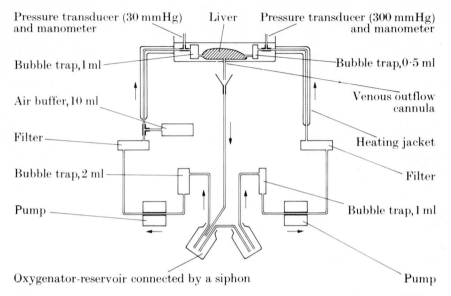

Pressure transducer (30 mmHg) and manometer

Liver

Pressure transducer (300 mmHg) and manometer

Bubble trap, 1 ml

Bubble trap, 0·5 ml

Air buffer, 10 ml

Venous outflow cannula

Filter

Heating jacket

Bubble trap, 2 ml

Filter

Pump

Bubble trap, 1 ml

Oxygenator-reservoir connected by a siphon

Pump

FIG. 3.14.    Apparatus for combined perfusion of portal vein and hepatic artery (Powis 1970).

In Powis's operation, a mid-abdominal incision is made, the intestines are displaced to the animal's left and the superior mesenteric artery (see Fig. 6.2) is located and cut between ligatures. The portal vein is located and all its branches apart from the splenic vein are tied and cut. This allows the intestine to be removed *en bloc*. The description by Powis of the continuation of the operation is as follows.

The spleen is displaced to expose the hepatic artery and nerve, which are dissected free and marked with a colour-coded loose ligature, the gastro-duodenal artery being cut between ties close to its origin with the hepatic artery. The bile duct is separated at this stage and a polyethylene cannula inserted. After dividing the left gastric artery the stomach and duodenum are removed. It has been found that the intravenous administration of 200 IU heparin at this point results in a much better quality of perfusion. The dorsal aorta is tied just below the origin of the coeliac axis. The right adrenal and renal veins are cut between ties, as are all tributaries entering the inferior vena cava below this point, leaving a stump suitable for cannulation. This leaves the liver and spleen attached to the body by ligaments only, which are cut, and by the vena cava and coeliac axis. The liver receives arterial blood through the hepatic artery and venous blood from the spleen. Two saline-filled cannulae are inserted into the stumps of the portal vein and vena cava. The thorax is opened and the thoracic vena cava tied, cut proximally, and dissected free with a small cuff of diaphragm. Simultaneously, the caval cannula is opened and perfusion with oxygenated blood started at a slow rate (2–3 ml/min) through the portal cannula. The splenic artery and vein can then be tied and the spleen removed. The tip of the arterial cannula is inserted into the coeliac axis through a slit made in the anti-mesenteric aspect of the aorta and arterial perfusion commenced at the rate of 0·5 ml/min. The whole liver is transferred to the perfusion chamber, covered with liquid paraffin and the perfusion pressure increased to 10 to 12 mmHg for the portal circuit and 80 to 120 mmHg for the arterial circuit.

### Characteristics of perfusion

The total blood flow is about 1·0 ml/min per g, the arterial circuit providing 15–20 per cent of the total flow The whole operative procedure takes 45 min and the liver is at no stage without a supply of oxygenated blood.

The total blood flow in this preparation is reported to be about 1·0 ml/min per g of wet weight. Perfusions have been continued for 3–4 h with a normal histological appearance after this time, while bile flow (0·04–0·07 $\mu$l/h per g), oxygen consumption (2·7 $\mu$mol/min per g in the 24-h starved rat liver) and the rate of glucose formation from lactate do not appear to differ from the results of portal perfusion alone. Provision for stimulation of the hepatic nerve has been made in this preparation, and there is evidence of release of noradrenaline into the perfusate. The rate of glucose synthesis from lactate is slightly stimulated by this manoeuvre (Powis 1970).

## Summary of the published techniques of liver perfusion

Of 494 references to 'perfusion in small animals' for 1968–9, 163 referred to the liver. These may be classified according to their authors' claim as to the method followed, viz. 41 followed Miller *et al.* (1951), 13 Brauer *et al.* (1951), 10 Schimassek (1963), 7 Mortimore (1961), and 3 Bücher, Schnitger, *et al.* (1965), with some 12 methods sufficiently different from the main stream to warrant separate description. The details of this classification are as follows.

### Miller (1951)

Liver perfused in isolated state, with diluted, heparinized rat blood, with additions of antibiotics and amino acids; gravity feed to portal vein and cannulation of inferior vena cava, the whole being conducted in a simple glass apparatus, sealed to allow gas collection and enclosed in a heated cabinet at 38–40°C. Oxygenation is by means of filming in a multibulb oxygenator. The bile duct is not cannulated as a rule and bile enters the perfusion medium.

### *Brauer* (1951)

Liver perfused in isolated state, with complex medium including rat blood (heparinized), bovine plasma albumin, salts, antibiotics, and glucose. Perfusion is by a mixture of gravity feed and direct pulsatile pressure, in a glass and rubber apparatus which is enclosed in a heated cabinet. The operative preparation is conducted 'aseptically' and includes cannulation of the bile duct.

### *Schimassek* (1963)

Liver is perfused in isolated state, after the general method of Miller. The operation and apparatus show the same origins, but in this, as in the introduction of a defined semi-synthetic medium, the author takes the first steps to a genuinely standard preparation with biochemical experimentation in mind. Standard conditions are defined as follows:

(i) Standard medium: washed bovine cells, 10 g % Hb. Tyrode solution, plus tetracycline. Bovine serum albumin, (purissima) 2·6 g %.

(ii) Flow rate 2–3 ml/min per g of liver in the first 20 min and 1·15 ml/min thereafter.

(iii) Perfusions last 3 h, and test substrates are given after 1 h.

(iv) Bile is collected (cf. Miller in original paper). This preparation was the first to be rigorously defined and standardized.

### *Mortimore* (1959, 1961)

Liver perfused *in situ*, a technique with theoretical disadvantages which have all been allayed in the event of application of this approach to many biochemical problems. Krebs *et al.* (1966–8) and Exton and Park (1967) have used this operative procedure, which is described in detail above, (p. 170, Method II). The method as described by Mortimore (1959) pumps medium directly into the portal vein, which is cannulated with the minimum of delay, perfusion beginning within 5 sec. A peristaltic, and later a valve and finger pump were used, to minimize haemolysis of the medium, which consisted of defibrinated rat blood diluted to a haematocrit of 22–24 per cent with Krebs–Henseleit buffer. Later buffers contained either heparinized rat blood in the place of the defibrinated form or bovine serum albumin (Armour) to a final concentration of 1·5–2·0 g % added with the buffer. This medium is therefore very similar to that of Schimassek (q.v.). It is gassed with 95% $O_2$ and 5% $CO_2$ in a rotating spherical flask, a very simple filming oxygenator, and the capacity of the apparatus is about 40 cm³ medium.

The author emphasizes the rapid non-traumatic preparation with a minimum anoxic period, and was able to demonstrate minor insulin effects which were not observed in the experiments of Miller *et al.* and have not been found by Cahill's group (Williamson, Garcia, Renold, and Cahill (1966), Garcia, Williamson, and Cahill (1966)). The apparatus was thought to cause no haemolysis, estimated visually and from the constancy of serum $K^+$ during 50-min liver-less perfusions. The most direct application of this method in biochemical studies is that described by Exton and Park and used too by Jefferson and Korner (1969). It is also the basis of the operation method of Hems *et al.*

### Exton and Park (1967)

The main modification of these workers has been to standardize the medium—washed rat erythrocytes, from fresh citrated rat blood, are suspended to a haematocrit of 22–24 per cent in Krebs–Henseleit buffer, with a final concentration of bovine serum albumin (Fraction V) of 2·35 g % (calculated from the published data on the basis that 3 per cent albumin in the KH buffer becomes $100 - 22/100 \times 3 = 2·35$ when this is diluted with cells to the prescribed haematocrit). The apparatus is contained compactly in a heated cabinet, using a standard seven-finger (Sigma) peristaltic pump in place of the valve pump of Mortimore, and the oxygenator is a simple rotating vessel not differing in essence from that of Mortimore and described in detail in Chapter 1 (Oxygenators). Once again, direct pumping of medium into the portal vein is the method employed and a total recirculating volume of medium of 55 ml is used.

### Schnitger, Bücher, Scholz, et al (1965)

These workers, in an attempt to define the experimental conditions still further, have devised a compact perfusion apparatus in which the isolated liver is perfused with buffer containing Dextran or albumin but no red cells. Direct pumping is used, and to overcome the problem of providing adequate oxygenation in the absence of erythrocytes a high flow rate (60 ml/min) and an artificial block to outflow was used. The result is that the liver swells and the quantity of medium within the organ at any one time is thereby increased. Recently this 'liver swelling' technique, which has been criticized, has been omitted with, the authors claim, no anoxia. This point is further discussed on p. 211.

### Williamson, Cahill, and others

These workers followed Bücher in using a red cell-free medium but employed perfusion conditions similar to those of Miller. The medium consists of 3 g % bovine albumin in Krebs–Henseleit buffer, to which glucose (180 mg) and heparin (50 mg) are added, perfused under a hydrostatic pressure of 15 cm $H_2O$. An additional modification of these workers has been to filter the medium (Millipore filters, see Chapter 1) before its use. Oxygenation is increased by the very high flow rates of medium attained in the absence of erythrocytes.

### Other methods

*Kvetina and Guaitani* (1969). These workers have devised a small-volume (15 ml) perfusion system, combining direct pulsatile inflow of diluted heparinized rat blood with a novel oxygenator. The whole is water-jacketed and has been applied to mouse liver and to lung perfusion in the rat (see Chapter 9). However it does not have any obvious advantages over simpler systems for biochemical work.

*Bohley et al.* (1966). In order to study cellular destruction in the rat liver, these investigators perfused in an apparatus similar to that of Miller, for periods of up to 117 h with a buffered saline (Ringer's minus $Ca^{++}$ and minus $Mg^{++}$). No special recommendation can be made for this approach to more general biochemical problems.

*Claret and Coraboeuf* (1968). In these experiments to determine membrane potential, the liver was perfused with simple buffers, presumably under hydrostatic pressure (the description is very brief). Tyrode and THAM were used to provide a pH range 7·45–9·07, and saccharose was used as the osmotic component. The perfusion time was only 15 min and in the light of the discussion of Schimassek and of others this is too short a period for 'equilibrium' to be established.

Further investigation is therefore needed to assess its value in general application.

*Heimberg, Fizette, and Klausner* (1964). These workers used a water-jacketed apparatus with a piston pump and falling film oxygenator to circulate a medium of heparinized rat blood, diluted 1:3 with Krebs–Henseleit buffer. Chambers, Georg, and Bass (1968) have used the same apparatus with defibrinated rat blood (40 ml) plus KH buffer (60 ml) with glucose and heparin added.

*Flock and Owen* (1965). This is basically the method of Brauer (q.v.), with a standard medium of washed rat erythrocytes in Krebs–Henseleit buffer and with 3 g % bovine serum albumin. The authors

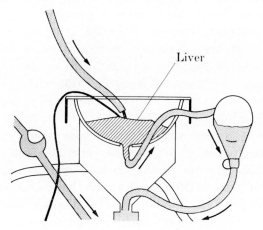

Liver

FIG. 3.15. 'Anti-siphon' device of Flock and Owen (1965). Detail from the circuit for perfusion of the liver; a levelling bulb is so placed that blood leaves the organ at the same pressure as it enters. 'Siphoning' of blood through the liver cannot occur. (For discussion see p. 203.)

are concerned about siphoning of medium through the liver if the outflow lies below the inflow. A simple modification is said to prevent this (see Fig. 3.15). This method is of interest in that it has been followed by Veneziale, Walter, Kneer, and Lardy (1967) in metabolic studies. However, there is no evidence of the deleterious effects of siphoning and the results of Veneziale *et al.* suggest that the liver is exactly as described by Schimassek, Exton, and Hems *et al.*

*Krieglstein* (1969). This investigator followed the method described in detail by Hems, Krebs, and others, but used a haemoglobin-free medium. By working at room temperature, the oxygen requirements of the liver are satisfied as detailed by Kessler and Schubotz (1968). The limitations of haemoglobin-free perfusion have been overcome in other ways (see, for example, the methods of Bücher and Scholz) or by restricting experiments to the use of small livers. By working at lower temperatures, new problems are encountered, reactions are slowed, and the changes expected during an experiment may be smaller or even absent when compared with their normothermic rates or those existing *in vivo*.

*Schreifers and Korus* (1960). This is the only group of workers since Trowell (1942) to publish biochemical studies using the reverse perfusion system of Hechter (q.v.). The medium, of citrated bovine blood in a glucose containing bicarbonate buffer, is passed through a test liver before the experiment, 'to remove emboli and vasoconstrictor substances'. A once-through system is used with inflow under gravity. Hence no pump is required. The oxygenator is a somewhat complicated one, based upon the design of Hooker (1915) with a rotating disc (cf. Kavin *et al.*, for intestine: Chapter 7) which throws medium against the walls of the containing cylinder.

The benefits of 'reverse' perfusion have been largely superseded by advances in the technique of forward perfusion, and an objective comparison using metabolic parameters has never been made.

The results of Trowell (1942) on urea synthesis might be compared with those of Hems *et al.* (1966) (see p. 179). The conditions offered by Trowell's reverse perfusion are clearly not optimal when expressed as the rate of urea synthesis from ammonium chloride, and workers proposing to follow Hechter or Schreifers would perhaps be best advised to test a metabolic function such as gluconeogenesis from lactate before embarking on new studies.

*Felts and Mayes* (1966). Based on Miller's technique, this method has been modified 'to make the preparation more suitable for metabolic studies, increase reproducibility and ensure successful perfusion at practically every attempt'. The only modification of significance is the introduction of defibrinated whole rat blood as perfusion medium, and the avoidance of heparin in both the medium and the rat during operation, since heparin is definitely implicated in lipoprotein lipase release, which promotes uptake and oxidation of triglyceride. The medium is prepared as follows. Fresh rat blood, collected from the

abdominal aorta (see p. 33) in ether-anaesthetized animals, is defibrinated by stirring with wooden applicator sticks. Ten rats are required to
provide 100–130 ml medium. The defibrinated blood is filtered through
glass wool, through cotton poplin and collected in a vessel cooled in ice.
Then it is dialysed at 4°C for 48 h, against Krebs–Henseleit saline,
containing both glucose (300 mg per cent for fed rat experiments,
100 mg per cent for starved) and amino acids (50 mg per cent Stuarts
amino acid mixture (Pasadena, U.S.A.)). The outside medium is
changed four times.

The apparatus is essentially that of Miller and is housed in a cabinet.
A Medtec pump (q.v.), two filters of the type described by Miller, a
bubble trap, and a humidifier are used. A gas flow (95% $O_2$:5% $CO_2$)
of 1 ft$^3$/h is provided, and two $CO_2$ traps collect outflowing gases. The
operation is not significantly different from that described by Miller,
but a good account is given in this reference. Flow is kept constant at
7 ml/min by adjusting the head of pressure, in contrast to the constant
pressure approach usually used.

*Burton, St. George, and Ishida* (1960). A small-volume modification of
Miller's method (25 ml circulating volume) has been used by these
workers, with special attention to the avoidance of haemolysis. In an
apparatus which includes a roller pump, a multibulb oxygenator, a
nylon mesh filter, and heparinized rat blood with tetracycline (25 mg/
ml), 6–10 per cent haemolysis is observed after 3 h perfusion through
the liver. Without the liver in the circuit, haemolysis reaches 6 per cent
after 100 h circulating in the apparatus. The apparatus has been more
recently modified (Burton, Mondon, and Ishida 1967).

*Long and Lyons* (1954). A very complex apparatus has been described
by these authors and used in liver perfusion studies. The plastic
apparatus is siliconized before use and may be sterilized. However, it
requires 2 hours to assemble and has not been widely used in biochemical
studies. Nevertheless, the pulsatile-valve pump described and illustrated has been used by Despopoulos (1966) for studies on hippurate
metabolism in perfused liver.

## Tests of liver function applicable to the perfused organ

Tests of function have been discussed after each of the methods
described, and the importance of this assessment has been discussed in

detail (see p. 127). Such tests may be used to assess the viability of the perfused organ and to determine its stability during the course of an experiment. In addition, modifications to the technique may be assessed objectively. In this section, an attempt is made to correlate the large amount of information available from various workers using one of the main methods outlined above. To this end, some of the experimental data has been examined for its value as a test of liver function.

*Ideals of test*

Any parameter of liver function which can be determined serially or continuously during perfusion may be used, be it a test applied to the perfusion medium or to the liver itself. It should be easily quantifiable. Thus, a continuous reaction should be linear with time, without a lag period, since this introduces difficulties of quantification of the rate of reaction. Alternatively, the magnitude of response to a given stimulus may be used as a test.

It is generally found that tests of endogenous unstimulated liver function are less reproducible than tests based on the introduction of a stimulus to produce a maximal response, for example, lactate added to produce the maximum rate of glucose synthesis. On the other hand, the tests introduced by Schimassek, in which stationary levels of crucial high energy intermediates and other metabolites in the liver are measured have proved extremely useful and will be evoked frequently as a basis of comparison between the liver perfused and *in vivo*, as well as to compare various techniques of perfusion. Simple physical parameters, pressure of perfusion, flow rate, wet weight/dry weight ratio of the liver, and the rate of bile flow are commonly determined and are important pre-requisites of normal liver function. None of them is in itself sufficiently critical to be a sensitive indicator of the state of the perfused organ. Clearance studies of chemicals such as bromsulphthalein (B.S.P.) or of particles such as carbon or bacteria are used but again appear to be relatively insensitive to the method of perfusion. B.S.P. clearance is discussed and illustrated on pp. 146 and 156. No truly comparative studies using this parameter have been conducted. The clearance of particles from the perfusion medium appears to be more closely related to the composition of the medium itself than to the state of the liver (see Jeunet *et al.* (1967) and Filkins *et al.* (1965)).

For biochemical experimentation, metabolic parameters of liver function would seem to be the most appropriate, and a variety of such tests can be suggested. Leakage of enzymes (Schimassek 1968) from the liver occurs when liver damage is apparent, but it is a late event resulting from extensive disruption of the cell membrane and is therefore not a sensitive indicator of the state of the liver during perfusion. Detailed experiments following the time course of loss of enzymes from the liver which is purposely damaged have been conducted in perfusion by Bohley *et al.* (1966). An objection to this as a measure of liver function is the non-linear rate of enzyme leakage with time (see Schmidt 1968), so that the results are at best semi-quantitative.

Leakage of metabolites, particularly glucose, and of ions, particularly $K^+$, has long been suggested as a sign of liver damage (Krebs *et al.* 1951) and tests based on determination of the loss of $K^+$ from the liver during perfusion have been widely applied (for example, by Frimmer, Gries, and Hegner 1967). Mortimore (1961) demonstrated that $K^+$ loss could be partially reversed by insulin added to the perfusion medium, and there seems to be no support from modern liver perfusion studies in the rat for the widely held view (Kestens 1964, Hardcastle and Ritchie 1968) based on work in the dog, that $K^+$ loss implies anoxia in the perfusion. The same experimental conditions under which glycogen breakdown and $K^+$ loss are demonstrated may be used to demonstrate oxygen consumption and energy requiring reactions at rates equal to or higher than those shown by the liver *in vivo* (Hems *et al.* 1966). It must be presumed that such potassium loss as occurs does not significantly interfere with these measures of liver function. More critical studies would help to resolve this apparent anomaly; studies analogous to those of Page *et al.* (1968) in the perfused heart or an extension of the studies of membrane potential in the isolated perfused rat liver (see, for example, Claret and Coraboeuf 1968) may show whether potassium loss during perfusion is a critical test of liver function.

Shedding of glucose by the liver is now well understood, and should not be considered as a sign of anoxia. Nor does its occurrence indicate liver failure (see, for instance, Kestens 1964)—rather the contrary—since it results, in fed rats, from the breakdown of glycogen and indicates the continuing activity of phosphorylase. Indeed, the stimulation of glucose shedding by glucagon, which is consistently observed in the perfused liver, has been itself suggested as a test of liver function (Miller 1961). When glycogen breakdown is prevented, by starting with a starved rat (Hems *et al.* 1966) or by adding glucose to the perfusion medium (Gordon 1963, Sokal *et al.* 1958), virtually no glucose is released into the medium. Glucose shedding *per se* is therefore no guide to liver function. (Note that glucose shedding (glycogenolysis) must be carefully distinguished from glucose synthesis from small molecules (gluconeogenesis), a process that may itself be a useful measure of liver function (Hems, *et al.*, see below).) Release of cholesterol and release of lipid both occur at a rate linear with time and might therefore be useful as tests of liver function, but it is not clear what proportion of each is newly synthesized and what is simply released from 'depots' (Heimberg, Dunkerly, and Brown (1966)). Synthesis is energy dependent, although to a lesser extent than in either urea or glucose synthesis, while release from depots is presumably not. Such complexity adds a source of variation that is undesirable in a test of function. In addition, lipid analysis is generally more difficult and time-consuming than the analysis of simpler carbohydrate precursors, and these tests have therefore not been widely applied.

*Synthetic reactions as test of function*

While earlier workers relied principally on the macroscopic and histological appearance of the liver, the oxygen consumption, the oxygen content of the outflowing blood, and the rate of bile production, Hems *et al.* (1966) discuss the application of tests of biosynthetic capacity as more telling criteria of function. Synthetic functions suffer very early when a tissue deteriorates and, in liver, the most exacting synthesis in terms of ATP requirements are gluconeogenesis and urea formation. Six moles of ATP are required for the synthesis of 1 mole of glucose from lactate, and 4 moles of ATP for the synthesis of 1 mole of urea. Since the rate of glucose and urea synthesis can each exceed 1 $\mu$mol/g fresh weight per min, the rates of the ATP requirement in the presence of lactate and ammonium salts are high enough to claim the greater part of the total energy supply of the liver.

Tests based on this observation are discussed below, and have been extended to the assessment of other organs in perfusion (see Chapter 4).

(i) *Protein synthesis.* The incorporation of [14]C lysine into protein has been observed and advocated as a test of liver function (Miller *et al.* 1951). The rates of reaction are relatively low, but are linear with time over the course of several hours.

(ii) *Glucose synthesis.* (a) *Endogenous.* A low rate of glucose synthesis is observed by the appearance of glucose in the medium of the perfused starved-rat liver. Although initially this is linear with time (up to

30 or 40 min) it falls thereafter, presumably due to lack of substrate. As a quantitative test of function therefore, it is of limited value. (b) *Substrate-stimulated maximal rates*. When lactate, for example, is added at saturating concentrations (10 mM and above), glucose synthesis occurs at a highly reproducible rate (1 $\mu$mol/min per g) for periods of up to 2 h, and this time can be extended by the further addition of substrate. The rate, calculated from the linear part of a curve of glucose content of the medium, versus time, and expressed per gramme wet weight/min has been very useful as an objective criterion in assessing modifications to the technique of perfusion (Hems *et al.* 1966).

(iii) *Urea synthesis*. (a) *Endogenous*. Miller has demonstrated an almost linear rate of urea synthesis without added substrate in livers perfused with whole rat blood. Obviously, the medium itself provides some substrate, and in experiments conducted with semi-synthetic media, the observed rate of 'endogenous' urea synthesis is somewhat lower. In Miller's hands, the rate of urea synthesis was reproducible when expressed per $cm^2$ surface area of rat, but not per g of liver. (b) *Substrate-stimulated maximal rates*. More reproducible is the maximal rate of urea synthesis that is obtained when a substrate (e.g. $NH_4Cl$) together with the catalyst, ornithine (Krebs and Henseleit 1932) are added.

For two reasons, the rate of glucose synthesis from lactate may be a more sensitive criterion of function than is urea synthesis, even in the presence of ornithine. Thus, although the rates of the two reactions are similar, gluconeogenesis requires 30 per cent more ATP, mole for mole, and involves a longer sequence of enzyme reactions than does urea synthesis (see Krebs (1964)). Further, when both reactions are allowed to proceed together, by the addition of lactate (10 mM) and $NH_4Cl$ (10 mM), urea synthesis is partially suppressed. Adding ornithine restores maximal rates, suggesting that the observed inhibition is not dependent upon lack of ATP.

Using some of the suggested tests of function which may be applied to preparations of perfused liver, some details of the results are included in the discussion that follows on assessing the effect of minor modification in technique.

## Choice of a technique for liver perfusion

A number of variations in technique has been introduced (see Methods), often for what seem arbitrary reasons. In the interests of

14

standardization of method (see Introduction) it is strongly suggested that one of the principal existing methods should be adopted and followed accurately, unless there is some pressing need for modification. The reasons for this are clear, and it has many advantages for the new investigator of perfused liver. Each innovation that is introduced should be tested by one or more parameters and compared precisely with the basic method. This is routine in simpler biochemical techniques but has often been neglected in the complexity of organ perfusion. It should now be possible to use the accumulated volume of published work on perfused rat liver to make an objective comparison of the several modifications introduced. In the following section an attempt is made to do this, using the 'tests of function' discussed in the preceding section.

### (a) *Comparison of isolated and* in situ *techniques*

At the outset, it should be said that there is a place for both methods in biochemical experimentation. The first essential is to establish any difference in metabolic function that might be attributed to the situation of the organ, either in the rat or in an organ chamber. For this purpose, the data of Miller *et al.* should be compared with that of Mortimore, in which a similar perfusion medium is used but the liver remains *in situ*, and this comparison is supplemented by one between the methods of Schimassek (or Menahan *et al.*) and Hems (or Exton and Park). These pairs of methods, both using semi-synthetic medium, differ in that the liver is isolated in one case and *in situ* in the other (see Methods). In such a retrospective study it has not been possible to find methods that differ only in this one feature, although some experiments by Menahan *et al.* (1968) and Hems *et al.* fall into this category. Such additional differences as occur are discussed separately in later paragraphs, and unless the contrary is specifically stated, they are not considered to contribute significantly to differences in results.

In some recent experiments, Menahan *et al.* (1968) used the medium proposed by Hems *et al.* (1966) for isolated liver perfusion and incidentally, examined some of the parameters of function investigated by the latter authors. Table 3.6 presents a comparison of data obtained by these two groups of workers, and shows that in most respects the results of *in situ* and isolated liver perfusion are similar. The differences, particularly the marked difference in the ratio $\beta$ hydroxybutyrate/ acetoacetate, may be related to the speed of preparing the perfusion, but this has not been conclusively established.

TABLE 3.6

*Some comparable data obtained in an isolated and an* in situ *liver perfusion: Menahan* et al. (1968) *compared with Hems* et al. (1966)

| Parameter | Method of perfusion: Isolated (ref.) (starved 24 h) | Method of perfusion: *In situ* (ref.) (starved 48 h) |
|---|---|---|
| Wt. of liver | 6 g (a) | 6–8 g (c) |
| Wt. of rat | 120–200 g (a) | 200–250 g (c) |
| Flow | 9·5 ml/min (a) | 15–25 ml/min (c) |
| Pressure | 8–12 cm $H_2O$ (a) | 20–30 cm $H_2O$ (c) |
| Bile flow | 0·5 ml/h (a) | 0·4 ml/h (c) |
| Glucose synthesis | | |
| (i) from lactate | 1·06 μmol/min/g (a) | 1·06 μmol/min/g (c) |
| (ii) from pyruvate | 0·7 μmol/min/g (b) | 1·02 μmol/min/g (d) |
| (iii) effect of glucagon | 2·64 μmol/min/g (a) | 1·86 μmol/min/g (d) |
| (iv) effect of oleate | 1·35 μmol/min/g (a) | 1·72 μmol/min/g (d) |
| Ketone synthesis (from 1 mM oleate) | 2·0 μmol/min/g (a) | 1·4 μmol/min/g (e) |
| Ratio β OHB/AcAc | | |
| (i) no substrate | 0·23 μmol/min/g (a) | 1·2 μmol/min/g (e) |
| (ii) with oleate | 1·03 μmol/min/g (a) | 3·4–2·7 μmol/min/g (e) |

References: (a) Menahan *et al.* (1968); (b) Menahan and Wieland (1969); (c) Hems *et al.* (1966); (d) Ross *et al.* (1967); Krebs (1968).

The quantitative and qualitative evidence appears overwhelmingly to lead to the conclusion that there are no significant differences between the liver perfused *in situ* and that perfused in isolation.

*General considerations.* The technique of liver perfusion *in situ* (Method II) has the following advantages.

(i) The operative technique is simpler and less liable to error. It may be readily performed by a single operator, since the tricky stage of transferring the liver from its situation in the rat to the organ bath is omitted.

(ii) Once set up, the preparation requires only a minimum of attention and may continue to function for the desired length of time without further adjustment. In contrast, the isolated liver appears to be much more susceptible to minor alterations in the lie of the liver, the position of cannulae, etc. and generally requires very close attention to ensure constancy of perfusion flow.

(iii) The liver is not handled and is therefore not exposed to mechanical damage. Haft and Miller (1958) comment that livers appear to fall into two groups, and explain the variable response of the perfused organ to insulin on this basis. Their explanation is that livers more carelessly handled do not respond to insulin. Mortimore, using the *in situ* technique, describes insulin effects that have not been satisfactorily reproduced in the isolated liver. Although there may be other explanations for these results, there do appear to be advantages in not handling the liver, as is the case when the *in situ* method is employed.

(iv) In the *in situ* technique, the bile duct need not be cannulated, since bile then drains outside the system into the duodenum. In the isolated liver, as has already been discussed, the bile duct must be cannulated for consistent results since bile entering the perfusion medium might conceivably have metabolic effects. The alternative, that of tying the bile duct (Abraham *et al.* 1968) would seem *a priori*, to be unsatisfactory since biliary obstruction has metabolic consequences for the liver (Lee and Haines, unpublished).

The isolated liver has the following advantages.

(i) The liver may be totally enclosed within an organ chamber, allowing total gas collections within the sealed system (see, for example, Morris (1960)). Loss of fluid by evaporation, and of gases by diffusion (Miller *et al.* (1951) cites Rostorfer *et al.* (1943)) are avoided in this way.

(ii) Sampling of liver lobes during perfusion is simplified if the liver is outside the rat. Not only are the individual lobes more accessible, but the ligation of the vascular pedicle, so vital in the liver *in situ* if perfusion is to continue without loss of medium, is not necessary. Medium flowing from the pedicle is collected in the organ chamber, and the flow to the remaining organ appears to be unaffected.

(iii) Freeze clamping for rapid sampling of liver (see Methods) is more easily achieved with the liver isolated and, if necessary, the whole organ may be clamped. This is rarely possible with the liver *in situ*.

(iv) Some of the surface measuring techniques (see Chapter 2) are dependent on the accurate placing and fixation of the liver during perfusion, and this may limit them to the liver in isolation.

*Conclusion.* For standard biochemical experimentation, the *in situ* technique is recommended. However, since no differences in metabolism between the two preparations has been found, the isolated liver should be used when frequent liver samples or surface measurements are to be the primary aim of perfusion.

(b) *Comparison of 'reverse' and normal portal vein perfusion* (Method III versus Methods I and II)

Some discussion of the method of reverse perfusion has been given on p. 179, following the description of the method. However, too little comparative data is available for a complete assessment of the method to be made. It is not recorded, for example, whether the liver perfused in this way produced bile, and the method might be dismissed but for the observation by Felts and Mayes of the 'uptake' of chylomicrons in reverse but not in 'forward' perfusion (Mayes and Felts 1967). One further advantage, suggested by the experiments of Arnold and Rutter (1963), is that in reverse perfusion the portal vein may be cut flush with the liver, ensuring that all pancreatic tissue is excluded from the organ chamber. Amylase production was 81 units as against 90 units in forward perfusion, which confirms the normal function of the liver perfused in the reverse direction and suggests that slight contamination with pancreatic amylase could occur in portal vein preparations. In only one study has oxygen consumption been determined in the reverse-perfused liver (Bristow and Kerly 1964). Using a simple saline medium, a figure of $0.62$–$0.67$ $\mu$mol $O_2$/min per g for the first 30 min of perfusion was observed. This is only 30 per cent of the figure found in forward perfusion (cf. Schnitger *et al.* 1965) and it is probable that the reverse preparation is functioning sub-optimally.

The standard portal perfusion methods are adequate for all biochemical purposes, and the technical difficulties which led Trowell to attempt reverse perfusion have been overcome.

(c) *Comparison of the principal perfusion media*

The almost limitless variation which has been applied to the composition of the perfusion medium is treated in some detail in Chapter 1. In the present discussion the media listed under each of the principal methods of perfusion will be considered, and the comparison is simplified if the conclusions of paragraph (a), that there is no significant difference between the metabolism of the liver perfused *in situ* or perfused in isolation, are accepted. Thus the following media are relevant:

Method 1A    *Medium* I. Heparinized rat blood, diluted by $\frac{1}{3}$ Ringer's solution and gassed with 100% $O_2$.
*Medium* II. Heparinized rat blood, 50% diluted and containing 35 mM glucose.
*Medium* III. As Medium II, with amino acid supplement.

Method 1B    *Medium* I. Schimassek q.v. (includes tetracycline).
*Medium* II. Menahan *et al.* q.v.
*Medium* III. As Schimassek, but haemoglobin-free (Williamson *et al.* 1966).

Method 2    *Medium* I. Hems *et al.* q.v. plus variations.

It was concluded from the discussion of the tests of liver function applied to perfusion that tests of synthetic reactions, with their high ATP requirement and dependence on a sequence of linked enzyme reactions, are a more sensitive indicator of the survival of hepatic cells than are the older physiological and physical parameters of function. Nevertheless, it is important to establish that the latter are within the expected limits for each of the perfusion media advocated. The media will be compared under the following headings: flow rate; perfusion pressure; duration of perfusion; macroscopic and microscopic appearance of the liver; bile flow; phagocytosis and metabolic functions including oxygen consumption, enzyme release and enzyme synthesis, from the liver.

(i) *Flow rate.* As expected, the use of more viscous perfusion media, such as whole blood, results in lower overall flow rates. The highest flow rates are obtained with haemoglobin-free media (e.g. 60 ml/min (Scholz 1968)); but in itself the flow rate is immaterial, provided that sufficient oxygen is supplied to cover the requirements of the liver under the conditions of the experiment. The minimal acceptable flow rates for some standard media at basal and maximally stimulated oxygen uptakes are included in Table 3.7.

Experiments in which lower flow rates are used should be carefully considered for the possibility of anoxia. In addition, the existence of 'shunts' postulated by Kessler and Schubotz (1968) (and see p. 122)

TABLE 3.7

*Flow rate of perfusion medium required for minimum oxygen transport*

| Medium | Hb content | Oxygen content (ml/ml) | (μmol/ml) | Minimum flow rate (ml/min) A | B |
|---|---|---|---|---|---|
| Miller | 3·5–5·5 g% | 0·07–0·1 | 3·2–4·4 | 0·57–0·78 | 1·8–2·5 |
| Schimassek | 10 g% | 0·16 | 7·0 | 0·36 | 1·14 |
| Hems | 2·5 g% | 0·058 | 2·57 | 0·98 | 3·1 |
| Williamson/Bücher | Nil | 0·024 | 1·04 | 2·4 | 7·7 |

*Notes.* Oxygen solubility is assumed to be 0·024 ml $O_2$/ml. Hb is assumed to bind 1·34 ml $O_2$/g, and 1 mol = 22·4 litres. Flow rate expressed as ml/min per g of liver. A: $O_2$ uptake basal, 2·5 μmol/min per g. (Schimassek 1963). B: $O_2$ uptake maximal (Hems *et al.* 1966; with lactate and ammonia+ornithine at saturating concentrations, 8 μmol $O_2$/min per g.

suggests that flow rate alone may not be a sufficient indicator of uniform provision of oxygen.

(ii) *Perfusion pressure.* In constant hydrostatic perfusion systems (e.g. Miller) most authors have found a perfusion pressure of 15–25 cm $H_2O$ to give satisfactory flow rates. In constant flow systems (Mortimore 1959) the pressure induced by direct pumping is of the same order. None of the media here under discussion causes the increased perfusion pressure (inflow or outflow block) discussed by earlier workers (Brauer *et al.* 1951).

(iii) *Duration of perfusion.* No systematic studies of the effect of medium composition on duration of perfusion are available. Authors using rat-blood media (Miller, Brauer, and others) obtained perfusion of from 6 to 9 h duration with apparently uniform function. Hems *et al.* suggest an experimental time of 130 min, although the liver does not 'fail' at this time. Schimassek records 3- and 4-h perfusions, but comments that the liver after 6 h shows an increased tendency to shed enzymes into the medium. However, it is not generally the purpose of perfusion for biochemical experiment to obtain very prolonged organ survival. Some component which allows prolongation of perfusion time may be missing from the semi-synthetic media and hence it is of special importance to confirm the viability of the organ for the required duration of an experiment in such systems. The remaining discussion is concerned particularly with short term experiment, up to 4 h, since there is very little information on longer experiments (see, however, Kim and Miller (1969)).

(iv) *Appearance of the liver.* Using media that contain red cells, the perfused liver assumes an appearance similar to the liver *in vivo*. Perfusion with haemoglobin-free media results in an organ of yellow colour, and Bücher's method of applying back pressure results in swelling, so that the external appearance of the liver is grossly abnormal.

*Microscopy.* Many authors report the 'normal' appearance of the perfused liver to light microscopy (Miller, Schimassek, and Hems) with no differences observed which might be attributable to differences in composition of the medium. However, there are no strictly comparable experiments of this type.

*Electron microscopy.* The appearances obtained by Jones, Ruderman, and Herrara (1966), using a haemoglobin-free medium to perfuse the liver with lipid, and those obtained by Hamilton (1967) in similar experiments but using a blood-containing medium as described by Mortimore, are so similar that without a direct comparative experiment this may be accepted as evidence of the identity of these two systems. In the face of this evidence, the minor but significant metabolic differences between the two systems suggest that electron-microscopic evidence alone (e.g. Slater, Sawyer, Delaney, and Bullock 1968) may be inadequate to distinguish normal from abnormal function.

(v) *Bile flow.* The effect on the constitution of the bile of variation in the composition of the perfusion medium is discussed by Brauer (1958). On the other hand, the rate of formation of bile appears to be similar for the various media presently under discussion (Table 3.8).

(vi) *Phagocytosis.* Phagocytosis of carbon (and presumably also of bacteria) requires a plasma factor (Filkins *et al.* 1965), and there is therefore a difference between this parameter determined in perfusions with Miller's medium and with Schimassek's. Since this difference is almost certainly related to a 'coating' of the particles to be engulfed, it cannot strictly be considered as a difference in liver function.

(vii) *Metabolic functions: oxygen consumption.* Comparison between experimental systems based on the determination of oxygen consumption under 'basal' conditions, i.e. without added substrates, may be misleading. Rat blood media contain sufficient substrate to stimulate

TABLE 3.8

*Bile formation and oxygen consumptions with various perfusion media*

| Medium | Bile flow ($\mu$l/min/g wet) | Oxygen consumption ($\mu$mol/min/g wet) |
|---|---|---|
| Miller | 0·25–0·5 | 3·0 (at 1 h, Forsander *et al.*) |
| Brauer | 1·0 | — |
| Schimassek | 1·0–1·5 | 2·2 |
| Hems | 1·0 | 2·2 (starved) |
| Williamson/Bücher | 0·085–0·05 in first hour | 2·0 |

respiration in the liver, and this is the most probable explanation for the observations of Schimassek, illustrated in Fig. 3.16. Oxygen consumption with a Miller-type medium is significantly higher than with the standard medium of Schimassek in the first 30 min of perfusion. Thereafter the results are very similar for the two media.

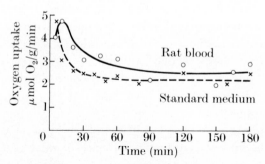

FIG. 3.16. Oxygen uptake by the perfused rat liver with whole rat blood or standard (synthetic) perfusion medium (Schimassek 1968).

Further comparisons between the results of workers using different media are included in Table 3.8. As discussed in Chapter 2, oxygen consumption is a useful, and possibly also a continuous, record of liver function. On the other hand, the results of Scholz and others demonstrate that these figures should not be considered in isolation. Uncoupling of oxidative phosphorylation may be the explanation for the apparently normal rate of oxygen uptake in their experiments, in which the rates of gluconeogenesis from various precursors are significantly lower than those observed by other workers using blood-containing media (Hems, Exton and Park, Menahan, and others).

(viii) *Release of enzymes.* Schimassek, Scholz, and Schmidt and Schmidt, among others, discuss the appearance of hepatic enzymes in the medium but no parallel data are available to enable a comparison of these results and those of perfusion with rat blood to be made. Moreover, it is unlikely that such information will be a more sensitive indicator of liver function than are the tests of synthetic reactions to be discussed below.

(ix) *Enzyme synthesis.* Only amylase synthesis has been systematically studied as a test of perfusion media. McGeachin *et al* (1967) give the following rates of synthesis for perfusion according to Method I (Brauer modification), with the following additional modifications: gassed with 100% $O_2$ (control); gassed with 100% $N_2$; livers damaged by exposure to N-nitrosodimethylamine; medium of washed red cells, suspended in buffer (cf. Schimassek):

$$38 \cdot 7; \quad 17 \cdot 8; \quad 8 \cdot 3; \quad 33 \cdot 6.$$

Thus, while liver damage is clearly reflected in a reduced rate of enzyme synthesis, the substitution of a non-blood medium does not significantly alter the rate from that of the control. Further parameters of function may be compared only on a qualitative or at best a semi-quantitative basis, in the absence of systematic studies of this point. Thus, protein synthesis, amino acid release, urea synthesis, and the metabolism of a wide range of small molecular substrates appears to be similar in preparations perfused with rat blood or with a semi-synthetic medium.

One outstanding difference lies in the observation of α-oxoglutarate uptake at a significant rate in the presence of rat blood (Miller *et al.* 1955), while Hems *et al.* have consistently failed to observe metabolism of this substrate except at a very low rate. Since the same workers also failed to observe uptake of α-oxoglutarate when rat blood, according to the description of Miller, was substituted for the medium of Hems, this apparent difference between the work of different laboratories must be of limited significance.

*Conclusions.* Such parameters of function as are available suggest that rat blood perfusion media (Miller) and semi-synthetic media of the type described in Methods IB and II are equally effective for metabolic studies in the perfused rat liver. The choice of medium remains with the investigator, and should be based upon the nature of the problem to be investigated.

### (d) *Minor variations in technique*

A number of minor technical details have been considered to be critical in the application of the methods of perfusion already discussed. These are as follows.

(1) *Constant pressure (hydrostatic supply) versus constant flow (direct pumping)*. Provided that the flow is adequate to oxygenate the tissue (Table 3.7) this modification has no effect upon liver metabolism. Comparison of the results of Exton and Park (1967) with those of Hems *et al.* (1966) shows the rate of glucose synthesis from lactate to be the same, while the tissue content of ATP, ADP and the ratio ATP/ADP are very similar for the two methods of perfusion.

(2) *Oxygenator, multibulb glass oxygenator versus rotating drum (Mortimore) or rotating mesh (Bücher)*. No differences can be attributed to the difference in oxygenator, as will be seen by comparing the rates of glucose synthesis, etc., i.e. the comparison between Exton and Park and Hems *et al.* discussed on p. 194. The Bücher and Scholz oxygenator is assessed by comparison with the similar haemoglobin-free perfusions of Williamson, Kreisberg, and Felts (1966) which use a multibulb oxygenator (see p. 201). There are therefore no differences that may be ascribed to these oxygenators but other considerations which influence the choice of an oxygenator are discussed in Chapter 1.

(3) *Siphoning*. Flock and Owen (1965) draw attention to the possibly deleterious effect of medium being siphoned through the liver, due to the arrangement of the venous cannula, and they introduce a special modification to counteract this phenomenon (see Fig. 3.15). Since their method has been used by Lardy for metabolic studies it is possible to state that there is no difference in the liver, as judged from its ability to form glucose from alanine or, less significantly, from fructose, at the same rate in the preparations of Hems *et al.* or Exton and Park and those reported by Veneziale *et al.* (1967). Flock and Owen's precaution is therefore probably unnecessary.

(4) *Pulsatile flow versus steady flow*. Although the portal flow is probably non-pulsatile *in vivo*, some authors recommend the use of a pulsatile inflow, for example, Kvetina and Guaitani (1969). This can be of two kinds.

(a) *Vascular pulsation*. The direct pumping method of Mortimore inevitably leads to a pulse wave in the portal vein (see Pumps).

Comparison of the results of Exton and Park with those of Hems *et al.* excludes any significant metabolic consequence of this difference in method.

(b) *Respiratory pulsation.* Ryoo and Tarver (1968) introduced a very complex method of liver perfusion, in which they claim that protein synthesis is continued during up to 12 h of perfusion. The modification allows the liver to be immersed in perfusion medium, and raised and lowered ten times per minute, simulating the movements of the liver due to respiration *in vivo.* Comparing the rate of protein synthesis in this system with that found by Miller is not directly possible, since these workers used [14]C leucine whereas Miller used [14]C lysine, and the results are expressed in different units. Nevertheless, both methods demonstrate a linear rate of protein synthesis by this method, provided substrate is available, and Miller's experiments have been conducted for as long as 12 h (Kim and Miller 1969). On qualitative grounds too, the two preparations are similar. Both Ryoo and Tarver (1968) and Tracht, Tallal, and Tracht (1967) using Miller's method, found the rate of protein synthesis to be inversely proportional to the osmotic pressure of the perfusion medium. It would appear to be unlikely that this rather elaborate modification of the original method is required for short-term biochemical experimentation. The general conclusion is that in the liver of the rat, as in many other organs, no case can be made out on metabolic grounds for the use of a pulsatile perfusion pressure.

(5) *Gas tension.* In the Introduction, the possible metabolic consequences of working with high oxygen tension rather than the 'normal' ranges of oxygen tension to which the tissue is exposed *in vivo* are discussed. In the liver there is evidence of damage to lysosomes which may be due to high oxygen tension in the perfusing medium. Abraham *et al.* (1968) in a perfusion system based on that of Felts and Mayes (q.v.) but with additional perfusion through the hepatic artery, found electron microscopic evidence of damage to lysosomes when the medium was gassed with 95% $O_2$. Gassing with air apparently resulted in no such evidence of damage. That lysosomes are disrupted during perfusion of the liver is evidenced by the detection of lysosomal enzymes in the perfusion medium from the beginning of perfusion by the method of Schimassek (Frimmer *et al.* 1967). The rate of release of enzymes is increased in these experiments when the standard medium of Schimassek is replaced by the haemoglobin-free medium of Bücher. Both experiments were carried out in the presence of 95% $O_2$, so that the oxygen

*content* of the latter medium is obviously lower. It is difficult to reconcile this observation on the enzyme release with the microscopic evidence of Abraham *et al.* and the question of oxygen toxicity remains open.

Despite the evidence of lysosomal damage, livers perfused with media with 100% oxygen retain function equal to or in excess of the rates observed *in vivo*. The occasional experiments conducted with air as the gas mixture do not appear to improve biochemical function (see, for example, Hems *et al.* (1966) and Ross *et al.* (1967)).

*Hyperbaric oxygen.* Similarly, there is no indication that providing an excess of oxygen by working under conditions of 2·5 atm pressure (Brauer *et al.* 1956) has any effect on liver metabolism.

*Oxygen lack.* On the other hand, oxygen lack has demonstrable metabolic consequences. Hoberman and Prosky (1967) demonstrate fumarate reduction, and a number of authors comment on the occurrence of anaerboic glycolysis in liver. Anoxia or hypoxia is to be avoided therefore, except for controlled experimental purposes.

(6) *Submersion of the liver.* Authors concerned about dessication of the isolated liver have devised perfusion systems in which the organ may be submerged in liquid paraffin (Gordon 1968, Powis 1970) or in 0·9% saline (Forsander *et al.* 1965). Comparison of the results of Menahan *et al.* and Powis, on the basis of glucose synthesis from lactate, suggest that immersion of the liver has no metabolic consequences. Discussion on this point, therefore, centres on the practicability of such immersion for biochemical experiment. Access to the liver for surface measurement or for sampling of lobes is seriously impeded, and the versatility of the preparation is thereby limited; on the other hand, in the carefully designed organ chambers normally used, dessication is not a serious problem and apparently does not lead to any loss of metabolic function.

(7) *Cannulation of outflow.* (a) *Portal vein perfusion.* Fisher and Kerly (1964) remark upon the presence of lymphatics along the upper inferior vena cava and suggest that tying a cannula into this vessel might obstruct these vessels, with metabolic consequence for the liver. The data obtained in their early experiments on amino acid release from the liver do not differ significantly from those of Schimassek, who cannulated the outflow in the usual way. In practical terms, cannulation of the outflow is important for clean sampling, and essential to the *in situ* technique. It appears to be without effect, provided that flow of medium is well maintained. (b) *Reverse perfusion.* Trowell (1942) comments, on the other hand, on the lymphatics that accompany the portal vein, which is the outflow in this preparation. Bollman's

description of the hepatic lymphatic vessels in the rat supports Trowell's view, but no direct study has been made of the importance or otherwise of free lymphatic drainage to liver function in perfusion. A comparison of forward and reverse perfusion methods (see above, p. 197) is too imprecise and does not reveal any differences that might be ascribed to this cause.

(8) *Antibiotics*. The problems of infection and the multiplication of bacteria in perfusion systems have been discussed in the Introduction. In liver perfusions it is common practice to include an antibiotic in the medium, and because this is rarely the case for other preparations *in vitro*, the effects of the additions must be examined in some detail.

(a) *Incidence of 'infection' in liver perfusion.* Several workers comment that samples of medium cultured during the course of perfusion show no, or minimal, bacterial growth. This includes workers who employ antibiotic additions in the medium (Schimassek and Brauer *et al.*) as well as those who do not (Hems *et al.* and Miller). No systematic study of this aspect has been published and experiments in which it is critical to have sterility are best conducted in special apparatus, for example, that of Carrell and Lindbergh (see p. 84).

Examination of data on biochemical experiments conducted in perfused liver have occasionally given rise to speculation that bacterial growth may have a contributory effect (see particularly Köblet (1968)) in that RNA 22s appears in the perfusion medium, a possible sign of bacterial multiplication (Bücher 1968).

An important cautionary note to the assessment of 'infection' by culture of the perfusion medium is the consistent observation that the perfused liver is capable of phagocytosis of bacteria, and that multiplication of phagocytosed organisms probably occurs thereafter within the Kupffer cells (Bonventre and Oxman 1965). The only sure test therefore, is, to culture both medium and liver.

(b) *Effects of antibiotics on perfused liver.* No systematic studies have been undertaken. Tetracycline is used by Schimassek; penicillin plus streptomycin by Brauer; ampicillin or tetracycline by Willms and Söling (1968) and Söling *et al.* (1966); and penicillin by Miller *et al.* 1951; while Hems *et al.* (1966) and Exton and Park (1967) use no antibiotic as a routine.

On comparing, therefore, the results of Schimassek, Söling, and others, with those of Hems or Exton and Park for glycolytic intermediates, adenine nucleotide content, and rate of glucose synthesis from lactate, tetracycline appears to have no effect on liver function in perfusion. Equally, the absence of an antibiotic does not appear to result in any defect in these parameters of liver function, and it is therefore unlikely that micro-organisms are contributing significantly to the results found. Especially reassuring is the linearity of glucose formation with time (Hems *et al.* 1966), since glycolysis by micro-organisms present in significant quantities might be expected to interfere with this test.

That antibiotics do have metabolic effects on the liver in perfusion has been amply demonstrated. Actinomycin D or puromycin are used, with the expected effect of inhibiting protein synthesis in a number of studies. Penicillin, in a dose of 1000 units/ml (the standard concentration in perfusion media is 100 units/ml, used, for example, by Brauer *et al* (1951)) results in a doubling of the rate of bile formation and a significant decrease in clearance of BSP and of $^{131}$I tetraiodothyronine (Gorman, Flock, and Owen 1967) in a perfusion similar to that of Miller. Under these circumstances penicillin is in fact secreted into the bile as well as being metabolized by the liver (Kind *et al.* 1968), and it might advisably be omitted from biochemical perfusion experiments for these reasons. In direct experiments to assess the possible effects of penicillin on protein synthesis, McGeachin *et al.* (1967) used the net synthesis of amylase, and Juchu *et al.* (1965) the induction of enzymes by 3:4 benzpyrine, to demonstrate that it had no such inhibitory effects as were clearly demonstrable with puromycin and actinomycin D respectively. The amylase test is more critical (see Choice of tests), and penicillin in the perfusion medium resulted in synthesis at a rate at the lower limit of the rather wide normal range. In the experiments based on the induction of enzymes by 3:4 benzpyrene, streptomycin was without inhibitory effect. In conclusion, for short-term perfusion of 3–4 h duration, antibiotic additions appear to be unnecessary. Tetracycline at the concentration suggested by Schimassek is not suspected of metabolic effects, however, and would be the antibiotic of choice. It is essential to assess the effects of this addition in each new experimental situation; in particular, there is no information about its capacity to influence protein synthesis. Nor is there yet convincing evidence that sterility is assured by this means; if guaranteed sterility is required, apparatus of the type designed by Lindberg and Carrell (see Chapter 1) might be the most suitable.

(9) *Anticoagulants* (see Introduction for general discussion). Hems *et al.* (1966) (Method II) inject heparin into the rat a few seconds before the operation begins, and no heparin is used in the medium. The resulting

perfusion is similar in physical and metabolic characteristics to that of Miller (q.v.), in which heparin is added to the perfusion medium but none is injected into the rat. No special study of the effects of anticoagulants has been made in this situation but their use is cited as one of the milestones in the successful application of perfusion to the rat liver (Miller, Shimassek).

On the other hand, perfusion without anticoagulants and using defibrinated blood which has been dialysed before use, has been successfully conducted by Felts and Mayes (1966). At a perfusion pressure of 14 cm blood, flow is at least 7 ml/min and satisfactory perfusion is continued for 4–5 h. No metabolic parameters are available on which to base a critical comparison of the two methods, and the central observation of Felts and Mayes remains controversial. These workers find that there is no uptake of chylomicrons by the rat liver perfused without heparin, and they conclude that activation of lipoprotein lipase by the anticoagulant explains the contrary observation by numerous earlier workers (Morris and French 1958; Hillyard *et al.* 1959; Heimberg *et al.* 1962; Rodbell, Scow, and Chernick 1964). Morris (1963*a*), the only worker apparently to perfuse with and without heparin in the perfusion medium found no difference between the two procedures.

Satisfactory liver perfusion may therefore be conducted without heparin, either in the medium or given to the animal. Its use remains routine, however, and there seems to be no difference between giving the animal heparin and adding it to the perfusion medium.

(10) *Supplements to the medium.* The argument in favour of a simpler defined perfusion medium is substantiated by comparing the results of Miller, Schimassek, and Bücher. (See p. 197.) Nevertheless, there are a wide range of supplements to the perfusion medium which have been used by different workers, and the effect of which may be assessed by comparing biochemical parameters of liver function. A number of the more important ones are considered here; further discussion of this important aspect of perfusion is contained in Chapter 1 (p. 17).

(i) *Glucose.* Glucose is often added to the medium for liver perfusion (for example, by Schimassek (1962)). Its presence does not apparently alter liver survival or influence the physical aspects of perfusion (compare Hems *et al.* with Schimassek). Addition of glucose to the medium maintains glycogen levels in the liver of fed rats (q.v.), but glucose uptake is minimal except under anoxic conditions or at very high concentrations. There is therefore no indication for the routine addition of glucose to the perfusion medium for the liver.

(ii) *Amino acid supplements.* For short-term biochemical experiments there is some tendency to supplement the medium with a mixture of amino acids in approximately their physiological concentrations (for example, by Bücher and Scholz, Korner and Jefferson, and Schimassek). Schimassek (1963) has demonstrated that the perfused liver sheds amino acids until the medium:liver ratio reaches a critical level for each amino acid. Both Miller and Jefferson and Korner (1969) have demonstrated that protein synthesis and enzyme induction are enhanced when extra amino acids are added to the perfusion medium. There is, however, no indication from tests of glucose synthesis of adenine nucleotide concentration or of glycolytic intermediates, that the liver perfusion without amino acid supplement differs from the liver *in vivo* (see Hems *et al.* 1966 and Exton and Park 1967, and compare with data from Schimassek 1962) and there is therefore no case for the routine addition of amino acids to the simple defined medium (Hems *et al.* 1966) or to whole blood perfusion media (Miller *et al.* 1951).

Studies in which maximal rates of enzyme induction and protein synthesis are of interest may, according to the report of Jefferson and Korner (1969) be best carried out using a medium in which amino acids at concentrations up to 200 times the normal are present. In fact, most experiments in which enzyme induction is to be studied have been conducted with a whole-blood perfusion medium (cf. Miller).

(iii) *Glucagon.* Schimassek comments on an abnormality in the concentrations of fructose-I-P and fructose-6-P in the perfused liver compared with those of the organ *in vivo*, concentrations that revert to 'normal' when glucagon is added to the medium. His suggestion is that glucagon may be required for normal liver function (Schimassek 1963). However, so many other effects of glucagon are now documented for the perfused rat liver (see Miller 1961 and Ross *et al.* 1967 for review) that its use as a routine addition cannot be supported.

(iv) *Plasma expanders.* Experiments in perfused liver give some guidance to the choice of agent required to provide the colloid osmotic pressure in semi-synthetic perfusion media (see Media). Bovine serum albumin, Fraction V, is used by most workers (see Schimassek; Hems; Exton and Park; and others). It may be dialysed before use (Hems *et al.*) freed from fat (Williamson, Browning, and Scholz 1969) or used directly from the manufacturer (Hems *et al.* 1966). No systematic study has been made, but the findings of Hems *et al.*, that acetate is present in the product from Armour Laboratory (see Media, Chapter 1) and of Menahan *et al.* (1968) that acetate has a stimulatory effect on

15

glucose synthesis from lactate, suggests that albumin is best dialysed before use in liver perfusion experiments. Comparing the results with commercial albumin and 'defatted' albumin shows no improvement in liver function, and this rather lengthy procedure may be reserved for special situations in which fatty acid metabolism is to be investigated (Williamson *et al.* 1969).

Albumin, Dextran, and Haemaccel are compared on the basis of glucose formation from lactate by Hems *et al.* (1966), with the following relative rates:

$$\text{Albumin} = 1.06,$$
$$\text{dialysed albumin} = 1.0,$$
$$\text{Dextran} = \text{'variable'},$$
$$\text{Haemaccel} = 0.90.$$

They are compared on the basis of the rate of release of enzymes from the liver into the medium, by Shimassek (1968), illustrated here by the mitochondrial enzyme, glutamate dehydrogenase:

| Addition | Albumin | PVP | Dextran | Gelatine |
|---|---|---|---|---|
| GluDH in medium after 4 h | 0 | 1200 | 1100 | 8000 |

Sucrose (0.3 M), used by Myers (1964) in place of a larger molecule, results in the loss of potassium from the liver, so that 70 per cent of the tissues', $K^+$ content appears in the medium within 30 min. This is vastly in excess of the figures suggested by Mortimore (1961) for rat blood perfusion—7 per cent per hour. Omission of a colloid osmotic agent is found to be deleterious to the function of perfused liver if the results of Bloxham (1967), who used this device, are compared with those of Schimassek (amino-acid release).

(11) *Optimum flow rate.* Flow rate in the perfused liver is usually higher than the flow rate *in vivo*, except when whole rat blood is used as the perfusate. Sufficient has been said of the metabolic similarity of these various methods for it to be clear that flow rate *per se* has no metabolic effect. Provided oxygen supply is sufficient, and the rate of substrate delivery to the organ is not the limiting factor, the results of Morris (1963*b*) confirm this. In his studies the rate of uptake of palmitate was found to be constant over a wide range of perfusion pressure and flow rate. A fall in the rate of palmitate uptake at low perfusion pressures (5 cm $H_2O$) (and hence low flow) can be attributed to the reduced oxygen supply and resultant anoxia of the liver.

Contrary results are found by Schreifers (1968) for the uptake of aldosterone and by Morris for chylomicrons, both of which are removed from the medium at a rate inversely proportional to the flow rate. The difference may be due to difference in method of 'uptake' in that physical phagocytosis is required, certainly for the uptake of chylomicrons, and this process might be expected to be relatively slow.

(12) *Recirculation versus once-through perfusion.* The advantages of recirculation in perfusion have been discussed (Introduction) and information from published work establishes that, at least for relatively short perfusions, accumulation of products does not influence liver function. Rates of glucose synthesis, of urea synthesis from ammonia, and of several other reactions remain linear so long as substrates are available (see Hems *et al.* (1966) and Miller *et al.* (1951)). Phagocytosis, which does fail after a period of perfusion with blood-containing media, is revived by changing the medium, and indicates the exhaustion of some component of the blood necessary for phagocytosis.

Workers wishing to use a recirculation method for the liver, but fearing that accumulation of products may damage it, have introduced continuous dialysis of the perfusion medium during an experiment as a means of keeping the composition of the medium constant (Abraham *et al.* 1968). The results of flow rate, the perfusion pressure, and the bile flow do not differ, and it is not possible to judge whether the metabolic function is different since no tests have been carried out in such a system.

(13) *Choice of perfusion volume.* In most perfusion systems devised for biochemical purposes, the proportions between mass of tissue and volume of medium are considerably different from the conditions *in vivo*, and it might be thought that this would influence the metabolic behaviour of the liver. No direct comparison has been made between large- and small-volume systems (see Kvetina and Guaitani 1969).

(14) *Back pressure.* Bücher and Scholz introduced this modification to ensure that the surface of the liver would be adequately supplied with medium, which in their system is haemoglobin-free. Oxygen consumption is similar to that observed in normal perfusion, as are the adenine nucleotide contents, but the rate of glucose synthesis from various precursors appears to be lower. In more recent papers these authors have abandoned this modification, which leads to loss of fluid from the liver surface. This does not occur in the haemoglobin-free perfusion conducted according to standard methods (for example, by Williamson *et al.* 1966), and is therefore attributable to the back

pressure. Alternative methods of dealing with perfusion with haemo-globin-free media are the use of low temperature (for example, Kriegl-stein 1969) or a very high flow rate (Williamson, Browning, and Scholz 1969) and there seems to be no need for the use of this device which damages the liver and impairs its metabolism.

(15) *Hepatic artery perfusion versus portal vein.* There is as yet no evidence of any metabolic consequences of perfusing the rat liver by the portal vein alone. Abraham *et al.* (1968) and Powis (1970) have published results in which the parameters of liver function are un-changed by perfusing both the hepatic artery and portal vein (see p. 183 for details); Powis's observation that glucose synthesis from lactate is not different in the liver perfused via the portal vein or by both routes is conclusive evidence that the hepatic artery may be ignored in this preparation.

(16) *Anaesthetic agent.* Little attention has been paid to the possible persistence of anaesthetic agents in the liver during perfusion, or of effects induced by anaesthesia during the preparation of the animal. No comparative studies have been done; the agents used are ether (Miller *et al.*), pentobarbitone i-p (Hems *et al.* and Jefferson and Korner), amytal i-p (Knox and Sharma 1968), chloralose plus phenobarbitone (Kvetina), urethane (Clodi and Schnack 1966), and 50% $CO_2$ (Miller *et al.* 1951), and perfusion is satisfactory in each case. There may of course be more subtle persistent effects such as, for instance, insulin release during the induction of pentobarbitone anaesthesia (Alberti and Biebuyck, personal communication). The effect of anaesthetic agents added to the perfusion medium is unlikely to be much guide to their function when given to the rat; hexobarbitone, for example, has a half-life of only 13 min in perfusion.

(17) *Temperature.* No difference is found between the results of Hems *et al.* and of Exton and Park which might be attributed to the small difference in temperature used by the two groups of workers. Hems *et al.* (1966) worked at 35°C, while Exton and Park (1967) used 37–38°C. Ideally, perfusion should be conducted at a temperature similar to that used for other methods *in vitro*, most usually at 37°C, and the observation by Hems *et al.* that the liver survives less well at this physiological temperature than at 35°C must be treated with reserve. Wider ranges of temperature may be compared by considering the data of Kalser *et al.* (1968) and of Krieglstein (1969). The first-named studies oxygen consumption at temperatures ranging from 15 to 38°C within one liver; the oxygen consumption is found to be

5 ml/100 g per min ($\equiv 2\cdot2\,\mu$mol/g per min) at 37°C; 2 ml/100 g per mm (0·88 $\mu$mol/g per min) at 25°C, and about 1 $\mu$mol/100 g per min (0·44 $\mu$mol/g per mm) at 17°C (the last two figures are not thought to be significantly different).

## SUMMARY

By the device of comparing published results from various research groups carrying out similar studies under slightly different conditions, it has been possible to suggest which of such differences are critical in liver perfusion. There are objections to such an approach, since there may be other, unrecorded differences in technique which contribute to the results published. However, the use of objective and quantitative tests of organ function helps to give the comparison some substance. The methods described in detail at the beginning of this chapter are considered to be those best characterized and most readily applicable to biochemical studies.

Finally, this approach may be of some value in assessing biochemical techniques of perfusion of organs apart from the liver. The literature is at present too scanty to undertake such a *post hoc* comparison, but early application of the principles discussed here for the liver may avoid some of the wasteful proliferation of methods of organ perfusion and allow reproducible studies to be undertaken by different research groups.

## REFERENCES

ABRAHAM, R., DAWSON, W., GRASSO, P., and GOLDBERG, L. (1968) Lysosomal changes associated with hyperoxia in the isolated perfused rat liver. *Exp. mol. Pathol.* **8**, 370.

AGARWAL, M. K., HOFFMAN, W. W., and ROSEN, F. (1969) The effect of endotoxin and thorotrast on inducible enzymes in the isolated, perfused rat liver. *Biochem. biophys. Acta* **177**, 250.

ARNOLD, M. and RUTTER, W. J. (1963) Liver amylase: III. synthesis by the perfused liver and secretion into the perfusion medium. *J. biol. Chem.* **238**, 2760.

BARNABEI, O. and SERENI, F. (1960) Induzione dell'attivita' tirosina-α-chetoglutarate transaminasica nel fegato isolato di ratto. *Boll. Soc. ital. Biol. sper.* **36**, 1656.

BAUER, W., DALE, H. H., POULSON, L. T., and RICHARDS, D. W. (1932) The control of circulation through the liver. *J. Physiol., Lond.* **74**, 343.

BEL, C. (1969) Conjugation of bromosulfonephthalein by isolated and perfused rat liver. *C. r. Seánc. Soc. Biol.* **162**, 1949.

BERNARD, C. (1855) Sur le mécanisme de la formation du sucre dans le foie. *C. r. hebd. Seánc. Acad. Sci., Paris* **41**, 461.

BLAKELEY, A. G. H., BROWN, G. L., and FERRY, C. B. (1963) Pharmacological experiments on the release of the sympathetic transmitter. *J. Physiol., Lond.* **167**, 505.

BLOXHAM, D. L. (1967) Effects of halothane, trichloroethylene, pentobarbitone, and thiopentone on amino-acid transport in the perfused rat liver. *Biochem. Pharmac.* **16**, 1848.

BOHLEY, P., KLEINE, R. FROHNE M. KIRSCHKE H., LANGNER, J., and HANSON, H. (1966) Langfristige Perfusionen autolysierender Rattenlebern. *Hoppe-Seyler's Z. physiol. Chem.* **344**, 55.

BOLLMAN, J. L., CAIN, J. C., and GRINDLAY, J. H. (1948) Technique for the collection of lymph from the liver, small intestine or thoracic duct of the rat. *J. Lab. clin. Med.* **33**, 1349.

BONVENTRE, P. F. and OXMAN, E. (1965) Phagocytosis and intracellular disposition of viable bacteria by the isolated perfused rat liver. *J. Reticuloendothel. Soc.* **2**, 313.

BRAUER, R. W. (1958) in *Liver function* (ed. R. W. Brauer), p. 113. Publication No. 4, American Institute of Biological Sciences, Washington, D.C.

—— LEONG, G. F., McELROY, R. F., and HOLLOWAY, R. J. (1956) Haemodynamics of the vascular tree of the isolated rat liver preparation *Am. J. Physiol.* **186**, 537.

—— —— and PESSOTTI, R. L. (1953) Vasomotor activity in the isolated perfused rat liver. *Am. J. Physiol.* **174**, 304.

—— PESSOTTI, R. L., and PIZZOLATO, P. (1951) Isolated rat liver preparation: bile production and other basic properties. *Proc. Soc. exp. Biol. Med.* **18**, 174.

BRISTOW, D. A. and KERLY, M. (1964) Transamination in perfused rat liver. *J. Physiol., Lond.* **170**, 318.

BÜCHER, TH. (1968) *Stoffwechsel der isoliert perfundierten Leber* (eds. W. Staib and R. Scholz), p. 11. Springer-Verlag, Berlin.

—— and KLINGENBERG, M. (1958) Wege des Wasserstoffs in der lebendigen organisation. *Angew Chem.* **70**, 552.

BURGI, H., KOPETZ, K., SCHWARTZ, K., and FROESCH, E. R. (1963) Fate of rat insulin in rat liver perfusion. *Lancet* (ii), 314.

BURTON, S. D., MONDON, C. E., and ISHIDA, T. (1967) Dissociation of potassium and glucose efflux in isolated perfused rat liver. *Am. J. Physiol.* **212**, 261.

—— ST. GEORGE, S., and ISHIDA, T. (1960) Small volume perfusion system of the isolated rat liver. *J. appl. Physiol.* **15**, 128.

CHAMBERS, J. W., GEORG, R. H., and BASS, A. D. (1968) Effects of catecholamines and glucagon on amino-acid transport in the liver. *Endocrinology* **83**, 1185.

CHANCE, B. and WILLIAMS, G. R. (1956) The respiratory chain and oxidative phosphorylation. *Adv. Enzymol.* **17**, 65.

CLARET, M. and CORABOEUF, E. (1968) Effets de variations du pH sur la polarisation membranaire du foie du rat isolé et perfusé. *C. r. hebd. Seánc. Acad. Sci., Paris* **267**, 642.

CLODI, P. H. and SCHNACK, H. (1966) Der Einfluss von Diuretika auf die Cholerese der Gallefistelratte und der perfundierten Rattenleber. *Wien klin. Wschr.* **78**, 774.

CRISP L. R. and DEBROSKE, J. M. F. (1952) A simple perfusion pump. *Rev. scient. Instrum.* **23**, 381.

DESPOPOULOS, A. (1966) Congruence of excretory functions in liver and kidney: Hippurates. *Am. J. Physiol.* **210**, 760.

ELIAS, H. and SHERRICK, J. C. (1969) *Morphology of the liver.* Academic Press, New York and London.

ENTEMNAN, C., HOLLOWAY, R. J., ALBRIGHT, M. L., and LEONG, G. F. (1968) Bile acids and lipid metabolism. 1. Stimulation of bile lipid excretion by various bile acids. *Proc. Soc. exp. Biol. Med.* **127**, 1003.

EXTON, J. H., CORBIN, J. G., and PARK, C. R. (1969) Control of gluconeogenesis in liver: IV Differential effects of fatty acids and glucagon on ketogenesis and gluconeogenesis in the perfused rat liver. *J. biol. Chem.* **244**, 4095.

—— and PARK, C. R. (1967) Control of gluconeogenesis in liver. I General features of gluconeogenesis in the perfused livers of rats. *J. biol. Chem.* **242**, 2622.

FELTS, P. A. and MAYES, J. M. (1966) Liver function studied by liver perfusion. *Proc. Eur. Soc. Drug. Toxicity* **7**, 16.

FILKINS, J. P., CHASE, R. E., and SMITH, J. J. (1965) Characteristics of a plasma factor governing carbon phagocytosis in the isolated perfused rat liver. *J. Reticuloendothel. Soc.* **2**, 287.

—— and SMITH, J. J. (1965) Plasma factor influencing carbon phagocytosis in the isolated perfused rat liver. *Proc. Soc. exp. Biol. Med.* **119**, 1181.

FISCHER, A. (1963) Dynamics of the circulation in the liver. *The liver* (ed. Ch. Rouiller), Vol. 1, p. 330. Academic Press, New York.

FISHER, M. M. and KERLY, M. (1964) Amino acid metabolism in the perfused rat liver. *J. Physiol., Lond.* **174**, 273.

FLOCK, E. V. and OWEN, C. A. (1965) Metabolism of thyroid hormones and some derivatives in isolated, perfused rat liver. *Am. J. Physiol.* **209**, 1039.

FORSANDER, O. A., RÄIHÄ, N., SALASPURO, M., and MÄENPÄÄ, P. (1965) Influence of ethanol on the liver metabolism of fed and starved rats. *Biochem. J.* **94**, 259.

FRIMMER, M., GRIES, J., and HEGNER, D. (1967) Untersuchungen zum Wirkungs mechanismus des Phalloidins: Freisetzung von Lysosomalen enzymen und von Kalium. *Naunyn-Schmiedebergs Arch. exp. Path. Pharmak.* **258**, 197.

GANS, H. and LOWMAN, J. T. (1967) The uptake of fibrin and fibrin-degradation products by the isolated perfused rat liver. *Blood* **29**, 526.

GARCIA, A., WILLIAMSON, J. R., and CAHILL, G. F. (1966) Studies in the perfused rat liver. II. Effect of glucagon on gluconeogenesis. *Diabetes* **15**, 188.

GERSHBEIN, L. and ELIAS, H. (1954) Observations on the anatomy of the rat liver. *Anat. Rec.* **120**, 85.

GOLDSTEIN L., STELLA, E. J., and KNOX, W. E. (1962) The effect of hydrocortisone on tyrosine and ketoglutarate transaminase and tryptophan pyrrolase activities in the isolated, perfused rat liver. *J. biol. Chem.* **237**, 1723.

GORDON, A. H. (1968) Perfusion of the rat liver under paraffin oil. *Stoffwechsel der isoliert perfundierten Leber* (eds. W. Staib and R. Scholz), p. 97. Springer-Verlag, Berlin.

GORDON, E. R. (1963) Glucose consumption by the perfused isolated rat liver. *Can. J. Biochem.* **41**, 1611.

GORMAN, C. A., FLOCK, E. V., and OWEN, C. A. (1967) Penicillin induced changes in function of the isolated perfused rat liver: alterations in thyroxine metabolism and sulfobromophthalein. *Endocrinology* **80**, 247.

GREEN, E. C. (1935) *Anatomy of the rat. Trans. Am. phil. Soc.* Hafner, New York 1963.

GREEN, M. and MILLER, L. L. (1960) Protein catabolism and protein synthesis in perfused livers of normal and alloxan-diabetic rats. *J. biol. Chem.* **235**, 3203.

HAFT, D. E. and MILLER, L. L. (1958) Alloxan diabetes and demonstrated direct action of insulin on metabolism of isolated perfused rat liver. *Am. J. Physiol.* **192**, 33.

HAGER, C. B. and KENNEY, F. T. (1968) Regulation of tyrosine α-ketoglutarate transaminase in rat liver. VII. Hormonal effects of synthesis in the isolated, perfused liver. *J. biol. Chem.* **243**, 3296.

HAMILTON, R. L., REGEN, D. M., GRAY, M. E., and LEQUIRE, V. S. (1967) Lipid transport in liver: I. Electron microscopic identification of very low density lipo-proteins in perfused rat liver. *Lab. Invest.* **16**, 305.

HARDCASTLE, J. D. and RITCHIE, H. D. (1968) The liver in shock: a comparison of some techniques used currently in therapy. *Br. J. Surg.* **55**, 365.

HASE, T. and BRIM, J. (1966) Observation on the microcirculatory architecture of the rat liver. *Anat. Rec.* **156**, 157.

HECHTER, O., SOLOMON, M. M., and CASPI, E. (1953) Corticosteroid metabolism in liver. Studies on perfused rat livers. *Endocrinology* **53**, 202.

—— ZAFFARONI, A., JACOBSEN, R. P., LEVY, H., HEANLOZ, R. W., SCHENKER, V., and PINCUS, G. (1951) The nature and the biogenesis of the adrenal secretory product. *Rec. Prog. Horm. Res.* **6**, 215.

HEIMBERG, M., DUNKERLY, A., and BROWN, T. O. (1966) Hepatic lipid metabolism in experimental diabetes. 1. Release and uptake of triglyceride by perfused livers from normal and alloxan-diabetic rats. *Biochem. biophys. Acta* **125**, 252.

—— FIZETTE, N. B., and KLAUSNER, H. (1964) The action of adrenal hormones on hepatic transport of triglycerides and fatty acids. *J. Am. Oil chem. Soc.* **41**, 774.

—— WEINSTEIN, I., DISHMON, G., and FRIED, M. (1965) Lipoprotein lipid transport by livers from normal and $CCl_4$-poisoned animals. *Am. J. Physiol.* **209**, 1053.

—— —— KLAUSNER, H., and WATKINS, M. L. (1962) Release and uptake of triglycerides by isolated perfused rat liver. *Am. J. Physiol.* **202**, 353.

HEMS, R., ROSS, B. D., BERRY, M. N., and KREBS, H. A. (1966) Gluconeogenesis in the perfused rat liver. *Biochem. J.* **101**, 284.

—— STUBBS, M., and KREBS, H. A. (1966) Gluconeogenesis in the perfused rat liver. *Biochem. J.* **101**, 284.

HERFATH CH. (1966) *Habil. Schrift der Med. Fak. d Philipps-Universitat.* Marburg.

HILLYARD L. A., CORNELIUS, C. E., and CHAIKOFF, I. L. (1959) Removal by the isolated rat liver of palmitate-1-$C^{14}$ and cholesterol-4-$C^{14}$ in chylomicrons from perfusion fluid. *J. biol. Chem.* **234**, 2240.

HOBERMAN, H. D. and PROSKY, L. (1967) Evidence of reduction of fumarate to succinate in perfused rat liver under conditions of reduced $O_2$ tension. *Biochem. biophys. Acta* **148**, 392.

HOHORST, H-J., KREUTZ, F. H., and BÜCHER, TH. (1959) Über Metabolitgehalte und Metabolit-Konzentrationen in der Leber der Ratte. *Biochem. Z.* **332**, 18.

HOOKER, D. R. (1915) The perfusion of the mammalian medulla: The effect of calcium and of potassium on the respiratory and cardiac centers. *Am. J. Physiol.* **38**, 201.

JEFFERSON, L. S. and KORNER, A. (1969) Influence of amino-acid supply on ribosomes and protein synthesis of perfused rat liver. *Biochem. J.* **111**, 703.

JEUNET, F. S., SHOEMAKER, W. J., and GOOD, R. A. (1967) Isolated double perfusion of the liver in the study of the reticuloendothelial system. *Nature, Lond.* **215**, 61.

JOHN, D. W. and MILLER, L. L. (1969) Regulation of net biosynthesis of serum albumin and acute phase plasma proteins (Induction of enhanced net synthesis of fibrinogen, acid glycoprotein, α2 (acute phase)—globulin and haptoglobulin by amino acids and hormones during perfusion of the isolated normal rat liver). *J. biol. Chem.* **244**, 6134.

JONES, A. L., RUDERMAN, N. E., and HERRERA, M. G. (1966) An electron-microscopic study of lipoprotein production and release by the isolated perfused rat liver. *Proc. Soc. exp. Biol. Med.* **123**, 4.

JUCHU, M. R., CRAM, R. L., PLAA, G. L., and FOUTS, J. R. (1965) The induction of benzpyrene hydroxylase in the isolated perfused rat liver. *Biochem. Pharmac.* **14**, 473.

KALSER, S. C., KELVINGTON, E. J., and RANDOLPH, M. M. (1968) Drug metabolism in hypothermia. Uptake, metabolism and excretion of $S_{35}$-sulphanilamide by the isolated perfused rat liver. *J. Pharmac. exp. Ther.* **159**, 389.

—— —— —— and SANTOMENNA, D. M. (1965) Drug metabolism in hypothermia. II. $C^{14}$ atropine uptake, metabolism and excretion by the isolated perfused rat liver. *J. Pharmac. exp. Ther.* **147**, 260.

KAVIN, H., LEVIN, N. W., and STANLEY, M. M. (1967) Isolated perfused, rat small bowel—technique, studies of viability, glucose absorption. *J. appl. Physiol.* **22**, 604.

KESSLER, M. and SCHUBOTZ, R. (1968) Die $O_2$-versorgung der hämoglobinfrei perfundierten Rattenleber. *Stoffwechsel der isoliert perfundierten Leber* (eds. W. Staib and R. Scholz). Springer-Verlag, Berlin.

KESTENS, P. J. (1964) *La perfusion du foie isolé (son application à l'étude de quelques problèmes de biologie)*. Edition Arscia, Bruxelles.

KIM, J. H. and MILLER, L. L. (1969) The functional significance of changes in activity of the enzymes, tryptophan pyrrolase and tyrosine transaminase after induction in intact rats and in the isolated perfused rat liver. *J. biol. Chem.* **244**, 1410.

KIND, A. C., BEATY, H. N., FENSTER, L. F., and KIRBY, W. M. (1968) Inactivation of penicillins by the isolated rat liver. *J. Lab. clin. Med.* **71**, 728.

KNOX, W. E. and SHARMA, C. (1968) Enzyme induction in perfused rat liver by glucagon and other agents. *Enzymol. biol. Clin.* **9**, 21.

KOBLET, H. (1968) Regulation der Ribonucleinsauresynthese in der isoliert perfundierten Rattenleber. *Stoffwechsel der isoliert perfundierten Leber* (eds. W. Staib and R. Scholz), p. 180. Springer-Verlag, Berlin.

KREBS, H. A. (1964) Croonian lecture 1963: Gluconeogenesis. *Proc. R. Soc.* **B159**, 545.

—— (1968) *Stoffwechsel der isoliert perfundierten Leber* (eds. W. Staib and R. Scholz), p. 129. Springer-Verlag, Berlin.

—— EGGLESTON, L. V., and TERNER, C. (1951) *In vitro* measurements of the turnover rate of potassium in brain and retina. *Biochem. J.* **48**, 530.

—— and HENSELEIT, K. (1932) Untersuchungen über die Harnstoffbildung im Tierkörper. *Hoppe-Seyler's Z. physiol. Chem.* **210**, 33.

KRIEGLSTEIN, J. (1969) Uber die Wechselbeziehung der plasma-und Gewebsproteinbinding von Promazine, untersucht an der isoliert perfundierten Rattenleber. *Naunyn-Schmiedebergs Arch. exp. Path. Pharmak.* **262**, 474.

KVETINA, J. and GUAITANI, A. (1969) A versatile method for the *in vitro* perfusion of isolated organs of rats and mice with particular reference to liver. *Pharmacology* **2**, 65.

LONG, J. A. and LYONS, W. R. (1954) A small perfusion apparatus for the study of surviving isolated organs. *J. Lab. clin. Med.* **44**, 614.

McGEACHIN, R. L., POTTER, B. A., and LINDSEY, A. C. (1964) Puromycin inhibition of amylase synthesis in the perfused rat liver. *Archs. Biochem.* **104**, 314.

—— —— and WILSON, C. W. (1967) Effect of nitrophenols and nitrobenzoic acids on amylase synthesis and transport in the perfused rat liver. *Archs. Biochem.* **122**, 265.

McGRAW, E. F. (1968) The effect of fasting on glucose utilisation in the isolated perfused rat liver. *Metabolism* **17**, 833.

MAYES, P. A. and FELTS, J. M. (1967) Lack of uptake and metabolism of the triglyceride of serum lipoproteins of density less than 1·006 by the perfused rat liver. *Biochem. J.* **105**, 180; personal communication.

MENAHAN, L. A., ROSS, B. D., and WIELAND, O. (1968) Studies in the mechanism of fatty acid and glucagon stimulated gluconeogenesis in the perfused rat liver. *Stoffowechsel der isoliert perfundierten Leber* (eds. W. Staib and R. Scholz), p. 142. Springer-Verlag, Berlin.

—— and WIELAND, O. (1969) The role of endogenous lipid in gluconeogenesis and ketogenesis of perfused rat liver. *Eur. J. Biochem.* **9**, 182.

MEYER, D. K. and SCAF, A. H. (1968) Inhibition of transport of D-tubocurarine from blood to bile by K-strophanthoside in the isolated, perfused rat liver. *Eur. J. Pharmac.* **4**, 343.

MILLER, L. L. (1961) Some direct actions of insulin, glucagon and hydrocortisone on the isolated perfused rat liver. *Rec. Prog. Horm. Res.* **17**, 539.

—— (1965) Direct actions of insulin, glucagon and epinephrine on the isolated perfused rat liver. *Fedn Proc. Fedn Am. Socs exp. Biol.* **24**, 737.

—— BLY, C. G., WATSON, M. L., and BALE, W. F. (1951) The dominant role of the liver in plasma protein synthesis. *J. exp. Med.* **94**, 431.

—— BURKE, W. T., and HAFT, D. E. (1955) Interrelations in amino-acid and carbohydrate metabolism. Studies of the nitrogen sparing action of carbohydrate with the isolated, perfused rat liver. *Fedn Proc. Fedn Am. Socs exp. Biol.* **14**, 707.

MORRIS, B. (1960) Some observations on the production of bile by the isolated perfused liver of the rat. *Aust. J. exp. Biol.* **38**, 99.

—— (1963a) The metabolism of free fatty acids and chylomicron triglycerides by the isolated perfused liver of the rat. *J. Physiol., Lond.* **168**, 564.

—— (1963b)

—— and FRENCH, J. E. (1958) The uptake and metabolism of $^{14}$C labelled chylomicron fat by the isolated perfused liver of the rat. *Q. Jl exp. Physiol.* **43**, 180.

MORTIMORE, G. (1961) Effect of insulin on potassium transfer in isolated rat liver. *Am. J, Physiol,* **200**, 1315.

—— and TIETZE, F. (1959) Studies on the mechanism of capture and degradation of insulin I$^{131}$ by the cyclically perfused rat liver. *Ann. N.Y. Acad. Sci.* **82**, 329.

MYERS, D. K. (1964) Potassium levels in regenerating rat liver. *Can. J. Biochem.* **42**, 1111.

PAGE, E., POWER, B., BORER, J. S., and KLEGERMAN, M. E. (1968) Rapid exchange of cellular or surface potassium in the rats heart. *Proc. natn. Acad. Sci. U.S.A.* **60**, 1323.

POWIS, G. (1970) Perfusion of rat's liver with blood: transmitter overflows and gluconeogenesis. *Proc. R. Soc.* **B174**, 503.

RODBELL, M., SCOW, R. O., and CHERNICK, S. S. (1964) Removal and metabolism of triglycerides by perfused liver. *J. biol. Chem.* **239**, 385.

ROSS, B. D., HEMS, R., and KREBS, H. A. (1967) The rate of gluconeogenesis from various precursors in the perfused rat liver. *Biochem. J.* **102**, 942.

ROSTORFER, H. H., EDWARDS, L. E., and MURLIN, J. R. (1943) Improved apparatus for liver perfusion. *Science, N.Y.* **97**, 291.

RUBARTH, S. (1958) Leber und Gallenwege. *Pathologie des Laboratoriumstiere* (eds. P. Cohrs, R. Jaffe, and E. Messen), Vol. 1, p. 155. Springer-Verlag, Berlin.

RYOO, H. and TARVER, H. (1968) Studies on plasma protein synthesis with a new liver perfusion apparatus. *Proc. Soc. exp. Biol. Med.* **128**, 760.

SCHIFF, L. (ed.) (1969) *Diseases of the liver*, 3rd edn. Lippincott, Philadelphia.

SCHIMASSEK, H. (1962) Perfusion of isolated rat liver with a semisynthetic medium and control of liver function. *Life Sci.* 629.

SCHIMASSEK, H. (1963) Metabolite des Kohlenhydratstoffwechsels der isoliert perfundierten Rattenleber. *Biochem. Z.* **336**, 460.

—— (1968) Möglichkeiten und Grenzen der Methodik. *Stoffwechsel der isoliert perfundierten Leber* (eds. W. Staib and R. Scholz), p. 1. Springer-Verlag, Berlin.

—— and GEROK, W. (1965) Control of the levels of free amino-acids in plasma by the liver. *Biochem. Z.* **343**, 407.

SCHMIDT, E. (1968) Austritt von Zell Enzymen ans der isolierten, perfundierten Rattenleber. *Stoffwechsel der isoliert perfundierten Leber* (eds. W. Staib and R. Scholz), p. 53. Springer-Verlag, Berlin.

—— SCHMIDT, F. W., and HERFARTH, C. (1966) Studien zum Austritt von Zell-Enzymen am Modell der isolierten, perfundierten Rattenleber. *Enzymol. biol. clin.* **7**, 53.

SCHNITGER, H., SCHOLZ, R., BÜCHER, TH., and LUBBERS, D. W. (1965) Comparative fluorometric studies on rat liver *in vivo* and on isolated, perfused, haemoglobin-free liver. *Biochem. Z.* **341**, 334.

SCHOLZ, R. (1968) Untersuchungen zur Redoxcompartmentierung bei der Hämoglobinfrei perfundierten Rattenleber. *Stoffwechsel der isoliert perfundierten Leber* (eds. W. Staib and R. Scholz), p. 25.

SCHREIFERS, H. (1968) Rattenleber perfusion als Methode zur Untersuchung des Stoffwechsels von Δ⁴-3-Ketosteroiden. *Stoffwechsel der isoliert perfundierten Leber* (eds. W. Staib and R. Scholz), p. 230. Springer-Verlag, Berlin.

—— and KORUS, W. (1960) Über die Kinetik des Stoffwechsels von Cortison, Prednison und 9α-Fluor-hydrocortison bei der Rattenleberperfusion. *Z. phys. Chem.* **318**, 239.

SLATER, T. F., SAWYER, B. C., DELANEY, V. B., and BULLOCK, G. (1968) The effect of sporidesmin and icterogenin on bile flow in the isolated perfused rat liver. *Biochem. J.* **110**, 15p.

SOKAL, J. E., MILLER, L. L., and SARCIONE, E. J. (1958) Glycogen metabolism in the isolated liver. *Am. J. Physiol.* **195**, 295.

SÖLING, H. D., KATTERMAN, R., SCHMIDT, H., and KNEER, P. (1966) The redox state of NAD⁺—NADH systems in rat liver during ketosis and the so-called 'triosephosphate block'. *Biochim. biophys. Acta* **115**, 1.

—— WILLMS, B., FRIEDRICHS, D., and KLEINEKE, J. (1968) Regulation of gluconeogenesis by fatty acid oxidation in isolated perfused livers of non-starved rats. *Eur. J. Bioehem.* **4**, 364.

SOLOMON, S. S., FENSTER, L. F., ENSINCK, J. W., and WILLIAMS, R. H. (1967) Clearance studies on insulin and non-suppressible insulin-like activity (NSILA) in the rat liver. *Proc. soc. exp. Biol. Med.* **126**, 166.

SPERBER, I. (1959) Secretion of organic anions in the formation of urine and bile. *Pharmac. Rev.* **11**, 109.

STAIB, W. and SCHOLZ, R. (eds.) (1968) *Stoffwechsel der isoliert perfundierten Leber*. Springer-Verlag, Berlin.

—— STAIB, R., HERRMANN, J., and MEIERS, H. G. (1968) Unterschungen über die Cortisol-glykoneogenese in der isoliert perfundierten Rattenleber. *Stoffwechsel der isoliert perfundierten Leber*, p. 155. Springer-Verlag, Berlin.

STRUCK, E., ASHMORE, J., and WIELAND, O. (1966) Effects of glucagon and long chain fatty acids on glucose production by isolated perfused rat liver. *Advances in enzyme regulation* (ed. G. Weber), Vol. 4, p. 219. Pergamon Press, New York.

THOMAS, H., GANZ, F. J., and BUSCHEMANN, E. (1968) Zur Biogenese von Homovanilloyl-glycin bei der Rattenleberperfusion. *Hoppe-Seyler's Z. physiol. Chem.* **349**, 1686.

TRACHT, M. E., TALLAL, L., and TRACHT, D. G. (1967) Intrinsic hepatic control of plasma albumin concentration. *Life Sci.* **6**, 2621.

TROWELL, O. A. (1942) Urea formation in the isolated perfused liver of the rat. *J. Physiol., Lond.* **100**, 432.

VENEZIALE, C. W., WALTER, P., KNEER, N., and LARDY, H. A. (1967) Influence of L-tryptophan and its metabolites on gluconeogenesis in the isolated, perfused liver. *Biochemistry* **6**, 2129.

WILLIAMSON, J. R., BROWNING, E. T., and SCHOLZ, R. (1969) Control mechanisms of gluconeogenesis and ketogenesis. I. Effects of oleate on gluconeogenesis in perfused rat liver. *J. biol. Chem.* **244**, 4607.

—— GARCIA, A., RENOLD, A. E., and CAHILL, G. F. (1966) Studies in the perfused rat liver. I. Effects of glucagon and insulin on glucose metabolism. *Diabetes* **15**, 183.

—— KREISBERG, R. A., and FELTS, P. W. (1966) Mechanism for the stimulation of gluconeogenesis by fatty acids in perfused rat liver. *Proc. natn. Acad. Sci.* **56**, 247.

WILLMS, B. and SÖLING, H. D. (1968) Regulation der Gluconeogenese durch Fettsäureoxydation in der isoliert perfundierten Rattenleber. *Stoffwechsel der isoliert perfundierten Leber* (eds. W. Staib and R. Scholz), p. 118. Springer-Verlag, Berlin.

# 4. Kidney

PERFUSION of the kidney has always been of considerable interest to physiologists and has presented great problems, partly because the kidney has unique properties of vascular control of flow rate (auto-regulation) and partly because close control of regional blood flow within the kidney is central to normal excretory function (Van Slyke, Rhoads, Hiller, and Alving 1934). The methods of analysis of renal function have reached such a high degree of perfection that defects are easily detected. All the preparations of which details have been already published have been criticized as unphysiological. Differences of a very few per cent in sodium reabsorption, for example, can be easily detected when magnified during the concentration of urine. Concentration of urine, in its turn, is normally so marked that small deviations from normal in the mechanisms involved result in a many-fold dilution of the urine produced. Such minor deviations from normal will permit substances not normally present to appear in the urine, such as glucose or protein.

Compared with the liver, in which a venous perfusion at corre-spondingly low pressures is sufficient to oxygenate and maintain normal biochemical and physiological functions, the kidney must be supplied by arterial medium at a pressure sufficient to pass through two capillary beds, that is, the glomeruli and the capillary plexus that surrounds the tubules and loops of Henle. It is not surprising, therefore, that this has proved difficult to reproduce under conditions of artificial perfusion.

On the other hand, much interesting biochemical and metabolic work has been conducted in less elaborate systems—tissue slices, homogenates, isolated tubules, isolated cells, or those with subcellular components (see Bennoy and Elliott (1937)). In none of these prepara-tions can excretion or reabsorption be said to function normally. The introduction of the perfused kidney, with defects in this very aspect of renal function, should therefore first be justified.

Substrate uptake, biosynthesis, and oxygen consumption, as well as excretion, reabsorption, and autoregulation of blood flow have been demonstrated in the various preparations of perfused kidney to be described. Where possible, comparisons are made between these findings in the isolated perfused organ and those found in studies made *in vitro* and *in vivo*. The preparation has many biochemical applications, even allowing for these defects that must, of course, be taken into consideration when extrapolating from the perfused kidney to its function *in vivo*. (For recent developments, see Addendum p. 257).

## Anatomy

The gross anatomy of the kidney of small laboratory animals is the same as that of other mammals, including man. The two kidneys lie one either side of the vertebral column in mid-abdomen. Their posterior surface is applied to the posterior abdominal wall, while the anterior surface is covered by a layer of peritoneum and surrounded by a layer of fat. In addition, there is a 'capsule' of connective tissue which is closely applied to the kidney. The kidney is bean-shaped, with the concavity (hilum) medially. The ureter, renal artery, and renal vein emerge and enter the kidney here. The adult kidney is not lobulated, but there may be a separate blood supply to the upper pole, the accessory renal artery. This is of some importance for perfusion since when present it arises from the aorta, and perfusion of the major renal artery as suggested by Weiss *et al.* (1959), and Nishiitsutsuji–Uwo *et al.* (1967) may leave some 20 per cent of the kidney un-perfused.

*Blood supply*. Each kidney is supplied principally by a single artery, arising from the aorta. The right kidney lies higher in the abdomen than the left, so that its lower pole is at the same level as the upper pole of the latter. Accordingly, the right renal artery arises from the aorta approximately 1 cm above the left, and lies opposite the origin of the superior mesenteric artery.

The usual, though not constant, arrangement is for the right supra-renal artery to arise from the renal artery rather than from the aorta and this is an important branch to find and to ligate during perfusion of the right kidney. On the left, the supra-renal artery is more commonly derived from the aorta itself. The renal veins are also single, and overlie the artery on each side to enter the inferior vena cava. Since the inferior vena cava is to the right of the aorta, the left renal vein crosses the aorta to reach it, while the renal artery is behind the inferior vena cava at its origin. The renal vein on the right has no tributaries, but on the left the spermatic vein enters its inferior aspect. For anatomical reasons (see below) most current methods perfuse the right kidney and this variation is unimportant; in any case the problem is solved by a single extra ligation (see Leary (1969)).

*Lymphatic vessels.* Lymphatics are not prominent in the kidney, although a network is present in the capsule and drains the area around the tubules. None are found in the glomerulus or in the medullary rays (Kriz (1967) suggests that they may be absent entirely from the medulla in the rat), and they drain together with the main blood-vessels. So far no account has been taken of the lymphatic system in any kidney perfusion studies.

*Nerves.* Sympathetic plexuses supply nerves to the renal artery which enters at the hilum. Their distribution in the rat kidney is not known in detail and so far their function has not been considered in perfusions.

## Microscopic structure

The basic unit of the kidney is the nephron, but organization has more recently been considered in relation to the blood vessels (Kriz 1967) so that a segmental structure is proposed upon which the physiological facts of counter-current concentration (Wirz, Hargitay, and Kuhn 1951) may be based.

From the diagram of Kriz (1967), Fig. 4.1, it will be seen that nephrons are arranged in an orderly fashion, with the glomerulus nearer the periphery. A convoluted and a straight portion of the proximal tubule descend into the kidney and become the loop of Henle, which lies in the medulla. The structure of this loop alters to form first the straight and then the convoluted portion of the distal tubule, both of which are once more in the cortex. Tubules enter collecting ducts by a short 'initial segment', and the collecting duct returns by a straight course to the renal papilla, where it empties into the pelvis of the kidney, to enter the ureter. It is the regular position in which the transition from one part of the nephron to the next occurs that gives the kidney its characteristic appearance in cross-section. One irregularity in this pattern is created by the existence of two sorts of nephron—the long, of which the loops of Henle turn in the inner medulla, and the short, in which the turn is in the outer medulla.

The outer cortex is clearly distinguishable from the medulla, and each layer is further subdivided (the divisions being visible to the naked eye or with the aid of a simple magnifying glass) into cortex, subcortical zone, and outer and inner medulla—the red and white medullae, respectively, of older biochemical literature; see Bloom and Fawcett (1968). Extensions of the medulla into the cortex—the medullary rays—can also be seen with the naked eye and are of importance in preparing kidney cortex slices (see below).

The cellular composition of each layer can be derived from Kriz's diagram. The outer and inner medullae are distinguished by the

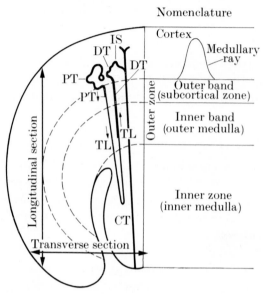

Fig. 4.1.   Nomenclature of kidney structure. The schema shows the structure of a kidney tubule in relation to the macroscopic divisions of the kidney as seen in section.

| PT | | proximal tubule | convoluted portion |
|----|--|-----------------|--------------------|
| PT | ↓ | | straight portion |
| TL | ↓ | | descending thin limb |
| TL | ↑ | Loop of Henle | ascending thin limb |
| DT | ↑ | | ascending thick limb (straight portion of distal tubule) |
| DT | | distal tubule | |
| IS | | initial segment of collecting tubule | |
| CT | | collecting tubule | |

presence of straight, distal tubules in the former; the existence of long and short nephrons ensures that both have loops of Henle. In a 200-g rat the proportions given for the various zones are $1:0.43:0.8:2.5$ for the cortex, subcortical zone, outer medulla, and inner medulla, respectively.

A detailed analysis of the anatomical information required for an understanding of the counter-current theory of urine concentration in the renal medulla is out of place here. Fig. 4.2, from Kriz, shows schematically his interpretation of the histological data, and suggests that the medulla is best understood if considered in connection with the known facts of its blood supply. Blood reaches the medulla in vasa

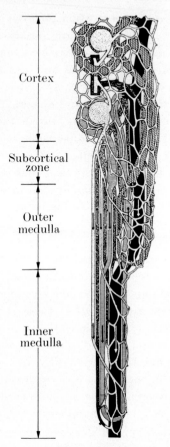

Cortex

Subcortical zone

Outer medulla

Inner medulla

FIG. 4.2. Schematic representation of the course of a long and a short nephron (from Kriz 1967). The direction of flow in a longitudinal section is indicated by the arrows. Short nephron—cross-hatched; long nephron—transverse stripes; collecting duct—black; vessels: arteries and capillaries—sparse dots; veins—dense dots.

recta arteries; these are formed by bifurcation of the efferent artery from the glomerulae adjacent to the medulla. Branches of these arteries supply two distinct capillary plexuses to the inner and outer medulla respectively. In each case, as already suggested, this is a second capillary plexus, since the blood has already traversed the glomerulus. The effect of this arrangement is that blood bathing the tubules of the cortex has first traversed the medulla and not vice versa.

### Physiological functions relevant to method of perfusion

*Autoregulation.* An early observation (Winton 1937, Ochwadt 1956) was that over a wide range of pressure variation in the renal artery the rate of flow through the kidney remained constant. This is clearly in contrast with many other organs and tissues, where increasing pressure increases flow and vice versa. It is described as 'autoregulation of blood flow' (Ochwadt 1956), and was originally considered to be due to a streaming effect of erythrocytes in the blood transversing the renal arterioles (Kinter and Pappenheimer 1956). However, from studies *in vivo* and *in vitro*, particularly those of Weiss *et al.* (1959) in the isolated perfused rat kidney, it is clear that the effect persists even in the absence of red cells. It is therefore an intrinsic property of the kidney and must be taken into account when planning perfusion experiments.

*Oxygen consumption and flow rate.* Oxygen consumption is proportional to the flow rate, a relationship that is again unique to the kidney. In the liver, for example, the oxygen consumed remains constant irrespective of the flow rate, provided there is an adequate supply. Kramer and Deetjen (1964) have shown that it is the linear relationship between oxygen consumption and sodium reabsorption that is the determining fact behind this observation. By virtue of the simple filtration that occurs in the glomerulus, as a constant fraction of the plasma flow rate, more sodium must be reabsorbed when the glomerular filtration rate increases. This fundamental relationship means that alteration in flow rate may, but does not necessarily, alter metabolic functions of the organ. Thus in the kidney, perfusion pressure and flow rate assume special importance in any attempt to standardize technique. Until more is known about the regulation of oxygen consumption in the perfused kidney, it would seem important to make these the central parameters of function.

### Experimental preparations other than perfusion

#### In vivo methods

Clearance studies, upon which most of the detailed information of renal function are based, are discussed in detail by Homer Smith

(1951), Pitts (1968), and others. Micro-puncture, which allows fluid from known segments of the kidney tubule to be sampled while the kidney retains its normal circulation *in vivo* is discussed by Richards (1938). The method is one requiring considerable skill but it has been applied, in addition, to the isolated perfused kidney (Bahlmann, Giebisch, Ochwadt, and Schoeppe 1967). An extension of the method, which has not yet been applied in perfused kidney, is the procedure of 'micro-tubular perfusion' (Sonnenberg and Deetjen 1964), whereby two micro-pipettes are inserted into the same tubule at varying distances along its length and standard solutions perfused between the two.

## In vitro methods

*Tissue slices*. By means of slices the cortex and medulla may be studied separately. Perfusion cannot replace this older technique in metabolic studies, since it is already clear that different metabolic processes occur in the two regions of the kidney (Krebs *et al.* 1963, György, Keller, and Brehme 1928). This is hardly surprising in view of the great differences in histology and physiological activity; cortex outweighs medulla by about 7:1 and its activities could therefore easily obscure those of the smaller component. However, there is much to be said for applying both experimental methods, i.e. slices and perfusion, to any problem; the similarities may help to confirm the physiological validity of the perfusion technique (Nishiitsutsuji–Uwo, Ross, and Krebs 1967) and differences may elucidate the metabolic basis of tubular excretory function.

*Homogenates*. Numerous experiments have been performed with homogenates of kidney, but usually without reference to the heterogeneity of the starting material, that is, that medulla and cortex are homogenized together.

*Cellular fractions*. Isolated kidney cells may be prepared by the method of trypsin digestion. A renal tubule preparation introduced by Burg and Orloff (1962) has found application in studies of gluconeogenesis (Güder and Wieland 1970, Nagata and Rasmussen 1970) and is clearly of very great interest in attempts to localize the various known metabolic functions of the whole organ.

*Sub-cellular fractions*. Mitochondria, nuclei, and other fractions prepared in well established techniques have been used to compare permeability and function of these organelles with those from the more commonly used liver fractions (for example, De Luca *et al.* 1957).

## METHODS OF PERFUSION

A small number of methods of perfusion of the rat kidney have been devised, and most workers have been concerned to study autoregulation (Weiss *et al.* 1959) and excretion-reabsorption functions of the organ (Bauman *et al.* 1963, and Bahlmann *et al.* 1967). More recently, studies of metabolism of the kidney have been made in this system (Nishiitsutsuji–Uwo *et al.* 1967, Turner, Neely, and Pitts 1968). Turner's apparatus, specially adapted to continuous oxygen uptake studies is illustrated and discussed in the context of oxygen measurement (Chapter 2). A considerable amount of careful observation has provided data by which the various approaches to the problem of perfusion of the rat kidney may be compared. Anatomically, there are few differences in approach available; the organ may be perfused via the aorta or directly via the renal artery. Uniform or pulsatile flow may be used, and the pressure may be exerted either with a pump or by means of a hydrostatic pressure head. As in other perfused organs, the most important variable has been the perfusion medium, and a considerable range has now been tested. Further comparative studies will, however, be required before a decision on an ideal perfusion medium can be taken with the sort of assurance that has now been achieved for the perfused liver.

A simple, if arbitrary classification of the methods published will be applied.

I. Isolated organ.
  A. Semi-synthetic medium.
     (i) Non-pulsatile flow (Nishiitsutsuji–Uwo *et al.*).
     (ii) Pulsatile flow (Weiss *et al.*, Nishiitsutsuji–Uwo modification).
  B. Blood medium (Bauman *et al.*).

II. *In situ* organ.
  A. Semi-synthetic medium.
     (i) Non-pulsatile flow (Fülgraff *et al.* 1968).
     (ii) Pulsatile flow (Bahlmann *et al.*).
  B. Blood medium (none described to date).

(*Note*. A further point of comparison could be made on the basis of recirculation versus non-recirculation methods, but in the absence of more extensive metabolic data this does not seem profitable at present.)

The operative technique of Nishiitsutsuji–Uwo *et al.* will be described in some detail, as that most likely to be applicable to the requirements of metabolic studies, and a variety of apparatus and media will be discussed.

## Method IA(i). Isolated kidney, perfused with semi-synthetic medium, with a constant perfusion pressure (Method of Nishiitsutsuji–Uwo, Ross, and Krebs (1967))

This technique was designed primarily for the study of renal metabolism but is also suitable for the study of some aspects of the secretory function. The rate of gluconeogenesis from a variety of precursors is rapid in this preparation and similar to that in kidney cortex slices, while the oxygen consumption of the perfused organ is about twice that of cortex slices, presumably because of the work done in secretion by the perfused kidney. In addition, there is some suggestion from this preparation that different substrates may be required for 'metabolic' and 'secretory' work.

### Method

The method is based on that of Weiss *et al.* (1959) for perfusion of rat kidney and on the liver perfusion techniques of Miller and of Schimassek (see Chapter 3), and Hems *et al.* (1966). Medium is perfused into the renal artery of the isolated right kidney from a hydrostatic head of 120 cm $H_2O$. This is determined by the overflow from the multibulb oxygenator and is therefore non-pulsatile. From the renal vein, the medium, which is a semi-synthetic one based upon that described by Schimassek (1962, see Media), flows to a reservoir from which samples may be taken. A pump then returns medium to the oxygenator, which is held some 120–140 cm above the kidney tray. The whole apparatus is enclosed in a warmed cabinet to control temperature.

### Apparatus

The original apparatus is a modification of the liver perfusion apparatus used by Hems *et al.* (1966) (see Chapter 3 and Figs. 1.12 and 4.3) and the modifications concern the following points.

(1) The distance between the bottom of the oxygenator and the 'kidney tray' is 120 cm, to obtain the required hydrostatic pressure.

(2) The kidney, unlike the liver, is not perfused *in situ* but is transferred to a kidney tray. This consists of a nylon mesh stretched

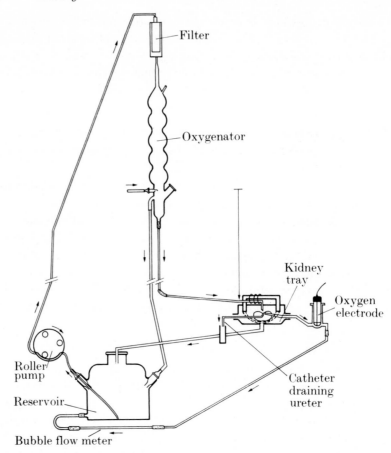

Fig. 4.3.    Circuit diagram of kidney perfusion (Nishiitsutsuji–Uwo *et al.* 1967).

over a ring (7·5 cm diam.). A stainless-steel strip is mounted about 2·5 cm above the tray to support the arterial cannula (Fig. 4.4).

(3) The temperature of the cabinet and the circulating medium is 38–40°C.

(4) The arterial tube from the oxygenator is mainly of glass to avoid oxygen loss because of the permeability of polythene to gases.

(5) The collecting vessel receives the venous outflow via a 2-ml graduated pipette, which also serves as a flow meter.

(6) The urine produced is returned to the collecting vessel to avoid loss of products of metabolism, except in experiments on the secretory function.

(7) The total volume of the perfusion medium is 100 ml.

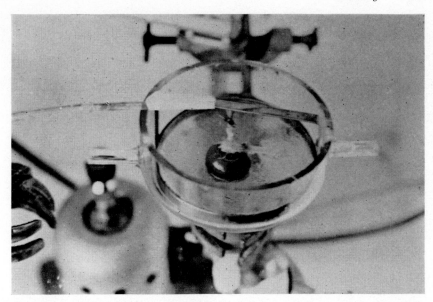

F IG . 4.4. Kidney support and chamber (Nishiitsutsuji–Uwo *et al.* 1967). The arterial cannula is bound to the horizontal metal limb of the kidney tray, allowing the organ to hang on to the nylon mesh base. The tray is contained in a glass funnel designed to conduct effluent medium and urine catheter through separate side-arms.

*Cannulae.* The arterial cannula is made of glass tubing (internal diameter 2·8 mm, external diameter 3·5 mm) drawn to a tip of 1·3 mm external diameter and 1·0 mm internal diameter. It is bent to a right angle 1·5 cm from the tip and the short limb of the cannula has little or no taper. The tip of the cannula is bevelled slightly to facilitate its insertion (Fig. 4.5). The venous cannula is placed in the inferior vena cava, and consists of a 3-cm length of Portex tubing (Portland Plastics Ltd.) size PP 270 (internal diameter 2 mm, external diameter 3 mm) cut off at an angle to form a sharp tip. With the cannula in position, the opening lies opposite the right renal vein. The other end of the cannula is temporarily closed by a loose plug of tissue paper. An array of arterial cannulae should be prepared since the diameter of the renal artery shows considerable variation; in addition, the bevelled tip is fragile, and failure in cannulation may be attributed to irregularities of the rim. Ureteric catheter (Portex) is size PP 10 (internal diameter 0·28 mm, external diameter 0·61 mm), cut to an acute angle to facilitate its insertion.

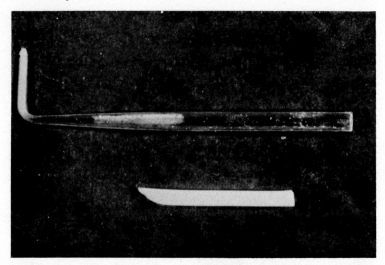

F I G . 4.5.  Cannulae for kidney perfusion (Nishiitsutsuji–Uwo *et al.* 1967). Above: glass cannula for renal artery; note right angle bend and the parallel sides and squared end of the cannula tip. Below: Polythene cannula for inferior vena cava.

## Perfusion medium

The standard medium is prepared as follows. Five grammes of bovine serum albumin (Fraction V; Armour Pharmaceutical Co. Ltd., Eastbourne, Sussex) is dissolved by stirring in about 70 ml of the saline of Krebs and Henseleit (1932). This solution is dialysed against 2 litres of the saline (five changes) for 48 h in a sealed vessel (to prevent loss of $CO_2$) at 4°C with shaking. After dialysis, the perfusion medium is brought to 111 ml with the saline. The final concentration is about 4·5% (wt./vol.). In the experimental series undertaken by Nishiitsutsuji–Uwo *et al.* (1967) it was found necessary to modify the medium used for liver perfusion to obtain optimum rates of gluconeogenesis from added substrates and low rates from endogenous sources. The modifications concern a higher concentration of albumin (which approaches the physiological value of the colloid osmotic pressure) and dialysis of the bovine albumin (which decreases glucose formation in the absence of added precursors). The presence of red blood cells proved unnecessary for full oxygenation.

## Operative procedure

Male Wistar rats weighing 350–450 g are starved for 42–45 h before use because higher rates of gluconeogenesis are obtained in the kidney of starved rats (Krebs *et al.* 1963). In addition, the operation is easier in the absence of perinephric fat. After anaesthetizing the rat with intraperitoneal Nembutal (0·1 ml of 6% wt./vol. solution per 100 g body wt.) an abdominal incision is made in the mid-line and extended laterally; and the intestine displaced to the animal's left. The right kidney is used for perfusion, because the mesenteric artery arises

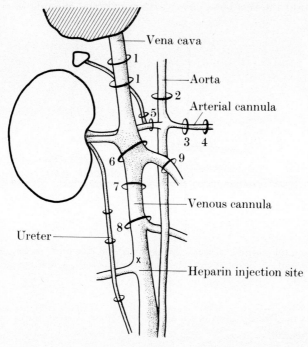

FIG. 4.6. Operative procedure for kidney perfusion (Nishiitsutsuji–Uwo *et al.* 1967). The right kidney and adjacent vessels are shown diagrammatically. The ligatures are numbered according to the description in the text. One possible site of heparin injection is indicated.

from the aorta at the same level as the right renal artery and a cannula can be passed from one to the other without blood loss and without stopping the blood flow to the kidney. To expose the major abdominal vessels and the right kidney, fat and perivascular tissue are cleared away by blunt dissection. The adrenal branch of the right renal artery is tied, and loose ligatures placed around the following blood-vessels (see Fig. 4.6): (1) inferior vena cava, just below the liver; (2) aorta, above the mesenteric artery; (3) mesenteric artery, near the aorta;

(4) mesenteric artery, further from the aorta than (3); (5) right renal artery, at its origin from the aorta; (6) inferior vena cava, between the left and right renal vein; (7) inferior vena cava, below the left renal vein; (8) inferior vena cava, more distally still; (9) left renal vein. The right kidney is mobilized as far as possible, leaving only its vascular and ureteric attachments. For the collection of urine, the ureter is cannulated (see Apparatus). Then heparin (0·2 ml, 200 units) is injected into the lower inferior vena cava; afterwards the opening in the wall of

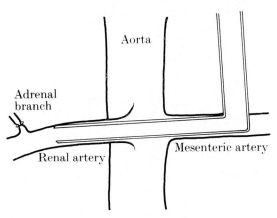

Fɪɢ. 4.7. Cannulation of renal artery via the superior mesenteric artery. The glass cannula is inserted into the mesenteric artery, across the lumen of the aorta, and into the renal artery.

the vein is closed by means of a ligature passed over the point of the injecting needle.

Next, ligature (9) is tied. The venous cannula is inserted in the inferior vena cava and tied in place (6) and (7). With the cannula in position, the opening lies opposite the right renal vein. The other end of the cannula is temporarily closed by a loose plug of tissue paper. The arterial cannula is filled to the tip with perfusion medium. The mesenteric artery is tied distally (4) and Spencer Wells forceps are used to maintain tension on the vessel. The artery is grasped at its origin with fine curved forceps and an incision made in the wall. The cannula is inserted and passed to meet the forceps, which are then removed. The tip of the cannula is advanced into the aorta and then into the renal artery opposite, (Fig. 4.7) allowing perfusion medium to flow to the kidney. The cannula is tied in place by ligatures (3) and (5). Table 4.1 outlines the main steps in the operation.

TABLE 4.1

*Perfusion of isolated kidney (operative procedure)*

---

(i) Ureter cannulated.
(ii) Left renal vein tied (lig. (9)).
(iii) Venous cannula inserted; tied at (6) and (7).
(iv) Mesenteric artery tied (lig. (4)); arterial cannula inserted and tied at (3) and (4), and perfusion begun.
(v) To remove the kidney, aorta tied at (2) and cut; inferior vena cava tied at (1) and cut.

Total operation time = 15 min.

---

In an alternative procedure the aorta is tied at (2) and the cannula inserted directly into the aorta and right renal artery. A short period of ischaemia of the kidney is unavoidable with this method of cannulation; it is therefore not the method of choice.

The aorta is tied at (2) and cut below the ligature. The inferior vena cava is tied at (1) and cut above the ligatures. The perfusion medium then flows down the venous cannula. The plug is removed and the first 8–10 ml of medium discarded. The kidney is transferred from the animal to the nylon-mesh tray. The whole operation takes about 15 min. The flow is 16–30 ml/min.

*Characteristics of perfusion*

The characteristics of this preparation are discussed by Nishiitsutsuji–Uwo *et al.* (1967) with additional information in papers by Weidemann *et al.* (1969). At the constant pressure (140 cm $H_2O$) selected for these experiments, flow rate varies between preparations, but is constant in any one preparation for from 1 to 2 h. This may be a good index of function, since oxygen consumption is linked to flow rate (q.v.).

The rates of flow observed (16–30 ml/min per g wet weight) exceed the normal plasma flow rate *in vivo*, which is about 2·75 ml/min per g (calculated from Spector (1956)) but this is partly a result of the necessary extra flow required to provide for the oxygen requirements of the kidney in the absence of erythrocytes.

*Physiological parameters.* Collection of urine in this preparation allows a number of excretory and secretory functions of the perfused organ to be examined and compared with the results *in vivo*.

*Urine flow.* The mean rate of urine flow in this preparation is 0·054 (0·02–0·115) ml/min per g wet weight of kidney, which agrees well with other data on the isolated perfused kidney (see below). These figures are about ten times higher than those in the normal rat, but may be equalled *in vivo* under conditions of diuresis induced by Mannitol. Both glucose and protein appear in the urine in small amounts in this preparation but both figures may be within normal limits for the rat *in vivo* (see Cori (1925), Spector (1956)).

*Glomerular filtration rate.* Creatinine clearance has been used as an estimate of glomerular filtration rate but in this preparation, although the rate of urine flow remains constant for 2 h, creatinine clearance always falls steadily with time. At low concentrations (0·1 mg/ml), creatinine is secreted in this preparation as it is *in vivo* (Fingl 1952). The mean rate of creatinine clearance described in this preparation is 0·136 ml/min, a figure lower than that obtained by most workers in the rat *in vivo* (Fingl 1952, Harvey and Malvin 1965). Despite these somewhat abnormal physiological parameters of function, biochemical tests demonstrate linear rates of reaction for well over the 2-hour perfusion period.

*Biochemical tests of functions.* The present authors describe the uptake of L-lactate and a variety of other substrates at high rates, which are generally linear with time. Glucose is synthesized from each of the precursors as expected, and at a rate similar to that found in tissue slices. Again the rates are linear with time and provide characteristic information on which to base a test of kidney function (see Chapter 2).

Table 4.2 shows a number of metabolic parameters based on the utilization of substrates by the perfused kidney. As in the case of liver, it is probable that the conversion of lactate or pyruvate to glucose is a more sensitive test of function than one based on another precursor, which may involve a smaller number of enzyme reactions and fewer regulatory steps.

*Oxygen consumption.* Oxygen consumption is linear in time and increases in response to any addition of a substrate. However the use of rate of oxygen consumption as a test of function in this preparation has limitations, due to its dependence upon flow rate. At a flow rate of 24 ml/min per g the basal oxygen consumption of the perfused kidney is

1600 $\mu$mol/h per g dry weight, a rate almost double that observed in kidney slices (Krebs, Hems, and Gascoyne 1963).

TABLE 4.2

*Metabolism of various substrates by perfused rat kidney*

(from Nishiitsutsuji–Uwo *et al.* (1967))

| Substrate added to medium 5 mM soln. | Amount of substrate removed ($\mu$mol/g dry wt./h) | Glucose formed ($\mu$mol/g dry wt./h) | Lactate formed ($\mu$mol/g dry wt./h) | Other products formed ($\mu$mol/g dry wt./h) | |
|---|---|---|---|---|---|
| None | | <3 (5) | <3 (5) | | |
| L-Lactate | 405±32 (10) | 82± 9 (10) | | Pyruvate | 89±20 (3) |
| Pyruvate | 1246±82 (10) | 304±30 (10) | 212±24 (10) | | |
| Glycerol | 282±13 (4) | 95± 7 (4) | 16± 6 (4) | | |
| Succinate | 1054±59 (5) | 310±50 (5) | 88±22 (5) | Malate | 77± 6 (2) |
| L-Glutamate | 518±49 (8) | 107± 9 (6) | <5 (5) | {Aspartate {Glutamine | 32± 4 (3) 196±14 (3) |
| L-Asparate | 687±59 (6) | 72± 7 (7) | <3 (5) | {Glutamate {Glutamine | 160±19 (5) 67± 3 (2) |
| L-Malate | 892±12 (4) | 206±16 (5) | 71±26 (4) | Fumarate irregular | (0–50) |
| Fumarate | 837±31 (5) | 265±26 (5) | 98±16 (5) | Malate | 150±16 (2) |
| Aceto-acetate | 146± 9 (3) | — | — | $\beta$ hydroxy-butyrate | 48± 4 (3) |

*Adenine nucleotides in perfused rat kidney.* Another measure of the functional integrity of the perfused organ is the stability of the adenine nucleotides and the ATP/ADP ratio. As shown in Table 4.3, the concentration of ATP falls and that of AMP increases at the end of the operation, presumably because of temporary anoxia during the preparation of the kidney for perfusion. Within 30 min the normal values obtained from non-perfused kidneys are restored. The total adenine nucleotide content, the ATP/ADP ratio and the concentrations of AMP are maintained rather more effectively than in the perfused liver (see Chapter 3, p. 174), irrespective of the addition of lactate or pyruvate. The decrease of the total adenine nucleotides after 2 h in the presence of lactate is 15 per cent. Although the addition of lactate

## TABLE 4.3

### Adenine nucleotides in perfused rat kidney

(from Nishiitsutsuji–Uwo et al. (1967))

| State of tissue | Duration of perfusion (min) | No. of observations | ATP (µmol/g dry wt.) | ADP (µmol/g dry wt.) | AMP (µmol/g dry wt.) | Ratio $\dfrac{\text{ATP}}{\text{ADP}}$ | Total adenine nucleotides |
|---|---|---|---|---|---|---|---|
| At end of operation | 0 | 4 | 4·81±0·33 | 2·83±0·22 | 2·71±0·44 | 1·71±0·14 | 10·84±0·26 |
| Perfused with lactate | 30 | 4 | 6·33±0·48 | 2·44±0·48 | 0·85±0·09 | 2·93±0·52 | 9·62±0·46 |
| Perfused with lactate | 60 | 4 | 6·87±0·59 | 2·85±0·18 | 0·84±0·12 | 2·41±0·06 | 10·56±0·65 |
| Perfused with lactate | 120 | 3 | 5·57±0·13 | 1·96±0·27 | 0·76±0·24 | 2·94±0·29 | 8·30±0·38 |
| Perfused with pyruvate | 30 | 3 | 6·48±0·70 | 2·32±0·29 | 0·66±0·25 | 2·80±0·10 | 9·67±0·79 |
| Perfused without substrate | 30 | 4 | 6·32±0·31 | 2·32±0·13 | 1·04±0·03 | 2·72±0·02 | 9·68±0·46 |
| Not perfused (control) | — | 4 | 7·20±0·38 | 2·66±0·07 | 0·78±0·10 | 2·72±0·08 | 10·69±0·37 |
| Not perfused (well-fed rat) | — | 4 | 6·18±0·90 | 2·30±0·37 | 0·68±0·13 | 2·74±0·15 | 9·13±1·36 |

Adenine nucleotides were determined in rapidly deep-cooled tissue. Rats were starved for 48 h except when stated otherwise. 'At end of operation' was the stage of removal of the kidney from the animal. Data refer to dry wt. The wet wt./dry wt. ratio of the normal kidney of starved rat, not perfused, was 3·94±0·14(8). Initial substrate concentration was 5 mM.

causes a fall in the concentration of ATP and in the ATP/ADP ratio in the liver, there are no significant changes in this direction in the perfused kidney. No significant difference is found between the adenine nucleotide content before perfusion of kidneys from well-fed and starved rats.

**Method IA(ii). Isolated kidney, perfused with semi-synthetic medium, with pulsatile flow** (Method of Weiss, Passow, and Rothstein (1959) and modification of Nishiitsutsuji–Uwo)

This preparation was devised to test the hypothesis that auto-regulation of blood flow in the kidney was independent of the presence of red cells in the perfusing medium (Pappenheimer effect q.v.). Since the original description in 1959, modifications have been made to allow pumpless perfusion (Leichtweiss *et al.* 1967), determination of pressure-flow relationships (Basar *et al.* 1968), of oxygen consumption and of tissue oxygen tension (Leichtweiss *et al.* 1969). A variety of perfusion media has been used (see Table 4.5), but all are haemoglobin-free.

Despite the range of method and function tested, no physiological parameters apart from flow and pressure, nor any metabolic parameters apart from oxygen consumption have been determined for this technique. There is clearly a need for such information before the preparation can be compared with the other techniques available.

*Apparatus*

A peristaltic pump (American Instrument Co. 5–8950) is used. The kidney is immersed in a thermostatic bath of physiological saline at 37°C. A mercury manometer records inflow pressure, and a bubble flow-meter is applied to the outflow.

*Pulse pressure.* Although described as pulsatile the pressure applied is in fact marked by its low pulse pressure. With a mean pressure of 140 mmHg (190 cmH$_2$O) a pulse pressure of 7 mmHg amplitude is applied with a frequency of 240/min. The mean pressure is also some 70 cm H$_2$O higher than that used by Nishiitsutsuji–Uwo *et al.* (Method IA (i)).

*Cannulae: arterial.* The specially designed arterial cannula, with its provision for removal of bubbles immediately before the medium enters the vessel, is illustrated in Fig. 4.8. It is the basis of the cannulae used in other methods described below. *Venous.* The venous cannula is of glass, 3·00 mm o.d. The last 0·75 cm of its length is perforated from all sides so that the venous outflow cannot become totally occluded by

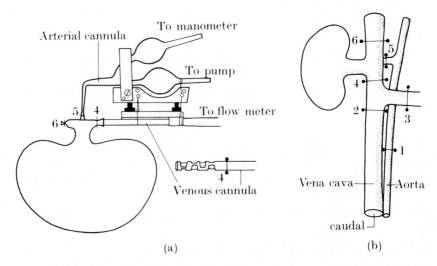

(a)    (b)

FIG. 4.8.    (a) Cannulae for kidney perfusion (Weiss *et al.* 1959). The diagram illustrates the excised kidney mounted for perfusion. The arterial cannula and venous cannula (see also inset) are held together in a clamp. Note the bubble trap and recording side-arm of arterial cannula. (b) Operative procedure for kidney perfusion according to Weiss *et al.* (1959). The right kidney and adjacent vessels are shown, with the position of ligatures. Compare Fig. 4.6, procedure of Nishiitsutsuji–Uwo *et al.* (1967).

kinking the vein. Back pressure from the vein is thought to be an important factor in failure of renal perfusion (Weiss *et al.* 1959).

*Ureter.* The ureter is not cannulated in this preparation, but clearly could be. (Leichtweiss *et al.* (1967) recommend glass for the ureteric cannula (diam. 0·6 mm) while exchanging the glass venous cannula for one of PVC.)

### Alternative apparatus

Pumpless perfusion may be performed with the apparatus diagrammatically portrayed in Fig. 4.9. A full description is contained in the writings of Leichtweiss *et al.* (1967, 1969) and of Basar *et al.* (1968).

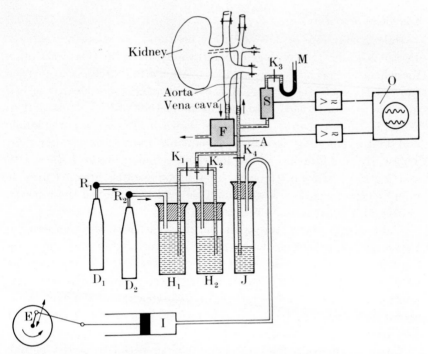

FIG. 4.9. Apparatus for pump-less perfusion of isolated kidney (Basar *et al.* 1968). $H_1$ and $H_2$ = medium reservoirs; $R_{1+2}$ pressure control valves; $D_{1+2}$ gas cylinders; A = arterial cannula; K = occlusive clamps; S = strain gauge (pressure record); F = electromagnetic flow-meter; O = oscilloscope recorder; E = eccentric drive, for I, injection to reservoir (to provide chosen pulse pressure wave); M = mercury manometer (for setting strain gauge before experiment). Note that the extra complexity of this apparatus allows studies on pressure of perfusion; the original apparatus described by Leichtweiss *et al.* (1967) is simplified by omission of the multiple pressure reservoirs, and is more suited to biochemical studies.

## *Medium*

Weiss, Passow, and Rothstein (1959) proposed the following medium: 6% Dextran (Plastan : Mead) dissolved in a salt medium, buffered with a mixture of phosphate ($pK_a$ 6·7) and acetate ($pK_a$ 4·73): (mM) NaCl, 160, Na acetate 5·0, $K_2HPO_4$ 5·0, $CaCl_2$ 1·2 $MgCl_2$ 0·5 mM, gassed with 100% $O_2$ and at pH 7·35. Glucose 5·0 mM is added. As already mentioned, a number of modifications have been made to this medium and Table 4·5 shows the medium that has been used for different experiments with this basic perfusion method (see Chapter 1, for discussion of details). One immediate observation is that while phosphate buffers in the required range (5·7–7·7), acetate does not (3·7–5·7).

*Operative procedure*

Male Wistar rats of 350–400 g are used and are anaesthetized with Pentobarbitone (130 mg/kg used intra-muscularly in this description, for which no special reason is given). After a mid-abdominal incision, the great vessels and origins of the renal arteries are cleared of connective tissue as already described in the method of Nishiitsutsuji–Uwo *et al.* Ligatures 1–6 are positioned as shown in Fig. 4.8, and the kidney and ureter are freed as much as possible from their retaining fat. This early mobilization allows for rapid removal of the kidney after starting the perfusion. Heparin, 1500 units/kg rat is injected into the inferior vena cava, low down, and a tie over the point of the injecting needle (cf. p. 234) is again required.

Eight minutes later, the operation proceeds, with the tightening of Ligatures 1–3, occluding the lower aorta, lower inferior vena cava, and the left renal vein. Venous cannulation is carried out first, (as described on p. 234). An incision is made in the inferior vena cava, below the point of entry of the left renal vein (already tied) and the cannula inserted for its opening to lie opposite the right renal vein. Ligature 4 is tied around it.

Aortic cannulation follows, and differs in detail from the method described by Nishiitsutsuji–Uwo. A bulldog clip occludes the aorta well below the left renal artery and an incision is made below this again. The cannula is already filled with medium from the prepared perfusion circuit and pulsatile flow is begun as the tip is inserted into the aorta. It is advanced beyond the bulldog clamp, which is removed without blood loss, since the cannula both fits well and is injecting medium in the reverse direction. The cannula is carefully advanced and manoeuvred into the right renal artery. Ligature 5 is tied and cannulation has thus been achieved without interruption of the renal blood supply. Ligature 6 is tied, diverting medium into the inferior vena cannula. The two cannulae are mounted on a rigid unit (see Fig. 4.8) to ensure their relative positions for optimum flow, and the ureter is cut as the kidney is lifted from the animal.

It is hardly necessary to compare the methods of renal artery cannulation introduced by Weiss *et al.* and that used by Nishiitsutsuji–Uwo *et al.* since the method used is merely a matter of personal choice. The Nishiitsutsuji–Uwo technique has the additional advantage of a 'second-go' via the aorta. The Weiss technique can, of course, be applied equally to the left and right kidney, while that of Nishiitsutsuji–Uwo *et al.* is restricted to using that on the right. Various figures for flow rate are given; about 8 ml/min per g seems to be the average (see below).

## Characteristics of perfusion

Autoregulation was detected in this preparation with a virtual linear relationship between pressure and flow, down to flow rates of 4 ml/min per g, at which rate it could no longer be detected (note that normal blood flow in the rat kidney is estimated to be no more than 5 ml/min per g). After about 2 hours (report in Leichtweiss *et al.* (1967)) the preparation fails, as indicated by the synchronous failure of autoregulation, falling urine flow, and increasing perfusion pressure. Despite higher perfusion pressures, the flow rate achieved was lower than that of Nishiitsutsuji–Uwo—8 ml/min against 16–30 ml/min. This may be attributable to differences in the perfusion medium. The non-pulsatile modification of the basic method, described in a later paper (Leichtweiss *et al.* 1969) has been used to determine oxygen consumption, glomerular filtration rate, flow, and tissue ATP content. The medium was varied, however, for some of these experiments (see Table 4.5) and it is particularly important that in the series concerned with the measurement of oxygen consumption the only substrate added was glucose; sodium acetate was omitted. With one additional difference, that Nishiitsutsuji–Uwo worked with 40 h starved rats while Weiss *et al.* used well-fed animals, the two series of results may be compared.

Flow rate is lower, although pressure was, in the main, higher. Figures of 50–180 mmHg (70–250 cmH$_2$O) are given. The creatinine clearance reported by Nishiitsutsuji–Uwo, 0·14 ml/min per g, is difficult to compare with that of 0–0·8 ml/min per g given by Weiss *et al.* but is clearly within this rather broad range.

At flow rates that are different, oxygen consumption cannot be compared (see p. 247). However, the publications of Weiss *et al.* give sufficient information (culled from several papers with minor differences in technique) for a graph of flow-versus-oxygen consumption to be constructed. The values for other perfused rat kidney preparations are entered on this curve and exhibit no major differences. The preparations compared are all perfused with haemoglobin-free media, at higher than physiological flow rates. Unfortunately, Bauman *et al.* (see below), the only workers to use rat-blood perfusion medium at physiological flow rates, do not record oxygen consumption. Nevertheless, it would appear that in this metabolic result, kidney perfusion has been successful in achieving inter-laboratory uniformity. Other points for comparison will be discussed at the end of this section.

**Method IB. Isolated kidney perfused with rat blood** (Method of Bauman, Clarkson, and Miles (1963))

This technique was devised to emulate as closely as possible the physiological conditions of the kidney *in vivo* and was used to determine its secretory and clearance characteristics. The method owes a great deal to the influence of Miller's liver perfusion method (Chapter 3). Direct pulsatile flow of oxygenated heparinized rat blood is pumped through the renal artery and collected via a venous cannula. The blood is oxygenated in a simple flask and recirculated with a peristaltic pump which provides a pulse pressure of 8–12 mmHg and a pulse rate of 300/min. The apparatus is enclosed in a warmed cabinet.

*Apparatus*

The apparatus, without the cabinet, is illustrated in Fig. 4.10. The kidney is contained in a small beaker, partly immersed in K–H saline (with dextrose 50 mg/100 ml). Outflowing medium flows to the reservoir, which is identical with that used by Miller *et al.* (Chapter 3) and lies 10–15 cm below the renal vein. This represents a negative venous pressure of this magnitude. The blood flows along a horizontal length of pipette, which may be used as a flow meter by the injection of a small air bubble the progress of which along the pipette is timed (cf. Nishiitsutsuji–Uwo, and Hems *et al.* for liver). The contents of the reservoir (70–110 ml) are stirred with a magnetic stirrer and delivered by means of the peristaltic pump into the arterial cannula. A manometer is attached to the side arm of the arterial cannula as suggested by Weiss *et al.*

*Pump.* A peristaltic pump of American Instrument Co. No. 5–9850 is used, and provides the pulse rate and pressure described above. The mean arterial pressure was within the range 100–140 mmHg.

*Cannulae.* These are as described by Weiss *et al.* (q.v.) and the shape of the arterial cannula is clearly shown in Fig. 4.10. *Ureter.* The ureter is cannulated in this preparation, usually with polythene tubing PP 10 (cf. Nishiitsutsuji–Uwo).

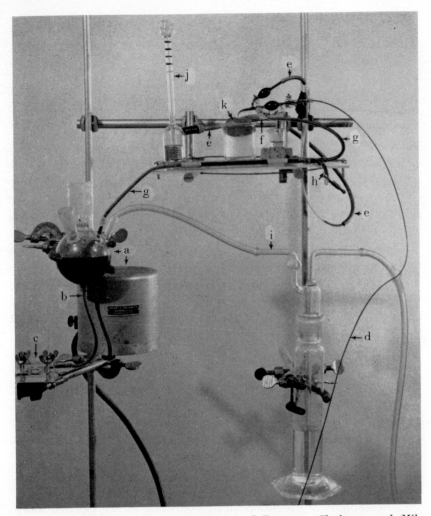

FIG. 4.10. Kidney perfusion apparatus of Bauman, Clarkson, and Miles (1963). Photograph of simulated perfusion system as it appears in the constant temperature cabinet. $a$, blood reservoir and magnetic stirrer; $b$, arterial-flow tract from reservoir to $c$, filter; the peristaltic pump, between $c$ and $d$, is not shown; $d$, arterial-flow tract (polyethylene tubing, P.E. 160) leads to double-bulbed arterial catheter and $k$, the kidney. Top bulb outlet connects via tubing, $e$, to mercury manometer, $j$, on the perfusion platform. $f$, ureteral catheter (P.E. 10); $g$, venous flow tract; $h$, air inlet in venous tract for measuring blood flow; $i$, flow tract for moistened 95% $O_2$–5% $CO_2$. The venous catheter and the humidifier are not shown. Pump tubing is vinyl Tygon plastic (American Instrument Co., No. 5-8966). All other parts of the blood-flow tracts are Silatube, i.d. $\frac{1}{8}$ in, o.d. $\frac{3}{16}$ in (Reiss Mfg. Co., Little Falls, N.J.). The arterial and venous catheters are fixed in apposition on a plastic yoke. The kidney, supported by the yoke and catheters, is suspended in warm Krebs–Henseleit medium. A clamp attached to the right upright bar extends to the arterial catheter.

*Cabinet.* No details are given, but that illustrated in the Appendix 1 would be suitable. The temperature for the present experiments was 38–40°C.

### Perfusion medium

Pooled cardiac aspirates from 8–10 fasted male Wistar rats of 300–350 g are collected, under pentobarbitone anaesthesia, intraperitoneally into 20-cm³, syringes which have been previously rinsed in K–H saline plus dextrose 50 mg per cent, followed by heparin in saline (200 units/ml, 0·9% NaCl). The blood is filtered through moist gauze. Additions are insulin 100 mg, sodium PAH 35 mg, and glucose 50 mg, all of which are dissolved in 4 ml K–H saline.

When made up as described, the final concentrations can vary, since the total volume of medium varies between 70 and 110 ml. It is a simple matter to standardize this by adding a proportion of the 4 ml K–H saline thus prepared according to the starting volume of rat blood in each perfusion. Heparin is also added, again to a variable extent, to give a final concentration of 3–6 units/ml. A further variable, which is of importance and is difficult to control, is the time that elapses between collecting the blood and beginning the perfusion. The authors allow between 70 and 280 min, during which time glucose content may fall and that of lactate rise, due to red-cell glycolysis. This leads to differences in the composition of the initial medium, which is to be avoided where possible (see p. 17). The ranges of glucose and lactate concentrations used are given by Bauman *et al.* (1963).

### Operative procedure

The technique of operation and preparation of the isolated kidney is exactly that described by Weiss *et al.* (see p. 242). The only additional manoeuvre is the cannulation of the ureter, which is carried out before the cannulation of the vessels, as described by Nishiitsutsuji–Uwo *et al.*

### Characteristics of perfusion

It is difficult from the published results to say how long 'successful' perfusion was continued with this preparation. Results obtained from 2 hours of perfusion are contained in the table of Bauman *et al.* and show that the initial flow rate was variable (2–3 ml/min) and was, in the main, at about half of that obtaining *in vivo* (c. 5·0 ml/min). This flow rate is, of course, very much lower than that obtained with blood-free media at similar perfusion pressure, a result that is not unexpected in view of the higher viscosity of blood. However, simple calculation shows that oxygen supply should be adequate, assuming the adequacy of oxgenation (q.v.), and assuming a figure for the oxygen requirement at this flow rate.

Unfortunately, no figure for oxygen consumption are given for perfusion with rat blood. It seems unlikely that the figure could fall on the curve constructed from the data of Weiss (see Fig. 4.11), since

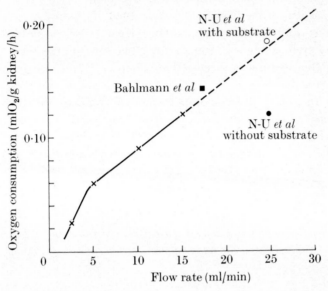

FIG. 4.11. Comparison of methods of kidney perfusion based on oxygen consumption. A curve (x—x) of oxygen consumption by the perfused rat kidney versus medium flow rate through the organ is based on data of Basar *et al.* (1968) and Leichtweiss *et al.* (1969), using an haemoglobin-free perfusion medium and the kidney of a fed animal. The oxygen consumption observed by Bahlmann *et al.* using a fed animal and rat blood medium, and by Nishiitsutsuji–Uwo (N–U) *et al.* using a 48-hour starved animal and haemoglobin-free medium with 5 mM sodium-pyruvate added as substrate, fall on this curve. In Nishiitsutsuji–Uwo's experiment, without the substrate, the oxygen consumption fell below the curve.

this lies below the figure for results obtained *in vivo*. This, in turn, brings into question the validity of Weiss's conclusions about auto-regulation, particularly that it no longer is to be observed at flow rates below 4 ml/min. A disturbing feature of the results presented here for rat-blood perfusion is the apparent inability of the kidney to remove lactate from the medium. On the contrary, its concentration tends to rise in the experiments documented. Quite rapid lactate uptake, with conversion to glucose, was described by Nishiitsutsuji–Uwo *et al.* in the haemoglobin-free perfusion (Method IA(i)) when the initial concentration was 5 mM lactate. No figures are given for the $K_m$ for the substrate, and it is possible that the concentrations present in Bauman's experiments were too low for uptake to occur. This may be unlikely, however, because of the appearance of considerable amounts of lactate in the urine under these conditions. Ability to remove lactate from the perfusion medium may be a critical test of function of the perfused kidney, as it is of the liver (Hems *et al.* 1966). Certainly, early attempts to perfuse kidneys by Nishiitsutsuji–Uwo *et al.* (unpublished) also failed to produce glucose from lactate. The rate of glycolysis of the red cells in the medium is discussed in Chapter 1 and also in Chapter 3 (Exton and Park).

It is therefore of some importance to compare the physiological functions of the kidney perfused with blood and that perfused with haemoglobin-free media (Table 4.4). The blood-free data is compiled from that of the foregoing preparations, with additional information from Bahlmann *et al.* (1967).

TABLE 4.4

*Comparison of physiological parameters of function in the blood-perfused and the haemoglobin-free perfused rat kidney*

| Function | Blood medium | Hb-free medium | *In vivo* |
|---|---|---|---|
| Medium-flow (ml/min) | | | |
| initially | Variable | Steady | |
| steady state | 3·7–2·1 | 10–25 | 5·0 |
| | (plasma 3·2) | | (plasma 3·3) |
| Pressure | Fixed | 70 or 85–105 mm | |
| | (100–140 mmHg) | Hg | |
| Urine flow (ml/min) | 0·028–0·073 | 0·046 or 0·055 | 0·0025–0·008 |
| increases | in 2 h | | ml/min |
| pAH clearance | 1·15 approx. | | |
| creatinine clearance | | 0·14 | |
| | | Nishiitsutsuji–Uwo | |
| Weight gain during | 'considerable' | 'considerable' | |
| perfusion | | | |
| Histology | gross tubular | tubular dilatation | |
| | dilatation and | (no ppt.) | |
| | protein precipitate | | |
| | in tubules | | |

The conclusion must be that physiologically there are no major differences attributable to the use of red cells in the medium, nor to the considerable differences in perfusion flow rate which result. At the present stage of information, therefore, it is probably no disadvantage to omit red cells (and, by extension, other blood constituents) from the medium, where their presence may interfere with the metabolic study in progress (see Chapter 1, Media). Certainly, it is neither the absence of red cells, nor the high medium flow rates that bring about the major physiological abnormalities reported. The recent micro-puncture studies of Bahlmann *et al.* (1967) in perfused rat kidney may offer an explanation for this.

**Method IIA(i). Kidney perfused *in situ* with semi-synthetic medium and non-pulsatile flow** (Method of Fülgraff *et al.* (1968))

For convenience, this method is described before that of Bahlmann *et al.* (Method IIA(ii)) on which it is based. It was used in this case to demonstrate the effect of diuretic drugs on the organ and to analyse the effect of substituting $Ca^{++}$ for $Na^+$ in the perfusion medium.

The kidney (and it is not quite clear from the description whether the left kidney only, or both kidneys are used), is perfused *in situ* beneath a layer of paraffin. Inflow is via the aorta, under a constant head of pressure provided by attaching a steady gas pressure to the sealed reservoir above the kidney. The blood-free medium can be sampled as it leaves the inferior vena cava, but it is not recirculated. A simple apparatus, similar to that already described by Leichtweiss *et al.* (q.v.) is used, together with a more complex semi-synthetic perfusion medium. Perfusions are conducted with apparent success for 3–4 h with this method.

*Apparatus*

Fig. 4.12 illustrates the apparatus required.

*Cannulae.* The cannulae are not described but the following are suitable. *Arterial.* A cannula suitable for the aorta is described by Sussmann *et al.* (Chapter 6 Pancreas). Alternatively, a simple polythene cannula of o.d. 2 mm may be used. *Venous.* The cannula used by Nishiitsutsuji–Uwo *et al.* (p. 231) or by Weiss (p. 239 Fig. 4.8(a)) is suitable. *Ureter.* PE10 polythene tubing is suitable, the end being cut at an acute angle with a razor blade.

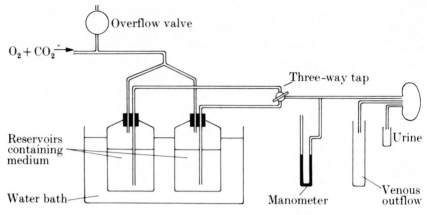

F I G .  4.12.   Perfusion of kidney without a pump. Diagram of the apparatus of Fülgraff *et al.* (1968). Pressure is provided by a gas cylinder, controlled by an overflow valve. Medium is contained in reservoirs warmed in a water bath. Alternative media are selected by using a three-way tap.

### Perfusion medium

The semi-synthetic medium used in this study is more complex than most. It is based on that of Weiss *et al.* (q.v.) with phosphate and acetate as buffers but, in addition, bicarbonate is added and pH controlled by the composition of the gas mixture. In other respects the medium resembles that of Bahlmann *et al.* with some arbitrary changes. These are (mM): $CaCl_2$ 2·5; NaCl 103·25; KCl 4·74; $MgSO_4$ 1·18; $KH_2PO_4$ 1·18; $NaHCO_3$ 25·0; ($Na^+$ (in Haemaccel) = 10·5); sodium acetate 5·0, sodium pyruvate 0·25, sodium lactate (?DorL) 3·5; glucose 6·2; urea 6·63 plus (in g per cent); Haemaccel, 2·7; bovine albumin (Mann) 0·051; insulin 0·05. Note that the form in which the lactate is added is not stated. Presumably the concentration is chosen in order to give a 'physiological' lactate–pyruvate ratio near to 10/1 (cf. Ruderman *et al.* (1971) Chapter 5 p. 310). It is suggested therefore that 2·5 mM sodium L-lactate should be used (see also Method IIA (ii)).

### Operative procedure

Rats of 250–300 g are used, anaesthetized with urethane (1·25 g/kg, intraperitoneally as 25% solution in normal saline = 0·5 ml/100 g rat). A mid-line abdominal incision with lateral extensions below clamps is made, exposing the aorta and inferior vena cava below the renal vessels. The ureter of the kidney to be perfused is cannulated as described (p. 234).

The aorta must be separated from the inferior vena cava and all its branches tied below the origin of the left renal artery. In addition, the spermatic and adrenal arteries are tied (these vessels are to be seen in Fig. 5.19) as they arise from the aorta. Ligatures are placed around the renal vessels on the left and right and the pair to the kidney not being perfused are tied. Heparin ($0 \cdot 1$ ml $= 100$ units is suggested) is injected into a large vein; perhaps the technique described by Nishiitsutsuji–Uwo (p. 234) is the most satisfactory.

Cannulation is achieved from below, the inferior vena cava being cannulated before the aorta. The methods are those described by Weiss, with the same precaution that flow through the aortic cannula is started before releasing the bulldog clip that occludes the vessel. In this preparation only the aorta is cannulated, no attempt being made to advance the cannula into the renal artery. Whatever results obtain, the method is to be criticized on the grounds of an unnecessarily complex operative procedure. The operation of Weiss or perhaps, more simply, that of Nishiitsutsuji–Uwo, involves the minimum of manipulation to achieve the same or better results.

## Characteristics of perfusion

Perfusion begins immediately, without any anoxic interval, with initial flow rates around 12 ml/min per g. Flow rate and glomerular filtration rate (inulin clearance) are similar to those of other preparations, but here both parameters fall within the first 40 min.

Fülgraff *et al.* also give additional information on the effects of diuretics (frusemide, 60 mg/l; chlorothiazide, 40 mg/l) on the excretion of water and sodium by this isolated preparation.

The experimental design chosen employs a 30-min pre-perfusion, to wash out blood and achieve 'equilibrium' but from the information available in this and the other preparations described, the kidney is failing from the start of perfusion and to ignore the first 30 min is to miss the time when the kidney is functioning nearest to its normal capacity *in vivo*. The ATP data of Nishiitsutsuji–Uwo do not help, unfortunately, since only post-operative and 30-min figures were determined (see Table 4.3). Clearly, the organ is abnormal immediately post-operatively, and this abnormality is corrected after 30 min. But it may well be quicker, as in the liver, which recovers a normal ATP content after 5 min (Hems *et al.* 1966). Urine flow and GFR, and medium flow rates are within the range described for similar methods in Table 4.4. The rate of failure of this preparation appears to be more rapid, but this cannot be attributed solely to the technique used *in situ* (cf. the following method).

**Method IIA(ii). Kidney perfused *in situ* with semi-synthetic medium and pulsatile arterial inflow** (Method of Bahlmann, Giebisch, Ochwadt, and Schoeppe (1967))

A more acceptable method of perfusion *in situ* is that on which Method IIA(i) is based. It combines a rapid operative procedure with a re-circulation circuit and pulsatile flow. The preparation has been used to study the mechanism in which all the rat renal perfusions so far described fail—the ability to concentrate urine.

The method is based on the standard liver perfusion techniques of Miller and of Schimassek and on the kidney perfusion technique of Bauman. The apparatus is housed in a heated cabinet at 39°C, and the kidney (and the rat) is held outside and above the cabinet, so that medium is pumped directly into the renal artery and falls from the renal vein into the multibulb oxygenator. This obviates the need for a second pump but has the disadvantage of allowing considerable desaturation of the arterial medium before it reaches the kidney from the bottom of the lung. This must then be taken into account in calculations of oxygen uptake by the kidney. A further modification of importance is the use of a more physiological pulse rate (100/min) than that used by Bauman or by Weiss (q.v.) although the pulse pressure remains similar, viz. 8–12 mmHg (10–16 cm $H_2O$).

*Apparatus*

As already stated, the apparatus is in a cabinet which is kept at a temperature of 39°. This allows the temperature of the medium leaving the arterial cannula, which is outside the cabinet, to be kept at 38°C. Fig. 1.4(d) shows the details of an apparatus which is very similar to that of Miller (see Liver): the multibulb glass oxygenator is 30 cm long with bulbs of 8 cm maximum diameter. This is close to the ideal suggested in Chapter 1. PVC tubing of internal diameter 1–2 mm and external diameter 6 mm is used throughout, and the oxygen perme-ability of this tubing creates a problem. Nishiitsutsuji–Uwo overcame this by making long stretches of the connections of glass with PVC only at the essential joints.

*Pump*. The details of the pressure and pulse rate and pressure provided by the pump have been discussed. The type of pump is not mentioned but see Chapter 1 for a suitable selection. The pressure recorded by a strain gauge on the side-arm of the arterial cannula can be adjusted by altering the pump setting (which also alters pulse rate) or by adjusting the clamp shown on the side-arm of the arterial inflow (see Fig. 1.4(d)). Finally, an electromagnetic flow meter on the venous outflow, and oxygen electrodes on the venous outflow and on the arterial by-pass, record flow and oxygen tension respectively.

*Cannulae.* Those of Weiss *et al.* are used.

## Medium

A semi-synthetic medium, similar to that of Schimassek is used. The main differences here are a higher albumin content (5·5 g per cent rather than 2·7 g per cent) and the addition of urea, which is thought to play a special role in the reabsorption of water in the distal tubule (see Pitts (1968)). Details of the medium are given in Table 4.5. For reasons that are not apparent, to 40–50 ml of this synthetic medium, pH 7·4 when gassed with 95% $O_2$, are added from 2–6 ml of blood from the donor rat. This gives a final haemoglobin content of 1·2 g per cent and inevitably complicates the interpretation of any biochemical data.

## Operative procedure

Inactin anaesthesia is used, followed by the intravenous injection of heparin, 20 mg/kg, and mannitol, 0·4 ml, 15% solution. The latter provokes an osmotic diuresis (see, for example, Harvey and Malvin (1965)) and it is in this state of maximum GFR that the kidney is perfused.

The operation is exactly as described in the account of Nishiitsutsuji–Uwo and of Weiss, with cannulation of the ureter and the Weiss technique of cannulation of the vena cava and the renal artery, via the aorta. Here the operations ends with the tightening of the ligatures numbered 1 and 2 in the Nishiitsutsuji–Uwo description. The kidney is perfused immediately, and remains *in situ*. Warm paraffin is poured into the open abdomen and maintains the temperature of the kidney. In addition, it facilitates the special technique of micro-puncture (see p. 227) which has been introduced into perfused rat kidney experiments by these workers.

## Characteristics of perfusion

Table 4.4 shows the main results on physiological function in this preparation, and they do not differ from other haemoglobin-free perfusions. A linear flow rate is maintained for 2 h while the glomerular filtration rate shows the expected progressive fall with time. Oxygen consumption lies on the curve constructed from Weiss's data in the first hour of perfusion, but is some 10 per cent lower in the second hour, indicating either exhaustion of lactate and pyruvate as substrates in the medium or failure of perfusion.

It is of particular interest that these workers have established that interstitial oedema, a constant feature of all the preparations described so far, may be attributable to three factors, all of which have become apparent as the result of micro-puncture studies: (i) inadequate compensatory adjustment of loop of Henle and collecting duct reabsorption of $H_2O$ and sodium, due to (ii) decreased tubular sodium transfer capacity, and (iii) decreased passage time of fluid along the proximal convoluted tubule, allowing less time for $Na^+$ reabsorption.

TABLE 4.5

*Composition of perfusion media used for kidney*

| Component | Medium | | | |
|---|---|---|---|---|
| | 1 | 2 | 3 | 4 |
| Bovine albumin | 5·0 | 5·54 | 0·051 | — |
| Haemaccel | — | — | 2·7 | — |
| Dextran (MW 60–90 × 10³) | | | | 3·0 |
| NaCl | 118·4(143) | 117·1 (147) | 103·3 (148) (+10) | 120 (145) |
| KCl | 4·74 | 4·74 | 4·74 | 5·0 |
| CaCl$_2$ | 2·54 | 2·54 | 2·5 | 0·8 |
| MgSO$_4$ | 1·18 | 1·18 | 1·18 | — |
| (MgCl$_2$) | — | — | — | 0·8 |
| NaHCO$_3$ | 25·0 | 25·0 | 25·0 | Nil (TRIS 10) |
| Na$_2$HPO$_4$ | — | — | — | 20·0 |
| NaH$_2$PO$_4$ | 1·18(K) | 1·18 | 1·18 | 5·0 |
| Na acetate | — | — | 5·0 | — |
| Na pyruvate | — | 0·25 | 0·25 | — |
| Na lactate | — | 2·75 (DL) | 3·5(?) | — |
| Glucose | — | 6·2 | — | — |
| Urea | — | 6·65 | 6·65 | — |
| Gas | 5% $CO_2$ | 5% $CO_2$ | 5% $CO_2$ | 100% $O_2$ |
| pH | 7·4 | 7·3–7·4 | 7·25–7·35 | |

Information concerns the standard medium described in each of the papers quoted. References are (1) Nishiitsutsuji–Uwo et al. (1967), (2) Bahlmann et al. (1967), (3) Fülgraff et al. (1968), and (4) Turner et al. (1968). Figures are $\mu$mol/ml (mM) unless otherwise stated. Total $Na^+$ is in parenthesis.

## GENERAL COMPARISON OF METHODS OF KIDNEY PERFUSION

It is clear that no method of rat kidney perfusion so far developed approaches the degree of similarity to the organ *in vivo* that has been achieved for the rat liver. The studies of Bahlmann point to the mechanical nature of the defect in tubular reabsorption by the isolated kidney, and this is not overcome either by using rat blood as perfusion medium or by imposing a pulsatile perfusion pressure.

Metabolic tests of renal function are, however, more reassuring, in that such preparations are viable for a few hours at least (see, for example, Nishiitsutsuji–Uwo *et al.* (1967) p. 236). The preparation maintains a normal ATP content, has at least the expected oxygen uptake, and metabolizes many substrates. In particular, high rates of glucose synthesis are observed, which is good evidence of the maintenance of cellular integrity in such preparations.

Of the methods devised for biochemical rather than physiological studies, that of Nishiitsutsuji–Uwo *et al.* has been investigated most fully and the tests of function suggested for the perfused liver have been applied. The preparation of Turner *et al.* (1968) (described in Chapter 2) is potentially of great interest since continuous oxygen and substrate uptakes may be followed. It should be possible to modify this apparatus for use at physiological temperature, where at present the oxygen content of the chamber limits the time available and the temperature at which experiments may be performed. The other methods described reinforce the considerable evidence of the physiological abnormality of the perfused rat kidney but point to the consistency of such a preparation. This is sufficient to justify its use in metabolic studies.

Apart from its direct value as a method studying metabolism in the presence of excretory function, perfusion may have a place in the assessment of function of organs stored for preservation and transplantation, using the metabolic tests discussed in this chapter.

### REFERENCES

BAHLMANN, J., GIEBISCH, G., OCHWADT, B., and SCHOEPPE, W. (1967) Micropuncture study of isolated perfused rat kidney. *Am. J. Physiol.* **212**, 77.

BASAR, E., TISCHNER, H., and WEISS, CH. (1968) Untersuchungen zur Dynamik druckinduzierter Änderungen des Strömungswiderstandes der autoregulierenden, isolierten Rattenniere. *Pflügers Arch. ges. Physiol.* **299**, 191.

BAUMAN, A. W., CLARKSON, T. W., and MILES, F. M. (1963) Functional evaluation of isolated perfused rat kidney. *J. appl. Physiol.* **18**, 1239.

BENNOY, M. P. and ELLIOTT, K. A. C. (1937) The metabolism of lactic and pyruvic acids in normal and tumour tissues. V. Synthesis of carbohydrate. *Biochem. J.* **31**, 1268.

BLOOM, W. and FAWCETT, D. W. (1968) *A textbook of histology.* Saunders, Philadelphia.

BURG, M. B. and ORLOFF, J. (1962) Oxygen consumption and active transport in separated renal tubules. *Am. J. Physiol.* **203**, 327.

CORI, C. F. (1925) The fate of sugars in the animal body. *J. biol. Chem.* **66**, 691.

DE LUCA, H. F., GRAN, F. C., STEENBOCK, H., and REISER, S. (1957) Vitamin D and citrate oxidation by kidney mitochondria. *J. biol. Chem.* **228**, 469.

FINGL, E. (1952) Tubular excretion of creatinine in the rat. *Am. J. Physiol.* **169**, 357.

FÜLGRAFF, G., HEHNE, H. J., SPARWALD, E., and HEIDENREICH, O. (1968) Die Wirkung von Calciumionen auf die glomeruläre Filtrationsrate und die Electrolyte—und Wasserauscheidung von Intakten und isoliert perfundierten Rattennieren. *Archs int. Pharmacodyn.* **172**, 49.

GÜDER, W. and WIELAND, O. (1970) The effect of 3′ 5′-cyclic AMP on glucose synthesis in isolated rat kidney tubules. *Gluconeogenesis* (ed. H. D. Söling), p. 240. Thieme-Verlag, Stuttgart.

GYÖRGY, P., KELLER, W., and BREHME, TH. (1928) Nierenstoffwechsel und Nierenentwicklung. *Biochem. Z.* **200**, 356.

HARVEY, A. M. and MALVIN, R. L. (1965) Comparison of creatinine and inulin clearances in male and female rats. *Am. J. Physiol.* **209**, 849.

HEMS, R., ROSS, B. D., BERRY, M. N., and KREBS, H. A. (1966) Gluconeogenesis in the perfused rat liver. *Biochem. J.* **101**, 284.

KINTER, W. B. and PAPPENHEIMER, J. R. (1956) Role of red blood corpuscles in regulation of renal blood flow and glomerular filtration rate. *Am. J. Physiol.* **185**, 399.

KRAMER, K. and DEETJEN, P. (1964) Oxygen consumption and sodium reabsorption in the mammalian kidney. I.U.B. Symposium Series, Vol. 31, *Oxygen in the animal organism* (eds. F. Dickens and E. Neil), p. 411. Pergamon Press, Oxford.

KREBS, H. A., BENNETT, D. A. H., DE GASQUETT, P., GASCOYNE, T., and YOSHIDA, T. (1963) Renal gluconeogenesis. *Biochem. J.* **86**, 22.

—— HEMS, R., and GASCOYNE, T. (1963) Renal gluconeogenesis IV. *Acta biol. med. germ.* **11**, 607.

KRIZ, W. (1967) Der architektonische und functionelle Aufbau der Rattenniere. *Z. Zellforsch. microsk. Anat.* **82**, 495.

LEARY, W. P. P. (1969) Aspects of the renin-angiotensin system. D.Phil., University of Oxford.

LEICHTWEISS, H-P., LÜBBERS, D. W., WEISS, CH., BAUMGÄRTL, H., and RESCHKE, W. (1969) The oxygen supply of the rat kidney: measurements of intrarenal $pO_2$. *Pflügers Arch. ges. Physiol.* **309**, 328.

—— SCHRÖDER, H., and WEISS, CH. (1967) Die Beziehung zwischen Perfusionsdruck und Perfusionsstromstärke an der mit Paraffinöl perfundierten isolierten Rattenniere. *Pflügers Arch. ges. Physiol.* **293**, 303.

NAGATA, N. and RASMUSSEN, H. (1970) Parathyroid hormone, 3′ 5′ AMP, $Ca^{++}$ and renal gluconeogenesis. *Proc. natn. Acad. Sci. U.S.A.* **65**, 368.

NISHIITSUTSUJI–UWO, J. M., ROSS, B. D., and KREBS, H. A. (1967) Metabolic activities of the isolated, perfused rat kidney. *Biochem. J.* **103**, 852.

OCHWADT, B. (1956) Zur Selbsteuerung des Nierenkreislaufes. *Pflügers Arch. ges Physiol.* **262**, 207.

PITTS, ROBERT F. (1968) Physiology of the kidney and body fluids. *Year book*, 2nd edn. Medical Year Book Publishers, Chicago.

RICHARDS A. N. (1938) Croonian Lecture: Processes of urine formation. *Proc. R. Soc.* **B126**, 398.

SMITH, HOMER W. (1951) *The kidney: structure and function in health and disease.* Oxford University Press, New York.

SONNENBERG, H., and DEETJEN, P. (1964) Methode zur Durchströmung einzelner Nephronabschnitte. *Pflügers Arch. ges. Physiol.* **278**, 669.

SPECTOR, W. S. (ed.) (1956) *Handbook of biological data*, p. 341. Saunders, Philadelphia.

TURNER, N. D., NEELY, W. A., and PITTS, H. R. (1968) Polarographic respirometer for the isolated perfused rat kidney. *J. appl. Physiol.* **24**, 102.

VAN SLYKE, D. D., RHOADS, C. P., HILLER, A., and ALVING, A. S. (1934) Relationships between urea excretion, renal blood flow, renal oxygen consumption and diuresis. The mechanism of urea excretion. *Am. J. Physiol.* **109**, 336.

WEIDEMANN, M. J., HEMS, D. A., and KREBS, H. A. (1969) Effects of added adenine nucleotides on renal carbohydrate metabolism. *Biochem. J.* **115**, 1.

WEISS, CH., BRAUN, W., and ZCHALER, W. (1967) Über die Einfluss von Änderungen des Urin-pH auf die Ausscheidung einige Aminosäureester ar der perfundierten Rattenniere. *Naunyn-Schmiedebergs Arch. exp. Path. Pharmak.* **258**, 83.

—— PASSOW, H., and ROTHSTEIN, A. (1959) Autoregulation of flow in isolated rat kidney, in the absence of red cells. *Am. J. Physiol.* **196**, 1115.

WINTON, F. R. (1937) Physical factors involved in the activities of the mammalian kidney. *Physiol. Rev.* **17**, 408.

WIRZ, H., HARGITAY, B., and KUHN, W. (1951) Lokalisation des Konzentrierungsprozessus in der Niere durch direkte Kryoskopi. *Helv. physiol. pharmac. Acta* **9**, 196.

## Addendum: Recent modifications in kidney perfusion technique

Method 1A (i) has been modified with very considerable improvement in the physiological parameters of function. With substrates, Na butyrate (2·5 mM) and D-glucose (5 mM) added to the perfusion medium, 97% sodium reabsorption is achieved, with a glomerular filtration rate (inulin clearance) of 0·6 ml/min (Ross, Leaf, and Epstein, unpublished results). The modification of the apparatus employs direct pumping in place of an hydrostatic perfusion (see Weidemann *et al.* p. 257); the apparatus and circuit are now similar to that used by Ruderman *et al.* (p. 309) for perfusion of the hind limb. Two Millipore filters, 14 μm pore size are mounted in parallel to allow continuous filtration of the perfusion medium. The perfusion medium is unchanged, but careful millipore filtration (0·5 μm pore size) into a sterile container is conducted immediately before perfusion. This preparation responds to antidiuretic hormone and to specific substrates, notably glucose, by producing a more concentrated urine, and offers greater scope than its predecessor for combined biochemical and physiological studies.

Recent work is reviewed by:

FRANKE, H., HULAND, H., WEISS, CH., UNSICKER, K. (1971) Improved net sodium transport of the isolated rat kidney. *Z. ges. exp. Med.* **156**, 268.

SCHUREK, H. J., LOHFERT, H., and HIERHOLZER, K. (1970) Na-resorption in der isoliert perfundierten Rattenniere (Abhangigkeit von Substratangebot und Na-load) *Pflügers Arch. ges. Physiol.* **319**, R 85.

# 5. Heart and Skeletal Muscle

## PERFUSION OF MUSCLE

FOR many years the main model for the study of biochemistry of muscle has been the heart. A number of perfused preparations are available and will be discussed in the first part of this chapter. Biochemically, skeletal muscle has been studied in homogenates, for example, in pigeon breast muscle, but for studies in 'intact' muscle it has been customary to use the diaphragm. This has been of particular use since, as a thin sheet of muscle, diffusion of oxygen is sufficient to allow incubation experiments to be conducted aerobically (see p. 300). There has been some suggestion that the central fibres of this preparation may be anoxic, and more recently a technique of retrograde perfusion has been introduced. This is discussed on p. 302.

For larger amounts of skeletal muscle it is necessary to use perfusion techniques if intact muscle preparations are required. A recent preparation of the perfused hind-limb of the rat has proved useful in metabolic studies, and this, together with some earlier preparations, will be discussed on p. 308.

Smooth muscle is not discussed in this section: it is an important component of the tissue perfused in preparations of blood-vessels (Chapter 8), intestine (Chapter 7), and uterus (Chapter 8), but a large number of incubation and superfusion techniques are given as an alternative in the pharmacological literature. These may be more appropriate to biochemical studies in smooth muscle than are the currently available perfusion methods.

## HEART MUSCLE

Heart has served as a model for study of metabolism in muscle, and perfusion has been the experimental method of choice since Langendorff. A succession of workers (Brodie 1903, Bleehen, and Fisher 1954, Morgan, Henderson, Regen, and Park 1961, and Opie 1965) has demonstrated the stability and versatility of the preparation.

The original method of Langendorff (1895), has survived with only minor modifications and remains the standard preparation upon which metabolic studies are performed (Bleehen and Fisher, Williamson and

Krebs (1961), Morgan *et al.*, etc.). However, this preparation of heart does little or no external work (see below) and the most important advance in recent times has been the 'working-heart' preparation of Morgan, Neely *et al.* 1967). This is an atrial perfusion in which considerable volumes of medium are pumped by the left ventricle.

The aortic perfusion of Langendorff survives however, for two reasons: it is technically very much simpler and requires a simpler apparatus. As will become clear from the discussion below, the type of 'work' that the heart performs, external work, by which is meant the pumping of blood (or of medium) against a peripheral resistance, or the alternative, which is simply the development of tension, without pumping blood out of the ventricle, does not appear to be reflected in major metabolic differences. The more elaborate provision of external cardiac work may therefore be quite unnecessary for standard biochemical studies.

In addition to the two basic techniques to be described, modifications in the operative preparation, which convert the Langendorff preparation into a 'closed' circuit, have been introduced by Arnold and Lochner (1965) and by Basar, Ruedas, Schwartzkopf, and Weiss (1968). This may have special application in physiological and biochemical studies and the methods are described below.

## Anatomy

The rat heart represents from 0·4 to 0·52 per cent of body weight in smaller animals (67–160 g) and a little less (0·32–0·37 per cent) in animals of from 224 to 289 g body weight. Donaldson (quoted by Jaffé and Garaller (1958)) gives the following figures:

| *Weight of rat* | *Weight of heart* | % |
|---|---|---|
| 15·0 g | 0·05 g | 0·33 |
| 100·0 g | 0·20 g | 0·20 |
| 200·0 g | 0·45 g | 0·23 |
| 250·0 g | 0·80 g | 0·32 |
| larger | *c*·1 g | — |

The dimensions of the rat heart are given as 1·2 cm long (0·8 in mouse, 2·0 in guinea-pig) and 2·5–3·0 cm circumference at the base (2 cm and 5–6 cm in mouse and guinea-pig respectively).

Fig. 5.1 shows schematically the arrangement of the four compartments of the mammalian heart and the circulation used in most heart perfusion studies is illustrated in Fig. 5.8.

The left ventricle reflects the extra load imposed upon it to provide the entire systemic circulation, by being some 3–4 times the weight of the right. Between the four compartments and the major vessels lie four valves, which prevent reversal of flow: right atrium → right ventricle = tricuspid valve; right ventricle → pulmonary artery = pulmonary valve; left atrium → left ventricle = mitral valve (two cusps); left ventricle → aorta = aortic valve. The Langendorff preparation depends upon the competence of the last-named valve; the

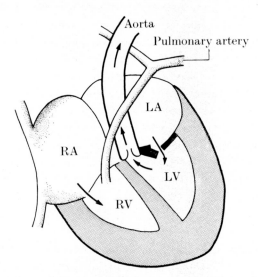

FIG. 5.1. Chambers of the heart. The heart is viewed in section: RA = right atrium; LA = left atrium; RV = right ventricle; LV = left ventricle. The pulmonary veins (not shown) enter the left atrium. Arrows demonstrate the direction of flow of blood in life.

coronary arteries, which supply the heart muscle, arise just above the aortic valve and are the only outflow when a hydrostatic pressure is exerted on the valve from above (Fig. 5.2).

In the rat, three vessels enter the right atrium; the two superior and one inferior venae cavae. Four vessels, the pulmonary arteries, enter the left atrium. Two arise from each lung and the four enter very close together on the posterior wall of the atrium.

*Coronary arteries.* As in other animals, in the rat there are two coronary arteries, a left and a right. The details of their distribution are similar to those in larger animals and, as in larger animals, they are 'end-arteries'. When one is tied, the sector of muscle supplied becomes ischaemic, since no anastomoses with adjacent branches exist. The coronary arteries supply all the muscle of the heart. The endothelium might be expected to receive some oxygen and substrates from blood

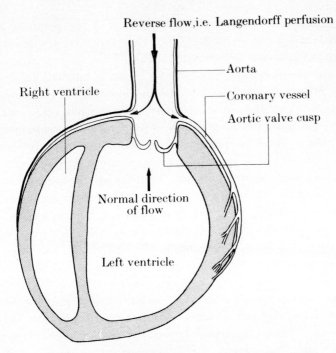

Reverse flow, i.e. Langendorff perfusion

Aorta

Right ventricle

Coronary vessel

Aortic valve cusp

Normal direction of flow

Left ventricle

FIG. 5.2. Basis of 'Langendorff' perfusion of heart. The aortic valve closes when pressure is exerted from above (reverse flow). Medium must pass into the coronary vessels to supply heart muscle, since it cannot now enter the left ventricle. The ventricle can empty only when intraventricular pressure rises above that in the aorta, to open the aortic valve.

(or medium) within the heart's cavities, but this represents only a very small fraction of the whole cardiac uptake.

*Venous drainage.* Coronary veins drain into the coronary sinus, which opens into the right atrium. Out-flowing medium in heart perfusion experiments is therefore normally allowed to drip from the right atrial orifices—the superior and inferior vena cava openings. However, in some preparations to be described a formal cannulation of the superior vena cava is made (Basar *et al.*) or the atrium is closed with appropriate ligatures, so that outflowing medium enters the right ventricle and thence the pulmonary artery (Arnold and Lochner 1965). In another, the simple aortic cannulation is performed as for Langendorff, but the heart is mounted apex uppermost, so that emerging medium does not come into contact with the outside of the heart (Rabitzsch 1968).

*Venae cordis minimae.* A small proportion of the coronary arterial flow (less than 5 per cent) does not enter the coronary sinus and is therefore not collected in the right atrium. The venae cordis minimae (Thebesian veins) drain independently, mostly into the right atrium but also into the ventricles. They become important in the Langendorff preparation. Thus Morgan *et al.* (1961) records that both ventricular and aortic pressures of the Langendorff preparation increased with each systole. Aortic diastolic pressure was found to equal the perfusion pressure, whereas peak systolic aortic pressure was from 4–10 mmHg greater than peak aortic pressure. Ventricular diastolic pressure increased from 2 to 12 mmHg over the range of perfusion pressures tested. These pressure changes suggest that ventricular filling occurs during diastole and that some fluid is ejected with systole. This was confirmed by inserting a stainless steel cannula through the mitral valve, when increasing amounts of fluid (1–6 ml/min) were pumped from the ventricle as perfusion pressure was raised.

Morgan *et al.* conclude that this medium arises from the aorta by incompetence of the aortic valve, and by drainage from the Thebesian veins (venae cordis minimae). Other authors, notably Arnold and Lochner (1965), have taken special precautions to drain the fluid by inserting a fine catheter into the left ventricle. The importance of changes in 'diastolic volume' in regulating the force of contraction is discussed below (p. 264).

**Microscopic anatomy** (Bloom and Fawcett (1968))

The heart wall is in three layers: an outer epicardium, which is a layer of connective tissue; an inner layer, the endocardium, one or more cells thick; and the predominant middle layer, the myocardium. Myocardium consists of cardiac muscle, with some blood-vessels, nerves, and connective tissue, but muscle cells so predominate that this is an almost unique opportunity to study the metabolism of a single cell-type in a perfusion system.

Anatomically, the muscle fibres are arranged as spiral or figure-of-eight bands which traverse both atria and ventricles and ensure that the heart contracts as a unit.

Cardiac muscle differs in several respects from skeletal muscle. Skeletal muscle fibres are arranged in parallel and show the well-known cross-striations. Cardiac muscle is also striated. However, (i) the fibres are not syncytial as was formerly thought, but are made up of separate cellular units, joined end to end by intercalated discs; these discs run transversely across the fibres; (ii) fibres are not simply cylindrical but bifurcate and connect with adjacent fibres to form a network; (iii) the nuclei are deep within the fibre, rather than at the surface as in skeletal muscle.

*Conducting tissue.* Specialized muscle cells are found in the conducting system of the heart. Small Purkinje cells form the sino-atrial node, or pace-maker, the rhythm of which determines the rhythmic contraction of the heart as a whole. Impulses from the pace-maker reach the atrio-ventricular node across apparently normal cardiac muscle, but spread from there to the ventricle in the specialized 'bundle of His'. The Purkinje cells here are particularly large, vacuolated, and contain high concentrations of glycogen (Bloom and Fawcett 1968). The physiological function of the Purkinje cell system in permitting synchronous contraction of the ventricle is clearly established. Their contribution to the metabolism of the heart has not been so clearly defined, but to study this problem in so small a heart as that of the rat would probably be unnecessarily difficult.

**Physiological function**

The heart beats rhythmically and ejects blood (or medium) during each contraction (systole). Relaxation (diastole) follows, during which the ventricles are refilled from the atria.

Cardiac output is the volume of blood put out by one half (left or right) of the heart in unit time. *In vivo*, this must obviously be the same for the left and right ventricles, since they are directly connected via the pulmonary vasculature. The 'work' of the left ventricle however, greatly exceeds that of the right, since a much greater circuit must be supplied. In consequence, the left ventricle is very much larger than the right (L:R = 3:1) and is more often used in biochemical studies (Neely, Liebermeister, Battersby, and Morgan 1967). Stroke volume is the output of one ventricle in a single contraction cycle, and so

stroke volume × heart rate = cardiac output.

The events of the cardiac cycle are best appreciated by considering the pressure diagrams of Wiggers (1952), which are reproduced in standard physiology texts. The subdivisions of the cycle are those proposed by Wiggers (1921). Systole begins with the rise in ventricular pressure and ends with release of tension in all muscle units. It is further divided into phases: isometric contraction, maximum ejection, and reduced ejection. The subsequent period of diastole is divided into phases of protodiastole, representing the time required for closure of the semi-lunar valves (aortic and pulmonary valves), isometric relaxation, rapid ventricular filling, diastasis, and filling by atrial contraction. Clearly this description applies only if the ventricle is allowed to fill, and of the perfusion systems to be described only the 'working' heart preparation of Morgan *et al.* (p. 279) makes provision for this.

The cardiac cycle modified by Langendorff perfusion is described by many authors (see, for instance, Neely *et al.* (1967)). However, as will be discussed later, the left ventricle is not empty in the standard Langendorff preparations of Morgan and of others, and, with medium to pump, tension is developed within the ventricle (Morgan *et al.* 1961), unless provision is made to empty the left ventricle continuously during perfusion (Jordan and Lochner 1962). The cardiac cycle in these preparations is therefore dependent on aortic pressure and on the duration of perfusion.

*Starling's law of the heart*

The energy of contraction is a function of the initial length of muscle fibres: in the case of the heart, the force of ventricular contraction is proportional to the volume of that chamber at the end of diastole. This is Starling's law of the heart and is the simple explanation of the extra work done by the perfused heart when medium is allowed to flow into the left ventricle from the left atrial cannula ('working-heart'). In the Langendorff preparation, little (Morgan) or no, (Arnold and Lochner) ventricular filling occurs.

*Work of heart contraction*

The work of the heart is approximately expressed as

$$\text{cardiac work} = \text{cardiac output} \times \text{arterial pressure},$$

but this neglect the factor of velocity of ejection, in the case of the working heart and, more important, does not describe the 'work' of isometric contraction. This occurs in the Langendorff preparation, where work is proportional to the peak systolic pressure (see, for example, Opie (1965)). Since intra-ventricular pressure is the critical parameter here, work in this preparation is predictably proportional to the pressure exerted to close the aortic valves, i.e. the perfusion pressure.

METHODS OF PERFUSION

Of the many different methods of perfusing the heart, the following are those most likely to have continuing value in biochemical studies.

I. Aortic perfusion: Langendorff preparation.
   (a) Original method: Langendorff (1895).
   (b) Recirculation method: Bleehen and Fisher (1954).
   (c) Improved recirculation method: Morgan *et al.* (1961).
II. Other modifications of Langendorff preparation.
   (a) Isovolumic preparation: Opie (1965).
   (b) Continuous emptying of the left ventricle: Arnold and Lochner (1965).
   (c) 'Inverted' heart preparation: Rabitzsch (1968).
III. Atrial perfusion: 'working heart' of Neely *et al.* (1967).
   (a) Original method.
   (b) Modifications: Chain, Mansford, and Opie (1969); Øye (1965).
IV. Closed circuit perfusions.
Method of Basar *et al.* (1968).

It is clear from the diagram (Fig. 5.2) that the Langendorff preparation perfuses the muscle of both ventricles, while only the left ventricle produces any tension, by contracting against the closed aortic valve. Similarly, in the 'working' heart preparations, coronary perfusion of both ventricles occurs, while only the left ventricle (and left atrium) perform external work.

The system modified by Basar *et al.* is a compromise, with partial filling of the left ventricle as in the Langendorff preparation and some right ventricular work to cope with the coronary flow entering from the closed right atrium. How this modification influences cardiac metabolism has yet to be analysed.

### Method I(a). Simple Langendorff preparation

The method originally described by Langendorff (1895) was applied to the dog and the rabbit. The heart was perfused through a cannula tied into the aorta and attached to the outflow of a Marriot flask.

Pressure for perfusion is exerted in this apparatus by running water from a tap into the connected reservoir flask. Langendorff's paper is illustrated with an engraving which is reproduced here (Fig. 5.3). Many modifications in apparatus have been introduced but in modern biochemical circles, aortic (Langendorff) perfusion is usually conducted in the rat heart by means of one or other of the apparatuses described below.

### Method I(b). Recirculation method of Bleehen and Fisher

This recirculation method employs a gas-lift system (see Chapter 1) and therefore requires no pump. However, the problems of frothing, especially if protein is present in the medium, are almost insurmountable. The apparatus is illustrated (Fig. 1.3), together with a device for limiting frothing. A perfusion pressure of 80 cm $H_2O$ is obtained by the gas lift and temperature control is by water-jacketing. No separate oxygenator is provided as this is thought to be unnecessary.

*Filter.* A Soxhlet filter is provided and probably accounts for the relative success of the method. Apparatuses lacking an effective filter result in perfusions in which steadily increasing perfusion pressure is required to maintain a constant flow (for example, that of Opie (1965)). Flow rate is determined on the inflow side in this apparatus (contrast Morgan's and most other heart perfusion systems, in which coronary outflow is monitored), by injecting a bubble of gas into the downflow through a fine polythene catheter. The bubble is allowed to escape into a side arm before reaching the heart.

*Perfusion medium.* Krebs–Ringer bicarbonate buffer, prepared as described by Umbreit (see p. 26), but containing only half the calcium concentration, was used by these authors.

*Operative procedure.* This was the same as that used by Langendorff and is described in the following section (see Morgan *et al.* (1961)).

### Characteristics of perfusion

Studies on insulin and glucose uptake were undertaken by the method's originators and metabolic studies have also been performed by Morgan (unpublished) who followed this method until their own improved method was devised. Williamson and Krebs (1961) used a

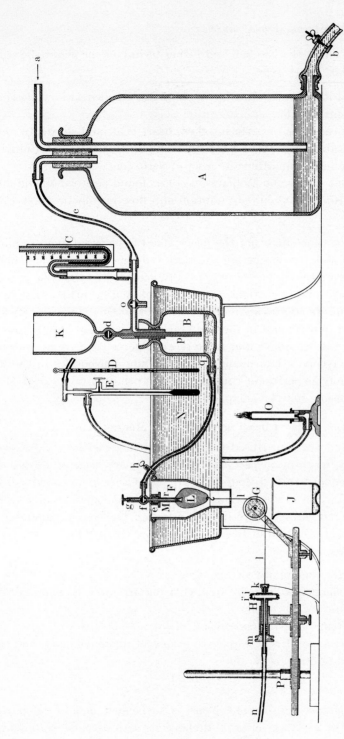

FIG. 5.3. Heart perfusion apparatus of Langendorff. A reproduction of the original engraving (1895).

268 *Heart and skeletal muscle*

modification of the Bleehen and Fisher technique for metabolic studies in the heart. They report

(i) that it is possible to omit the anti-froth device without detriment to the preparation;

(ii) Linear rates of perfusion flow, heart beat, and substrate uptake are obtained for up to 2 h with this preparation if determinations are begun 15 min after the onset of perfusion;

(iii) that Fisher and Williamson (1961) found between 10 and 20 per cent drop in oxygen consumption and flow rate in this first 15-min period.

It must be concluded that the heart, like the perfused liver, requires a period of stabilization before constant experimental conditions obtain. In the liver, 40 min has been generally accepted as the time required, on the basis of observations by Schimassek and by others that *in vivo* levels of intermediates are regained within that time. Less attention has been paid to this point in the preparations of perfused heart. However, the observations above and more recent figures concerning concentrations of glycolytic intermediates and adenine nucleotides in the heart *in vivo* and in perfusion confirm that the shorter interval of 15 min 'pre-perfusion' may be adequate.

### Method I(c). Isolated heart preparation of Morgan *et al.*

The rat heart is perfused through a cannula tied into the stump of the aorta, resulting in perfusion of the coronary arteries, as described by Langendorff (1895). Medium is supplied either under gravity from a head of constant hydrostatic pressure or, when a small volume recirculation circuit is used, directly from a pump. The special advantages of this method over that of Bleehen and Fisher (I(b)), which it has largely superseded, are

(i) better control of flow and pressure;

(ii) a smaller volume of recirculating medium is required (5 ml minimum);

(iii) Mixing is faster (as a result of (ii));

(iv) the apparatus is simpler to construct, more compact, and easily cleaned (Morgan *et al.* 1961).

*Apparatus* (Fig. 5.4)

*Recirculation system.* The heart is suspended in a water-jacketed chamber of $2 \times 20$ cm internal dimensions with a coarse sintered-glass filter disc sealed into the lower portion. The aortic cannula is mounted in a rubber (Teflon would be better; see p. 81) stopper, through which pass the gas inlet tube and an outlet vent for the excess gases. The inflowing gas is delivered by means of a fine plastic tube extending into

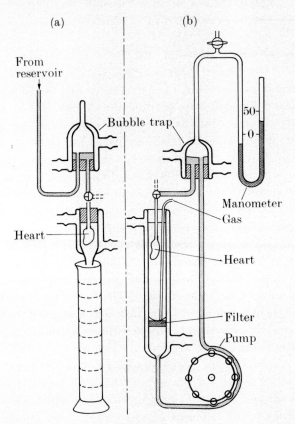

(a)                          (b)

From reservoir

Bubble trap

50
0

Manometer
Gas

Heart

Heart

Filter
Pump

FIG. 5.4.   Apparatus for 'Langendorff' perfusion of heart (Morgan *et al.* 1961). (a) Once-through perfusion. Medium enters the aorta after passing from a reservoir circa 100 cm above a bubble trap. Outflowing medium is collected in a measuring cylinder. (b) Recirculation perfusion. In a similar circuit, a pressure of 150 mmHg is provided by a roller pump. Pressure is recorded by a side-arm manometer and medium recirculates from a reservoir below the heart.

a small pool of perfusion medium which collects on the surface of the sintered-glass filter. After passing through the filter, medium is recirculated by a peristaltic pump model No. 5-8950, American Instrument Co. Inc., or any other suitable pump (see Chapter 1, Pumps). However, the wave form applied to the heart is considerably damped by passing first through a bubble trap containing 1–2 ml of air, so that the characteristics of the pump wave form may be unimportant.

*Perfusion medium.* Krebs–Henseleit bicarbonate buffer pH 7·4, with 95% $O_2$:5% $CO_2$ as the gas mixture is used in the standard preparation. 95% $N_2$:5% $CO_2$ is substituted in experiments on anoxia.

*Operative procedure.* Rats of 250–300 g weight were used by the originators of the method (heart size is of some importance in standardizing the technique; see Morgan *et al.* (1961)), and fasted for 18 h before use. Heparin is injected intra-peritoneally (5 mg) 1 h before killing the rat (or intravenously into the femoral vein, immediately before an experiment in an anaesthetized animal). The rats were killed by decapitation. Within 20 s the thorax is widely opened by incisions to remove the anterior wall (see Fig. 5.10). This is described on p. 285. The heart is grasped with the left thumb and forefinger and lifted ventrally. A scissor-cut across the great vessels about 5 mm from the heart allows it to be removed and immersed in ice-cold saline.

After 1–2 min cooling (this is not necessarily desirable and greater speed may, in theory, reduce the likelihood of anoxic damage to the heart), the heart is mounted on the aortic cannula as described on p. 288 and aortic flow begun at once. The first 8 ml of medium, which contains all the blood washed from the heart, is allowed to run to waste, and in these early experiments no other time was allowed for 'equilibration'. Before mounting the heart on its cannula the volume of medium required for recirculation, plus the 8 ml wash-out medium, is added to the reservoir. Turning on the pump, the level in the bubble trap is raised to 2 cm and a rate of pumping is chosen which gives a pressure in the manometer of 150 mmHg (aortic cannula clamped). In-flowing gas is equilibrated with warmed water at 37°C before entering the apparatus, to prevent loss of volume by evaporation.

## Methods of sampling in heart perfusion studies

### Coronary outflow samples and coronary flow rate

Medium drips from the heart and runs over the surface of the ventricles before collecting in the bottom of the heart chamber (oxygenator). In the once-through system, the apparatus design allows medium to be collected directly into a measuring cylinder and returned to the reservoir when flow-rate determinations are complete. This is unsuitable for determination of effluent oxygen tension, for which the modification described by Opie (p. 274) is ideal. Medium may then be run directly into a chamber containing an oxygen electrode or collected under paraffin for example, in preparation for a chemical gas analysis. In Morgan's recirculation circuit, the heart chamber is also the oxygenator. The heart must either be lifted from the chamber by removing the bung with the aortic cannula before each sample or, with a simple blown-glass heart chamber (as illustrated), the heart and oxygenator may be separated. Samples and flow rates are then determined as described above. Refinements to allow the use of a flow meter or continuous sample device are available (see Rabitzch 1968; and Chapter 1, Flow meters).

### Heart muscle samples

The perfused heart lends itself to the rapid freezing technique of tissue sampling devised by Wollenberger, Ristau, and Schoffa (1960), and this was introduced by Morgan *et al.* (1961) and by Newsholme and Randle (1964). With the heart perfusing on the cannula, the surrounding chamber is lowered and the heart rapidly crushed between the plates of the 'Wollenberger clamps'. In the same movement the heart is detached from the cannula and immersed in liquid nitrogen while still held in the clamps. To prepare a powder, which is an essential preliminary to any extraction procedure with this technique, Morgan introduced a useful device, the percussion mortar. The original apparatus is illustrated in Fig. 2.4, and its use described in detail in Chapter 2. Extraction procedures are discussed by Newsholme and Randle (1964) and by Enser *et al.* (1967). Preparation of frozen powder for glycogen determination or for enzyme assays (when deproteinization is avoided) is also discussed.

Methods of determining the physiological parameters of cardiac function are described below (see Method II(a)).

### Characteristics of the preparation

Contractions start very shortly after perfusion begins (see Method II for description). Perfusion pressure is 35–50 mmHg (50–70 cm $H_2O$) with a coronary flow rate of 7–10 ml/min for at least 1 hour. Heart rate (200–240 beats/min) is similarly steady for 1 hour.

The oxygen tension of the medium flowing into the heart reaches 550 mmHg (i.e. approximately 80 per cent of the expected figure for saturation with 95% $O_2$ at the appropriate temperature and water vapour pressure) but is quite adequate to the purpose. At the flow rates given, effluent oxygen tension, determined with an oxygen electrode, gives an oxygen consumption of $0.1$ ml/min per g ($4.5$ $\mu$mole/min per g) and this figure is constant for 1 hour.

Finally, the rate of glucose uptake and the sorbitol and water space of the perfused heart remained constant for 1 hour's perfusion.

## Microscopy

As might be expected in a perfusion with medium that contains no added colloid (see p. 37), interstitial oedema is seen in hearts fixed after perfusion with Krebs–Henseleit buffer containing glucose (Brown, Cristian, and Paradise 1968). Comparing incubated heart muscle with heart perfused under these conditions, Brown *et al.* report loss of cross-striations, and the presence of vacuoles in the incubated muscle only, while both preparations were abnormal in showing extensive interstitial oedema.

A similar study with hearts perfused under these conditions, but with 5 per cent bovine albumin added to the perfusion medium showed no interstitial oedema (Weissler *et al.* 1968). Unfortunately, no single investigator has compared the two experimental conditions. There is a very extensive literature on the metabolic activities of the heart perfused in Morgan's apparatus (see Opie (1968) for review) and some of this information may be of use in systematically assessing the viability of the preparation.

## Method II(a). Isovolumic recirculation method of Opie

In an attempt to standardize the Langendorff preparation adapted by Morgan (I(c)), Opie introduced the following modifications, and the resulting preparation is sufficiently different from that of Morgan to warrant a separate description.

(i) Variation in perfusion pressure is possible, either in a once-through system, or by recirculation (cf. Morgan).

(ii) Ventricular size is kept constant by the introduction of a fluid-filled latex balloon into the cavity of the left ventricle.

(iii) Drainage of the right ventricle, to prevent the development of right ventricular pressure, is accomplished by an incision into the base of the pulmonary artery (cf. p. 291, Method III(b)).

(iv) Modifications which permit the synchronous determination of electrical activity (e.c.g.), force of contraction (ventricular balloon) and coronary flow (syphon). The last modification also allows analysis of accurate oxygen content, since coronary outflow is collected without contact with the outside atmosphere.

*Apparatus*

The perfusion apparatus proper is identical with that of Morgan *et al.* (The criticism of Neely *et al.* (1967) that Opie's preparation showed signs of failure, recognizable by the increasing aortic pressure required to maintain coronary flow rate, is attributed by that author to an inefficient sintered-glass filter, and is easily overcome.)

*Perfusion medium*

The medium is a modified Krebs–Henseleit bicarbonate buffer, with half concentrations of $Ca^{2+}$ and $Mg^{2+}$, and gassed with 95% $O_2$, 5% $CO_2$.

*Operative procedure*

The Morgan technique (q.v.) is applied, with arrest of the heart induced by immersing it in ice-cold saline. No medium is perfused after mounting the heart on the aortic cannula; instead, a balloon is inserted into the stationary left ventricle, through a small incision in the left atrium.

The balloon, of fine rubber (later latex was introduced (Opie 1965)), has a volume when filled, some 30 per cent greater than that of the left ventricle, but contains only 0·04 ml $H_2O$ (Kadas and Opie 1963).

A water-filled polythene cannula (internal diameter 0·6 mm) connects the balloon to a transducer pressure-recorder. Heart contraction is restarted by perfusing with standard medium at 35°C. Note that a low temperature is used which has the advantage of prolonging effective perfusion but the disadvantage of being an unphysiological temperature. The pressure in the balloon is adjusted to atmospheric pressure, by injecting water from a micrometer-syringe. The system is then sealed, and henceforward ventricular contraction against the balloon, is isovolumic. Variation in pressure within the balloon (15–20 cm $H_2O$) occurs with each systole, but the degree of distention of the balloon must be kept constant in any series of experiments. From peak systolic pressure, the work-load may be calculated (Opie 1965).

*Electrocardiogram.* Two platinum plates touch opposite sides of the ventricle; the other lead is attached to the (metal) aortic cannula and a continuous trace is possible.

*Coronary flow.* The usual approach to coronary flow collection and measurements may be modified as follows when oxygen consumption is to be determined. The heart is enclosed in a small bottle (4 ml volume), which is sealed to the under surface of the stopper holding the heart cannula. The outflow collects in the bottle and a syphon tube leads back, past a tap, to the perfusion chamber. The heart is thus immersed in perfusion medium and out-flow samples can be collected without exposure of the medium to the air. The immersion of the heart in this way is said not to alter its mechanical or biochemical function (Opie 1965), however, this is not true for the 'working' heart preparation which, not surprisingly, stops beating when enclosed.

*Characteristics of perfusion*

The crucial observations that Opie was able to make in this modified Langendorff system, concern the variation in oxygen and substrate uptake by the heart, with changes in the aortic perfusion pressure. Perfusion pressure affects peak-systolic pressure, as recorded from the intra-ventricular balloon. This observation is discussed on p. 294, and the conclusion must be that for linear and stable Langendorff preparations, the perfusion pressure, and with it the coronary flow rate, must be kept constant. A suitable way of ensuring this is to adopt the modifications indicated here. One objection remains, and that is largely dealt with in Method II(b). Filling of the left ventricle from an incompetent aortic valve, or from drainage of venae cordis minimae (q.v.) still occurs and contributes a non-linear factor in Opie's modified method. Fluid accumulation in the left ventricle may, however, be less than expected since the mitral valve is now incompetent as a result of the introduction of the left ventricular balloon.

**Method II(b). Aortic perfusion with drainage of left ventricle: Arnold and Lochner**

The problem of filling of the left ventricle during standard Langendorff perfusion has been discussed. It results in a progressive increase in the diastolic length of muscle and must therefore influence the 'work' of the heart (Morgan *et al.* 1961). The present preparation incorporates the following modifications: (i) a fine catheter inserted into the left ventricle and to which suction is applied, keeps the chamber empty; (ii) the left atrium is opened, both to allow insertion of the catheter (i) and to prevent filling of this chamber; (iii) the right atrium is cannulated via the inferior vena cava to collect coronary venous samples before the medium has been in contact with the exterior. This is a potent source of error in oxygen uptake studies in the other preparations described where medium dripping from the right atrium is collected without special precautions. (This particular problem may be better solved in the 'closed' preparation of Basar *et al.* to be described (Method IV).)

## Method

Unlike other preparations, with the exception of that of Basar *et al.*, cannulation of the aorta is carried out with the heart *in situ*. When perfusion is in progress and after the additional cannulation of the right atrium via the inferior vena cava, the heart and lungs are excised together. The lungs are then excluded from the circulation by means of ligatures around each hilum, and medium outflow is ensured by incising the pulmonary artery.

## Apparatus

This is a once-through perfusion, under gravity but the operative procedure to be described can equally be used with minor modifications of standard apparatus (Morgan *et al.* 1961, etc.).

## Medium

Tyrode solution is offered as a suitable perfusion medium in the original method, but should probably be replaced by Krebs–Henseleit bicarbonate buffer if stable conditions of pH are to be achieved (see Chapter 1, Media).

## Operative procedure

Rats of 350–500 g weight are anaesthetized with ether. The chest is opened as described on p. 285 and the aorta is exposed in order to position a ligature around it just above the heart (see Fig. 9.1 for details). Care is required to avoid including the pulmonary artery, which lies to the *left* of the aorta in this situation.

The aorta is cannulated through a transverse incision (complete transection presents difficulties in cannulation) and the cannula is tied in place, with the tip above the aortic valve. Perfusion begins at once, under 60 mm $H_2O$ hydrostatic pressure.

An incision is made into the tip of the left atrium, for later insertion of the ventricular drainage catheter. A ligature is passed around the inferior vena cava, below the heart, and a fine polythene catheter is inserted through a small incision in the vessel below the ligature. The catheter is advanced into the right atrium and tied in place.

The heart and lungs are now excised as a block and mounted in the apparatus. The hilum of each lung is tied and a longitudinal incision into the pulmonary trunk allows perfusion medium to escape from the right ventricle. Suction applied to the right atrial chamber draws coronary venous blood directly into a mixing chamber containing an oxygen electrode, and is equally useful as a source of coronary venous samples.

*Ventricular drainage.* Two fine tubes of polythene, one inside the other, are inserted through the left atrial incision into the left ventricle. Using, for example, Portex PP 10 within PP 60 with respective o. d. 0·61 and 1·22 mm, i.d. 0·38 and 0·76 mm; suction on a tube of this diameter permits flows of up to 6 ml/min. Suction is applied to the inner tube, while the outer tube prevents its occlusion against the wall of the heart. No figures are given for the volume of fluid drawn off in this way, but from Morgan's discussion (Morgan *et al.* 1961) it may be between 2 and 6 ml per min.

## Characteristics of perfusion

Little of relevance can be said of cardiac function in this preparation since the work published concerns function in the potassium-arrested heart (Arnold and Lochner 1965) of rat and guinea-pig. It is to be expected that the problems of non-linearity of cardiac work encountered by Morgan *et al.* (1961) would disappear; but a systematic comparison would be necessary to show that this more elaborate preparation had any advantages in metabolic studies over the standard Langendorff technique.

One important application of Arnold and Lochner's preparation is that described by Huhmann, Niesel, and Grote (1967). The arrangement of the apparatus, in this case, incorporating a membrane oxygenator of special design (see Oxygenators) is shown in Fig. 5.5. In combination with a technique for measuring surface spectroscopy (p. 123), this membrane oxygenator allows the perfused heart to be studied so as to distinguish between the oxidation of myoglobin and cytochrome c.

Working at 25°C, in what is essentially Arnold and Lochner's preparation, Huhmann *et al.* found coronary flow rates of 8–15 ml/min after 10–15 min perfusion, and these were maintained for 1–2 hours. Oxygen consumption was linear with time (4·5 ml $O_2$/100 ml tissue per min $\equiv$ 3 $\mu$mol/min per g) and, from analysis of their data, they conclude that there is no arterio-venous shunting of oxygen in the perfused heart.

This is in contrast to the finding of Kessler and others in liver and kidney perfusions. In those organs, it is concluded that the presence of oxygen in the venous effluent of haemoglobin-free perfusions is not evidence of adequate oxygenation of the entire tissue. It is reassuring that in the heart, which is almost invariably perfused with haemoglobin-free medium, these limitations do not apply. Finally, these authors give histological data, which in contrast to that of similar perfusions conducted according to Morgan's technique, show no interstitial oedema. It

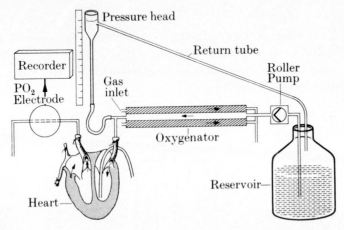

FIG. 5.5.    Heart perfusion in a closed circuit: method of Basar *et al.* (1968). The heart is cannulated both on inflow (aorta) and outflow (pulmonary artery). For operative preparation see Fig. 5.15. A pressure head is established by means of a pump and oxygenation by the membrane 'dialyser' described by Röskenbleck (see Fig. 1.9(b)).

is difficult to attribute this to the very low albumin concentration (1 g/l) added to the medium for the present series of experiments, and it might rather be a reflection of the fact that no anoxic period occurs during the preparation of the heart for perfusion by the method of Arnold and Lochner.

## Method II(c). Aortic perfusion with an inverted heart; continuous coronary flow determination (Rabitzsch)

Offered as a method specially suited to studies with labelled substrates, this preparation incorporates the following features, considered by its author, to be lacking in earlier preparations: (i) the simplest possible circulating system, with optimal condition of the heart for upwards of 2 h; (ii) small volume with rapid flow of fully oxygenated medium, in an apparatus which allows variation of recirculating volume and perfusion pressure; (iii) continuous coronary flow measurements (probably no better in this than they are in other preparations); (iv) avoidance of the usual contact between outflowing medium and the surface of the heart. The latter is seen as a possible source of contamination of the perfusion medium with cardiac proteins, fat, isotopes or with surface exudate. It is only in this point that the method is significantly different from other 'Langendorff' preparations already described.

*Apparatus*

Fig. 5.6 illustrates the apparatus used in this perfusion, with a reservoir of reducing diameter which allows either a larger or a smaller perfusion volume. The cannula mounting is such that the heart can be inverted after the pre-perfusion phase.

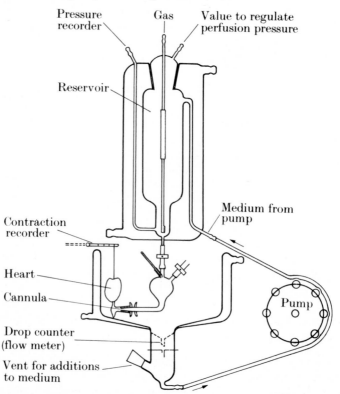

FIG. 5.6. Perfusion of inverted heart: Rabitzsch (1968) apparatus. The heart is mounted upside down on the aortic cannula and the cannula fits by a ground-glass joint onto the holder. Perfusion pressure is provided by pumping into a closed reservoir system.

*Operative procedure*

The standard operation is performed and 50 ml of medium is perfused (to waste) with the heart in the orthodox position. Recirculation
is established and the aortic cannula is then rotated through 180°. In
this new position, with the apex uppermost, coronary flow is constant
for 2 hours at a perfusion pressure of 70 mmHg (95 cm $H_2O$). The gas
flow to the reservoir, which contributes to this perfusion pressure, is at
a rate of 2–5 litres/min. The perfusion volume can be adjusted to equal
the maximum coronary flow in ml/min so that rapid mixing is assured.
No metabolic data are available for the preparation, and some objective
information is required to confirm the necessity of the extra precautions
introduced.

### Method III(a). Atrial perfusion: 'working heart' method of Morgan (Neely *et al.*)

This important preparation is described in the original publication of
Neely, Liebermeister, Battersby, and Morgan (1967) and is a description
of such detail and clarity that it is difficult to improve upon it here.
Despite the completeness of the description, however, workers with
experience in the field have had many difficulties in establishing the
preparation to Morgan's specifications and an attempt is made in the
following description to concentrate upon the difficulties.

Fig. 5.7 illustrates the apparatus of Morgan *et al.* Modifications introduced by Chain, Mansford, and Opie (1969) and Mansford (1969) are
minor, but are thought by these workers to be critical. These modifications are, in turn, based on modifications first applied by Øye (1965).
In each case attention to detail in construction of the apparatus,
operation, and heart-mounting techniques is critical to successful cardiac performance. No other organ perfusion technique offers such
immediate and visible evidence of 'failure'.

### Method III(b). Incorporating the modifications of Chain, Mansford, and Opie

*Circuit diagram for heart perfusion* (Fig. 5.8)

The circuit is simplified to show, in the dashed line, the 'Langendorff'
system for non-working heart. The flow from an oxygenated reservoir
enters the aortic cannula after passing through a bubble trap (B.T.I.).
The effluent in this system is from the coronary sinus only and passes
through the oxygen electrode chamber before collection. The solid line

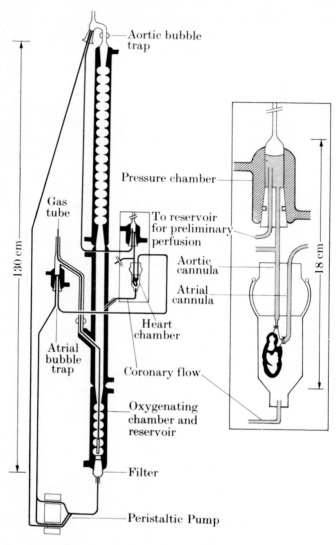

FIG. 5.7.   Apparatus for atrial perfusion of heart (Neely *et al.* (1967)). A perfusion apparatus for 'working' rat heart. The complete apparatus is shown with an enlargement of the heart chamber, cannula assembly, and pressure chamber (inset). For details see Fig. 5.9.

shows the circuit for the 'working-heart', in this case the system used by Opie and Mansford.

Oxygenated medium is pumped from the base of the mixing column through a filter into a bubble trap whence it enters the left atrium. The left atrial cannula is supplied under gravity at a hydrostatic pressure which can be adjusted. Contraction of the left ventricle follows and

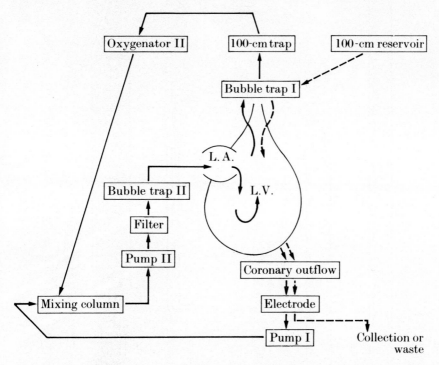

Fɪɢ. 5.8. Simplified circuit diagram of 'working' heart apparatus. The diagram refers to Fig. 5.9(b). LA = left atrium. LV = left ventricle. The circuit in (----) is the path followed by medium in Langendorff pre-perfusion, and that in (——) during 'work'.

medium is ejected through the aorta. The stream divides, the major flow, about 66 per cent, being propelled through a bubble trap (II) and beyond to a height of 100 cm where the overflow from a second bubble trap feeds the oxygenator (II). The smaller fraction of the aortic stream enters the coronary arteries and follows these vessels, providing the immediate oxygen and substrate supply to the contracting heart. Coronary outflow in this circuit is collected and pumped back to the mixing column. In the diagram, the relation in height between the various components is indicated in cm, the heart being taken as zero.

Fig. 5.9 shows the complete apparatus set up for Langendorff pre-perfusion, recirculating 'non-working' heart perfusion, and recirculating 'working' heart perfusion. In addition, the water-jacketing with a circulating system for temperature control is shown.

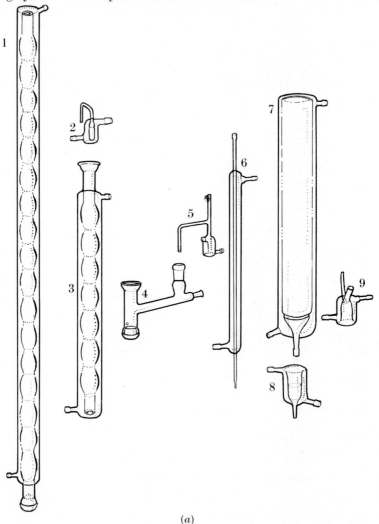

(*a*)

FIG. 5.9. (a) Glass components of the 'working heart' perfusion apparatus (Chain, Mansford, and Opie 1969). All parts are jacketed, borosilicate glass. 1, Oxygenator II. 2, 100-cm bubble trap. 3, Mixing column. 4, Connecting piece. 5, Bubble-trap II. 6, Condenser. 7, 100-cm reservoir. 8, Heart chamber. 9, Bubble-trap I. (b) Apparatus connected for use. ═ = Perfusion circuit. ⋯ = Water-warming circuit. Note particularly the positioning of roller clips to facilitate rapid conversion from 'Langendorff' to 'working' heart circuits.

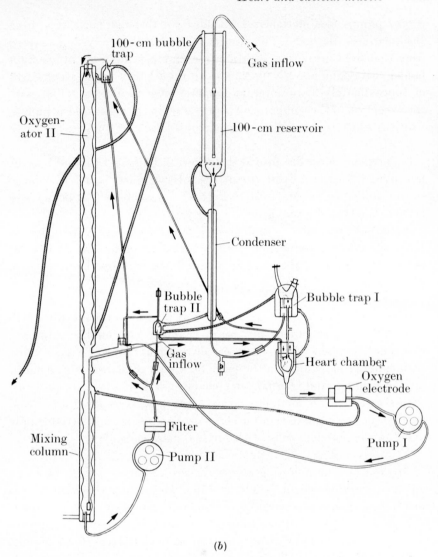

(b)

Fig. 5.9.    (*continued*.)

In a standard arrangement two complete apparatuses are mounted on a frame and provided with a linked thermo-regulatory system. Beneath the table on which the apparatus stands are housed water bath, gas cylinders, and control panel. To the right are the oxygen electrode read-out meters (Beckman) and a U-V recorder attached to the pressure transducer (q.v.). The pumps are Watson–Marlow (MHRE)

rotary pumps (see Methods), electrodes are Bishop–Clarke type from
Beckman (see Methods). The filter is a closed Millipore system adapted
from that used in routine filtration (for example, Millipore in-line filter
holder; catalogue no. XX30 02500). All tubing is of transparent vinyl,
of approximately 3 mm internal diameter (such as Portex NT/6), with
the exception of the compression tubing used in the rollers of the pumps,
which is siliconized tubing as, for example, Esco TC156.

*Connections.* Most are provided by Portex Plastics Ltd. and may be
improvised. In the present circuit the following are used: 'Y' connections (Catalogue No. M/71; M/70 Portex), 'connectors' (Catalogue Nos.
M 630, M 634, polypropylene).

Note that roller clips from disposable transfusion sets (Capon Heaton
Ltd.) are used in connection with Y pieces in preference to two- or
three-way taps since they can be manipulated with one hand if necessary.

*Glassware.* None of this apparatus is at present commercially available. The separate items are heart chamber (1), bubble traps (3),
electrode chambers (2), open reservoir of about 750-cm$^3$ capacity and
with sintered-glass filter in the base, 30–40 cm condenser, oxygenator,
mixing column, and thermometer chamber. All of these are provided
with a water jacket for temperature control. Also required are a connecting piece with side-arm for coronary-flow return and atrial supply
overflow and gassing 'rods' with sintered-glass ends.

*Accessories.* The following accessories are required: a table of which
an ideal working height is 3 ft, a frame with adjustable clamps, gas
manifold to provide separately controllable gas supply to each reservoir, bungs of Tygon, rubber tubing for water-jacket circuit, thermostat
for water bath, a circulating pump of sufficient strength to provide
2 litres/min, which is the minimum required to maintain a double
circuit such as that illustrated here, at 10–15°C above ambient temperature, and a thermometer marked in tenths °C.

*Cannulae.* Whereas a simple Langendorff perfusion can be achieved
without special care, for the 'working-heart' to perform satisfactorily,
the inflow and outflow cannulae are of specific dimensions and material,
i.e. stainless steel, o.d. = 0·134 inches (atrial), bent to angle of 120°,
and o.d. = 0·11 inches (aortic).

*Principle of the 'working-heart' preparation*

The medium is introduced into the left atrium under a pressure between 5 and 20 cm $H_2O$. This is sufficient to fill the left ventricle and induce a forcible contraction. After entering the left ventricle, out-flowing medium is forced into the aorta by ventricular contraction under a pressure of at least 100 cm water. Two cannulae are required, an atrial and an aortic, and the operation takes place in three stages: (1) removal of the heart from the anaesthetized animal and its transfer to ice-cold saline; (2) the aorta is cannulated and the heart perfused with medium at 37°C, in a retrograde manner (after Langendorff); (3) the left atrium is cannulated via the conjoined entrance of the two major pulmonary veins (see Fig. 5.13). When the second cannulation is complete, the preparation is tested for leaks and for its capacity for work and then the usual preparatory Langendorff, non-working per-fusion is continued. After a variable interval, 15 or 30 min in most laboratories, 'work' is begun by stopping the aortic inflow and allowing medium to enter the left atrium. The direction of flow in the aorta is reversed and 'work' begins when the ventricle contracts.

*Perfusion medium*

Krebs–Henseleit medium$+5$ mM glucose is filtered before use through a Millipore (0·5 $\mu$m), and gassed with 5% $O_2$:5% $CO_2$ in the usual way.

*Operative procedure*

The rat, usually 250 g and upwards in weight, is lightly anaesthe-tized with ether (see Methods of anaesthesia). Without preparation of the skin, the first incision is made with large pointed scissors along the lower margin of the ribs. The first cut is made boldly across the upper abdomen, with care not to injure the liver. This cut is extended later-ally, and for the purpose the rat may be held in the left hand. Then the point of the scissors is directed towards the head and a single cut made through all the layers of the thorax including the ribs, up to the clavicle. This is done first on the left side and then the right side and finally the whole of the free flap of the anterior chest wall is removed with a cut transversely at the level of the second rib (Fig. 5.10).

The lungs collapse and expose the heart, which continues to beat vigorously. The next incision is critical. The following description follows the method of Opie (personal communication) and has been found satisfactory. The heart is grasped firmly between the thumb and fore-finger of the left hand and lifted, drawing it to the animal's right. This exposes the pulmonary veins and their site of entry into the left atrium. The point of the scissors is passed horizontally and behind the left atrium, the pulmonary veins, and the aortic arch. Before closing the blades, the instrument is drawn well over to the animal's left to leave the maximum length of pulmonary vein in continuity with the atrium

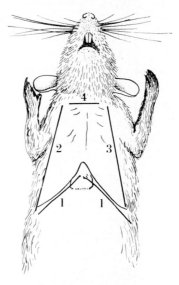

FIG. 5.10.   Operative procedure for perfusion of the heart. I. Incisions for rapid exposure of the heart and great vessels. The incisions, made with heavy scissors, are made in numerical order, following the description in the text (p. 285).

(Fig. 5.11). With this cut, and the scissors pointing downwards and away from the operator, the aorta is cut at about 4–5 mm from the ventricle. This length is adequate for cannulation. If the cut is made too near to the heart a hole is made in the left atrium, rather than through the pulmonary veins, and this makes cannulation difficult. Even if successful, the risks of leakage are high. If the cut is made with the scissors not in the vertical plane, the tip of the left auricle may be inadvertently removed, and the atrium will fail to fill when tested for leaks. If the cut is made too far from the heart then lung is included. This is erring on the side of safety, as the lung and pulmonary vessels can be cleared when the heart is mounted on the aortic cannula. Finally, if the point of the scissors is not directed downwards, the aorta may be cut too short for satisfactory cannulation.

The heart thus removed is transferred rapidly to ice-cold Krebs–Henseleit medium (usually substrate free) and thence, with minimum delay, to the aortic cannula, to start perfusion. Delay is critical since the organ is anoxic. If for some reason a delay of over 3–4 min occurs, it

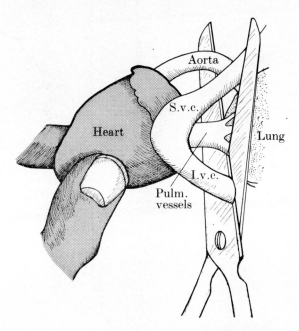

F I G. 5.11.   Operative procedure for perfusion of the heart. II. Removal of heart from thorax. The heart of the rat is grasped between thumb and index finger and drawn to the animal's right. One scissor blade is passed carefully behind all the major vessels. For details see text (p. 286). IVC, inferior vena cava; SVC, superior vena cava.

is probably wise to abandon the heart at this stage. This is based on un-controlled observations and may be unnecessary, but much time may be saved by starting again.

Some accounts (Neely *et al.* 1967, Mansford 1969) suggest 'cleaning' the heart at this stage, to remove thymus, lung, and other tissues. This is clearly undesirable at this time as it prolongs the period of anoxia. It can more easily be accomplished after aortic cannulation; see below.

*Aortic cannulation.* Although aortic cannulation can be carried out single-handed, for the sake of speed it is useful to have an assistant. The cannula is filled to the tip with medium. It is a useful preparation first to run 5–10 ml of medium through the aortic cannula; cold and desaturated medium is thereby replaced by warm oxygenated medium and the heart beats without delay when flow begins. The aorta is grasped from

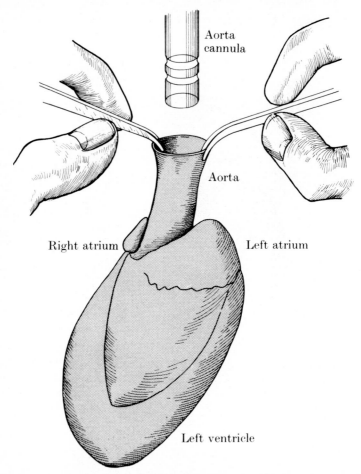

FIG. 5.12. Operative procedure for perfusion of the heart. III. Aortic cannulation. The cut margin of the aorta is grasped in two pairs of forceps and lifted onto the fluid-filled cannula.

opposite sides in two pairs of fine curved forceps and gently lifted over the straight metal cannula (Fig. 5.12). The assistant ties a single ligature of 3/0 thread around the aorta. It is important to watch that the tip of the cannula does not impinge upon the aortic valve by being too far down the aorta. There is no danger of this happening if the original cut was directed downwards as described, when the aorta will be longer than the free length of cannula. The aorta may be torn longitudinally during mounting and usually the heart must be rejected when this occurs. It may be saved if this damage is recognized early, and a second ligature placed around the vessel.

Single-handed aortic cannulation is accomplished by use of a bull-dog arterial clip, as described in the original paper by Morgan *et al.* (1961). The aorta is slipped over the cannula as described and held in position by applying the clip. This leaves both hands free to position and tie a ligature.

Before tightening and fixing this ligature, but after flow has begun, the heart should be rotated about the axis of the aortic cannula so that the atrial cannula and left atrium are in proximity.

A flow of medium at the maximal rate is begun as soon as the ligature is tied, and this ends the period of operative anoxia. Unless some damage has been done to the heart during operation, or there is some error in the composition of the perfusion medium, the heart begins to beat within 10–15 s of starting the perfusion. Even this delay may be due to a temperature effect. Thus, in the apparatus described here, there may be a considerable drop in temperature of medium in the cannula and adjacent tubes during the period of stasis required for operation and cannulation. This is a design fault but its effect may be minimized by running 3–5 ml of medium from tap 'X' to waste immediately before opening the supply to the aortic cannula. In this way 'cold' medium is discarded (see above).

*Atrial cannulation.* Atrial cannulation may be conducted at leisure since the heart is fully perfused via the aorta and is no longer at risk from anoxia. It takes some practice to find the pulmonary veins and distinguish this orifice from that entering the right atrium, and the superior and inferior venae cavae. Using a curved forceps, the two may be distinguished, since the points go to the left and emerge in the left auricular appendage if correctly inserted into the left atrial orifice. If the cut has been well performed a hole about 3 mm in diameter is present in the left atrium where the two pulmonary veins entered together. As often as not, some stump of the two veins is present. Some cleaning up is necessary. Fragments of attached lung should also be removed at this time. With a curved forceps the orifice is defined and distinguished from the right atrial orifice (see above).

Two pairs of fine curved forceps are needed for the cannulation. The atrial wall is grasped as it lies on either side of the cannula and by turning the wrists outwards the wall is slightly everted. Then the atrium is drawn over the cannula itself (see Fig. 5.13). When the cut is 'right',

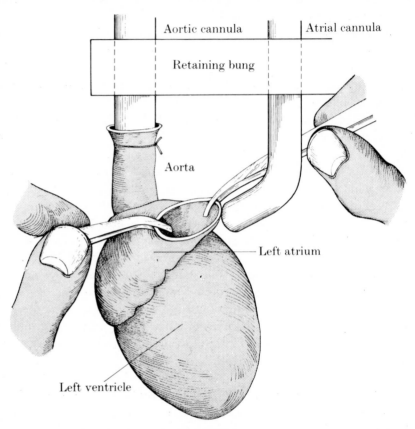

FIG. 5.13.    Operative procedure for perfusion of the heart. IV. Atrial cannulation. The heart is rotated about the aortic cannula until the left atrial orifice lies opposite the specially curved cannula. Two pairs of forceps are passed through a prepared loose ligature (not shown) to grasp the cut margins of the atrium. The atrium is drawn a short distance onto the cannula before tightening the ligature.

the heart remains in this position when released and a length of 3/0 thread may be passed round the atrium. Cannulation can thus be achieved entirely single-handed. More usually it is necessary for an assistant to prepare a loop of thread which is passed over the atrial cannula after the pulmonary vein has been identified and the heart rotated into the correct position. The forceps are passed through this loop to grasp the edges of the left atrial orifice and after the atrium has been drawn over the cannula the prepared loop is drawn tight, its position being maintained with the forceps.

After the first loop has been tied the atrium can be adjusted by pulling on the threads, locating the cannula in such a way that its end is not occluded against the posterior wall of the atrium. Tap 'Y' is opened for a 'second' to watch the left auricle fill, which it does only if the cannula is correctly placed and the atrium has not been punctured during the preparation. The first two or three heart beats may be observed to confirm that the preparation can 'work'. Then the atrial ligature is tied tight round one of the notches on the cannula.

*Incision into pulmonary trunk and right ventricle.* The final step is to incise the right ventricle and its pulmonary trunk. This allows the escape of medium accumulated in the right side of the heart from minor coronary veins, a factor that would otherwise result in poor cardiac function and apparent loss of circulating medium. The incision is made with pointed scissors through the base of the ventricle and the origin of the pulmonary trunk, i.e. across the valve (Fig. 5.14).

There may be difficulties in atrial cannulation and adjustments to the heart. Before any attempt is made to introduce the atrial cannula, the heart should be rotated about the aortic cannula—the two free ends of the ligature applied to the aorta may be pulled in opposite directions to achieve this. When the left atrial orifice lies opposite the atrial cannula, it may still be necessary to raise or lower the aortic cannula (by moving its position within the bung) or to rotate the angulated atrial cannula to lie as close to the atrium as possible. Once mounted (q.v.), the heart may still not perfuse well. The two cannulae should be gently moved so that no tension exists between them. Between the two cannulae, and overlying the base of the heart, there may be a residual condensation of pericardium and clearing this by means of a single incision with scissors, parallel to the aorta may, without risk to the heart, result in greater ventricular output.

Continued failure of the heart to work may then be due to leakage from the left atrium (the auricle is sometimes incised during the preparation) or to the occlusion of the tip of the atrial cannula against the opposite wall of the atrium. Needless to say, errors in the composition of the perfusion medium, inadequate gassing, residual detergent, and the like may still prevent heart work and it is only by attention to detail that these are overcome.

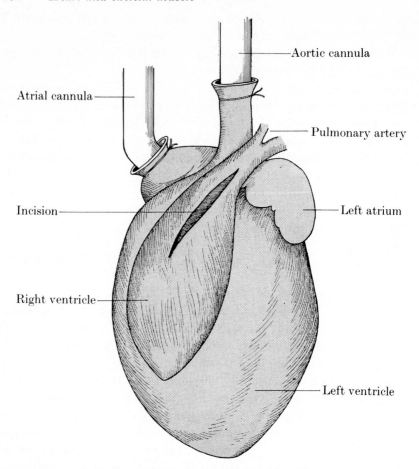

FIG. 5.14.   Operative procedure for heart perfusion. V. Incision into pulmonary tract. The heart is viewed from in front. A single incision is made with scissors, along the long axis of the pulmonary artery at its base, to extend into the right ventricle.

### Coronary flow determination

The heart chamber is separated from the oxygenator in this apparatus and there is no difficulty in collecting medium issuing from the right atrium/pulmonary artery. Medium is collected into a measuring cylinder and returned to the circulation through side-arm $X$. The method is primitive but effective for most purposes, since coronary flow rate in this preparation is remarkably constant.

The modification of Opie, for the collection of samples without exposure to the air, cannot be applied (see p. 274). The alternative arrangement, illustrated in Fig. 5.9 allows samples to be drawn from the heart chamber into an oxygen electrode chamber—inevitably, the medium is exposed to gas in the heart chamber and to gas losses during passage through the tubing that leads to the electrode (see Chapter 1). By turning the tap, samples of coronary outflow may be collected.

A continuous flow meter of simple design would have the advantage that the apparatus need not be opened each time a sample is to be taken and returned to the circuit. Loss of volume must attend each such manoeuvre, with loss of $^{14}CO_2$ and inequality of gas tension and of temperature as a further hazard.

### Measurement of aortic output

In the present arrangement, the average investigator must ascend a step-ladder to sample the aortic outflow and to determine the cardiac output of the preparation. The upper bung with its attached tubes is removed from the oxygenator and held at the correct height, while a graduated tube collects the outflow for a measured time. For most purposes this is an adequate technique, but a sampling method that allows the chamber to remain closed and the aortic pressure head constant would be preferable.

### Method III(c). Incorporating the modification of Øye

In early experiments with Morgan and others, Øye perfused the rat heart via the left atrium in the standard apparatus and using the standard operative technique (q.v.). A complex medium, which included rat erythrocytes and a mixture of substrates was used, in contrast to the standard use of Krebs–Henseleit buffer with glucose (Neely *et al.* 1967).

### Perfusion medium

Krebs–Henseleit buffer (25 mM bicarbonate) contained 1 per cent serum albumin and 10 per cent rat erythrocytes plus the following substrates: glucose, 4·1 mM; plus sodium pyruvate, glutamate, and fumarate, each at 1·54 mM. Experiments were conducted at 34°C. It is not clear why erythrocytes were used in this study, and they were omitted from the medium in the later studies of Øye and his group, with the observation that the oxygen content of the coronary arterial medium was in excess of that found *in vivo*.

*Performance of atrial perfusion of heart with erythrocytes in the medium*

Perfusing with a left atrial pressure of 10 mmHg ($=$13·6 cm $H_2O$) against a column of aortic hydrostatic pressure of 81·5 cm $H_2O$, the heart produced an aortic output of 20 ml/min and a coronary flow of 10 ml/min. The preparation was stable, by these parameters, for 1·5 h. However, the total flow through the heart is only half that obtained for the heart perfused with Krebs–Henseleit saline plus glucose. Morgan reports the following figures obtained at the same left atrial perfusion pressure: aortic output $=$ 43 ml/min per g wet wt.; coronary flow $=$ 15·4 ml/min per g wet wt.

Later reports from Øye's laboratory (Landmark, Glomstein, and Øye 1969) contain disconcertingly low values for the aortic output—8 ml/min for a preparation in which the standard apparatus was used with a perfusion medium containing 10 mg/ml albumin (Fraction V Sigma) and 1·8 mg/ml glucose. These are accepted as minimum normal criteria of function in the preparation, but are so much lower than the values found by either Neely *et al.* (1967), or by Mansford, or even by Øye's group itself in earlier experiments (Øye 1965) that they must be suspect.

## Comparison of biochemical findings in Langendorff and atrial ('working') heart perfusions

The extra complexity introduced by the use of a 'working' heart preparation is such, and technical difficulties experienced by almost all workers using this method for the first time so frequent, that it is important to offer a justification for its use in place of the Langendorff technique. Morgan *et al.* and Mansford and Opie have published such comparisons, and Opie (1965) contributed important observations concerning the dependence of oxygen consumption on perfusion pressure in the Langendorff perfused heart. At the outset, a misconception appears to have entered the hypotheses of biochemists working on this problem (Neely *et al.* 1967). External work as such is seen as a vital function of the heart *in vivo* which is lacking from the aortic perfusion technique of Langendorff. It was felt that development of tension in isometric contraction differed fundamentally from the 'work' of pumping medium to a height of 100 cm: Neely *et al.* describe the need for the 'working' heart preparation (despite the observations of Opie) . . . 'Opie's preparation does not pump fluid or do external work, preventing studies of the effects of these variables on oxygen consumption and substrate utilisation.' The information on biochemical parameters Morgan; Chain, Mansford, and Opie) largely confirms what physiologists would anyway have predicted, namely, that 'tension' and 'work' are identical in terms of the mechanics of muscle contraction, and would be anticipated to have the same requirements for substrates, oxygen, and ATP. Neely *et al.* (1967) and the data provided by Chain, Mansford, and Opie, and by Mansford (1969) confirm the observation, originally made by Opie (1965), that oxygen consumption and perfusion pressure are proportional in the Langendorff preparation. Oxygen consumption increased three fold when perfusion pressure was raised from 40 to 120 mmHg (55–165 cm $H_2O$), using the pump to increase the perfusion pressure in these experiments. The highest oxygen consumption observed in this 'non-working' heart was the same as that observed in the atrial perfusion ('working') at a left atrial perfusion pressure of 10 cm $H_2O$. The latter heart pumps almost 45 ml medium/min, to a height of 100 cm as well as providing a coronary flow of 18 ml/min. Chain, Mansford, and Opie (1969) extended these comparisons to include the rate of substrate uptake in the two preparations. Quantitative differences occur, but in many features, the 'working' and non-working heart preparations are similar. Tissue concentrations of glycolytic intermediates and of adenine nucleotides are similar in the two preparations, and similar to the concentrations found *in vivo* (Kraupp *et al.* 1967). No particular benefit has been incurred in these studies by the use of the elaborate and time-consuming 'working' heart preparation. One considerable advantage in the 'working' preparation is the ease and accuracy with which performance may be quantified. In the Langendorff preparation only coronary flow and heart rate are available as physical parameters of function. As an example, the observation by Opie that increasing perfusion pressure is required to maintain constancy of coronary perfusion with time, may be taken as a sign that the preparation is failing. Morgan *et al.* (1961) and Opie, in later experiments, overcame this objection by introducing efficient filters into the circuit. The 'working' heart, on the other hand, has aortic output as a quantifiable parameter of function, although this is dependent upon the heart achieving an aortic pressure of at least 100 cm $H_2O$. Failure is indicated very early by reduction in output, and later by the inability of the heart to sustain a column of 100 cm fluid. Anoxia, lack of substrate, poisons, and drugs may all be quantitatively tested on this basis. Linearity of oxygen and substrate uptake is an additional means of assessing function in this preparation and, together with the easily observed abnormalities of cardiac rhythm which may also indicate failure of the preparation (see, for example, Enser *et al.* (1967)), the 'working' heart preparation is easily and accurately assessed. Nevertheless, it is doubtful if it will be necessary to test all biochemical hypotheses on the heart in both Langendorff and 'working' preparations and, since the former is so much simpler, its use will continue for some time to come.

## Method IV. Complete-circuit Langendorff preparation of Basar *et al.*

As in the method of Arnold and Lochner (Method II(b)), it is appreciated by the authors of this method that existing Langendorff perfusion, with effluent medium dripping from the right atrium or the pulmonary artery over the surface of the heart, and on into a reservoir from which samples may be taken, contains too many uncertainties for the accurate assessment of oxygen consumption. Basar *et al.* (1968) devised a special operative procedure to study the flow and resistance characteristics of the coronary circulation. Since the resulting preparation is a well-controlled closed-circuit perfusion of the entire coronary system, it may have important advantages for metabolic and biochemical studies.

### *Method*

The coronary circulation of an isolated rat heart is perfused with a synthetic medium, saturated with oxygen. The aorta is cannulated *in situ* (contrast Langendorff method) and the perfusion pressure determined by the pressure of gas above the reservoir of medium (see, for example, Leichtweiss *et al.* 1967 see Chapter 4). Outflow from the coronary circulation is collected via the left superior vena cava. The circuit and the anatomy of the dissection are illustrated in Fig. 5.15.

### *Apparatus*

No description of the apparatus is given but the reader is referred to the kidney perfusion apparatus (Fig. 4.9) used by the same authors and incorporating the same pressure device for perfusion. A closed reservoir receives the gas supply (100% $O_2$) and perfusion pressure is determined by the combination of gravity and gas pressure. A re-circulation system would be suitable for biochemical studies, as discussed in Chapter 1. Alternatively, a cabinet perfusion system, with high-pressure adjustments as, for example, that used by Nishiitsutsuji–Uwo for kidney) might be applied, and would allow this preparation to be handled just as other organ-perfusion methods.

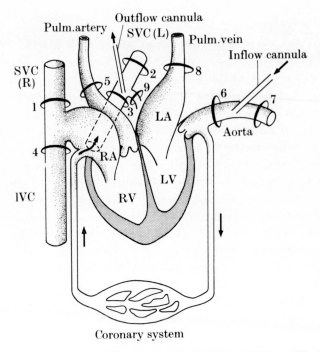

FIG. 5.15. Operative procedure for closed-circuit perfusion of heart (Basar *et al.* 1968). Perfusion is via the aorta, as in the Langendorff preparation, but cannulation is performed with the heart *in situ*. The ligatures required for this cannulation, and for outflow cannulation of the superior vena cava are illustrated (see description in text). LV, left ventricle. RV, right ventricle. RA, right atrium. LA, left atrium. SVC (R) and (L), superior vena cava right and left. IVC, inferior vena cava.

*Cannulae.* Both arterial and venous cannulae are of glass of internal diameter 1 mm.

*Perfusion medium*

Again, full details are not given. The medium is described simply as 'isotonic solution, with 3·5% Haemaccel' (see Media for details). In addition, 300 mg% glucose (16·7 mM) and 0·4 Int. units insulin/100 ml were added for the experiments recorded, with the occasional extra addition of papaverine, $5 \cdot 10^{-5}$ g/ml, to prevent constriction of the coronary vessels.

Assuming that the medium is similar to that described by this group for kidney perfusion, then see p. 254. But it is probable that any of the media already described for Langendorff or 'working' preparations would be suitable for use in this modified method.

*Operative procedure*

Rats of 250–300 g weight are anaesthetized with Inactin 150 mg/kg body weight, by intraperitoneal injection. Tracheotomy is performed and the animal is artificially ventilated as described in Chapter 9 for the duration of the operative preparation. The chest is opened by two parasternal incisions, removing the intervening sternum and anterior chest wall after making the necessary transverse incisions (see Chapter 9 for lung perfusion method). A clamp is placed across the upper sternum before completing its removal, to avoid blood loss into the chest. The layers of loose connective tissue overlying the heart and great vessels (pericardium) are cleared. Ligatures (1)–(8) (Fig. 5.15) are positioned; in almost all respects, these are the same as those used in the perfusion *in situ* of lung (Leary and Ledingham, see Chapter 9, Method I). Ligature (5) presents some difficulties, since the pulmonary artery is bound to the ascending aorta and is also very fragile. Ligatures (2) and (3) are emplaced after displacing the heart to the animal's right, and an additional ligature (9) around the very substantial azygos branch (see Fig. 5.15) prevents loss of medium by that route.

*Venous cannulation.* The left superior vena cava is grasped in curved forceps and an incision is made below Ligature (2). The cannula, filled with medium to the tip, is inserted and fixed in place with Ligature (3). Ligature (2) is tied round the cannula after tying the vessel and helps to steady it. Ligatures (1) and (4) are tied, followed by (5), so that no build-up of pressure occurs within the right atrium. Aortic cannulation is performed *in situ* as already described (p. 275, Arnold and Lochner), and perfusion is begun at once. Finally, the pulmonary vein ligature (8) is tied, and ligatures (10) and (11) are tied around the two auricles, restricting the flow of medium to the coronary vessels.

*Performance of perfusion*

Flow rates of approximately 10 ml/min are reported (Basar *et al.* 1968) and diagrams illustrating the pressure: flow responses of the heart perfused in this way are given. Otherwise the characteristics of the preparation are unknown and metabolic studies, particularly of oxygen consumption, and perhaps of arterio–venous differences, might be rewarding in such a preparation.

One defect is noted by the authors. During the course of perfusion some 3–5 per cent of the medium is lost 'through the ventricular walls' (times are not given). This is significant for metabolic studies in a recirculation system and should perhaps be investigated further before further recommending this preparation. A possible route for loss of medium is that already discussed by Morgan and partly solved by Arnold and Lochner, namely, leakage of the aortic valve under pressure. Alternatively, the Thebesian veins (venae cordis minimae, p. 262) may be an important means of loss.

### Choice of perfusion method for metabolic studies in the heart

The Langendorff preparation of perfused rat heart has, particularly in the hands of Morgan *et al.* and of other investigators using the apparatus they designed, contributed a great deal of biochemical information (see Opie (1968) for review). It has been criticized as 'unsatisfactory' in that no external work is performed. But this concept has now been tested in practice (atrial 'working' perfusion), and the theoretical basis of the 'similarity' of isotonic and isometric contraction confirmed, viz. no qualitative biochemical difference is to be anticipated for the two 'types' of work. The simple Langendorff preparation, with variation in the work load brought about by varying the perfusion pressure, therefore retains its place as the standard tool in biochemical investigation on the heart. The 'atrial' perfusion provides much higher systolic pressure (up to 200 cm $H_2O$) and such pressures cannot be achieved in the Langendorff preparation. Accordingly the 'working' heart should be used if maximal rates of uptake are required. Of the more recent developments, that of Arnold and Lochner seems to offer advantages. No anoxic period occurs during the preparation, the ventricle is drained continuously so that the force of contraction remains constant, and coronary outflow is collected directly from the right atrium. At present, however, the only objective evidence of the benefits incurred with this preparation are that no oedema is seen on histology, after perfusion.

A new type of perfused heart preparation (new for the rat, at least) which conforms more closely to the criteria of a successful perfusion technique outlined in Chapter 1 is that of Basar *et al.* and is described here. It has yet to be evaluated in biochemical and functional terms and has the disadvantage of operative complexity in place of the very simple single cannulation of the Langendorff preparation. A notable omission in the field of heart perfusion is any detailed analysis of the effects of variations of perfusion medium on the biochemical parameters of cardiac function. Sporadic studies have been made, both in the

Langendorff (Gamble *et al.* 1970) and in the 'working-heart' preparation, but much of the information is available only by comparing the work of several groups (see Enser *et al.* (1967), Chain, Mansford, and Opie (1969), and Gamble *et al.* (1970); and compare with Kraupp *et al.* (1967)). Indirect comparisons of this sort suggest that there are no great differences between the saline-(or other medium) perfused heart and the heart *in vivo*.

STRIATED MUSCLE

## Diaphragm

The rat diaphragm in incubation has been a standard preparation in biochemical and pharmacological experiments and has been taken as representative of striated muscle in this animal. The special advantage of diaphragm over other striated muscles is its relative accessibility to oxygen from the surface. Its average thickness is only 500 $\mu$m (Creese, Hashish, and Scholes 1958), which brings it into the range where oxygen diffusion is adequate to reach the innermost fibres. This is certainly not the case for whole undamaged striated muscle from other sources, which must be sectioned, homogenized, or 'teased' to give a preparation that can be adequately oxygenated by simple incubation.

If the rat diaphragm is thin enough for oxygen to diffuse to the centre, is there any need to perfuse it through vascular channels? The arguments presented by Rowlands (1969), in conjunction with the results of experiments to compare the perfused diaphragm with incubated preparations, suggests that there is a place for the perfused diaphragm in biochemical studies. Thus, she suggests that perfusion has advantages: (1) diffusion pathways are reduced; (2) anoxia, which occurs in the central fibres and can, in fact, be detected by ionic changes (Creese 1954, 1958) should be avoided; and (3) the metabolic and diffusion effects of the mesothelial and pleural surfaces are eliminated.

The preparation devised and developed by a succession of workers (see below) has recently been shown to have some advantages over the suspended preparation of Bülbring (1946) and it will therefore be described.

*Anatomy*

The diaphragm is a dome-shaped musculo-membranous structure separating the thoracic and abdominal cavities. Its peripheral portion consists of muscle fibres (striated muscle) originating at the sternal, costal, and vertebral margins of the thoracic outlet and converging to an insertion in a central tendon. Developmentally, the central tendon arises from an embryonic layer of cells separating the heart from the liver, while the muscle derives from muscle of the surrounding body wall.

*Blood supply.* The main arteries are left and right inferior phrenic vessels arising variably (in man) from the aorta, below the diaphragm, or from the coeliac artery (left). Additional arteries derived from the inter-costal vessels supply the diaphragm at the periphery. Burgen,

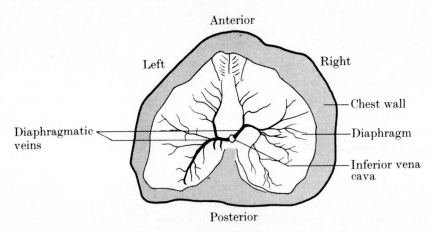

FIG. 5.16. Venous drainage of the rat diaphragm (from Burgen, Dickens, and Zatman 1949). Injected preparation of the venous drainage of the rat diaphragm. The large vein draining the left half of the diaphragm and the main right lateral vein can readily be seen.

Dickens, and Zatman (1949) comment that the arterial supply is complex and unsuitable to the purpose of close intra-arterial injection. The venous drainage, however, was demonstrated to be ideal for the same purpose. Practically the whole of each dome of the diaphragm drains into one large vein, which empties into the inferior vena cava as it passes through the tendinous arch of the diaphragm (Fig. 5.16). It is upon this observation that the method of retrograde perfusion of the diaphragm (Rowlands and others) is based.

*Nerve supply.* The nerve supply to the diaphragm is by means of two phrenic nerves, one left and one right, which descend through the thorax on the lateral surfaces of the pericardium and spread over the superior surface of the diaphragm. Electrical stimulation of the nerve causes contraction; direct stimulation of the muscle sheet has a similar effect and is the method chosen for use with the perfused diaphragm preparation.

*Lymphatics.* Lymph vessels penetrate the diaphragm (in man), but are not accessible to study in the present system.

*Histology.* Essentially, the diaphragm is a layer of striated muscle, some 500 $\mu$m thick (Creese *et al.* 1958), with a layer of mesothelium or pleura on each of its surfaces. The central tendon has a high collagen content and is usually excluded from the perfusion.

*Method of perfusion* (Brownlee and Straughan 1957: modified according to Rowlands, 1967, 1969)

Following the observation of the practical value of the venous drainage of the diaphragm for injection of drugs (Burgen *et al.* 1949), Brownlee and Straughan developed a technique of retrograde perfusion of a segment of the tissue using eserinized K–H buffer as perfusing medium. However, oxygenation was still by way of the surrounding bath medium in this preparation and the present truly vascular (but still retrograde) perfusion is that developed by Whaler (1960) and described in detail by Rowlands (1969). Simple metabolic studies have been carried out, and a comparison with other diaphragm preparations on this basis will be discussed. Oxygenated medium is perfused into the thoracic inferior vena cava and thence into the main diaphragmatic veins. The segment of diaphragm so perfused is isolated from tendon and rib-cage before mounting it in an organ bath. Medium is run into the vein under gravity and no attempt has been made at recirculation-perfusion in the present preparation (Rowlands 1967, 1969).

*Apparatus*

The water-jacketed system consists of a reservoir in which medium may be gassed by means of a bubbler, a valve-pump, with a bubble trap and 'depulsator' to reduce the wave oscillations, and an organ bath. Most of the connecting tubing is polythene, but in the original apparatus a short length of stainless-steel tubing is immersed in the water bath that surrounds the organ chamber, to act as a heating coil. A special ⊥-shaped frame is provided to hold the diaphragm and the attachments of electrical leads for stimulation of the muscle during a perfusion. Fig. 5.17, reproduced from Rowlands (1967), illustrates this

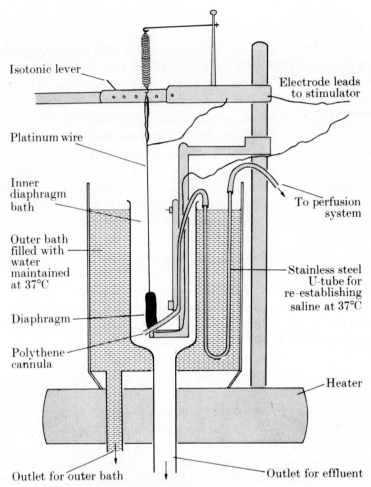

Isotonic lever

Electrode leads
to stimulator

Platinum wire

Inner
diaphragm
bath

To perfusion
system

Outer bath
filled with
water
maintained
at 37°C

Stainless steel
U-tube for
re-establishing
saline at 37°C

Diaphragm

Polythene
cannula

Heater

Outlet for outer bath

Outlet for effluent

FIG. 5.17.  Apparatus for perfusion of rat diaphragm (Rowlands 1967). The diaphragm is held in an ⊥-shaped metal support and suspended from a platinum wire within an organ bath. Electrical stimulation of the muscle is achieved by applying a potential difference across these electrodes. For details see p. 302.

arrangement. In a more recent modification, Rookledge *et al.* (personal communication) report the use of the Watson–Marlow pump (see Chapter 1) for perfusion under otherwise identical conditions.

*Perfusion medium*

Krebs–Henseleit buffer, containing 0·1 per cent glucose and gassed with 95% $O_2$:5% $CO_2$ is used.

*Operative procedure*

Fed rats of 2–300 g weight are anaesthetized with 20 mg/kg Nembutal (intra-peritoneal). The ventral aspect of the animal is shaved, and the abdomen is opened through a transverse incision, beneath the xiphi-sternum. Cutting the falciform ligament (which lies in the mid-line, beneath the xiphisternum and diaphragm, but above the liver) allows a ligature to be placed around the inferior vena cava, between the liver and diaphragm. This is tied, and the ends are left long.

The thorax is opened by means of a transverse incision just above the xiphisternum, extended upwards on either side. This allows the anterior chest wall to be removed, exposing the heart, the lungs, and the phrenic nerves. The nerves, left and right, are located and cut, and the thoracic inferior vena cava is cannulated directly, in the direction of the dia-phragm. (An improvement is to cannulate the vessel through the right atrium, as described for *in-situ* liver perfusion (p. 171, Fig. 3.11)). A slow flow of saline medium begins at once and is increased to 2 ml/min by adjusting the pump. The liver is removed, and the oesophagus cut, exposing the abdominal aspect of the diaphragm. The right hemi-diaphragm is to be perfused in this preparation and the main vessel to this area, now perfusing with saline, is located; its branches are ligated, using 3/0 thread on a curved needle, thereby isolating the segment of diaphragm. A ⊥-shaped platinum electrode is sewn to the peripheral edge of this area, with care not to damage the diaphragmatic muscle, or to include intercostal muscle in the preparation. The perfusing segment of diaphragm is now cut from the rib-cage and from the adjoining muscle, in a direction parallel to the muscle fibres. Finally, the cannula is separated from the heart, cutting through the upper end of the vena cava (or through the atrium), and the preparation is mounted on the ⊥-shaped holder. Thus, one end (the peripheral) of the diaphragm is held on a platinum electrode, the other, the central tendon, is attached to a platinum wire, and this becomes the second electrode. The other end of the wire is attached, in the original preparation, to an isotonic recording lever, which writes on a smoked drum. More sophisticated recording techniques may be applied.

Rowlands describes the following method of stimulation and record-ing. Stimulation of nerve endings is achieved with a pulse of 50 $\mu$s duration, strength 10 V. 'Direct' stimulation of muscle occurs when a pulse of 500 $\mu$s duration and 80 V is applied. Downward displacement of the lever on the smoked drum represents shortening of muscle.

As a test for complete perfusion, the sensitivity of nerve stimulation to the inhibitory effect of tubocurarine is suggested. The test, carried out at the beginning of perfusion, consists of the injection of 30 $\mu$g (in 0·15 ml 0·9% NaCl) of d-tubocurarine into the tubing immediately before the cannula. In a well-perfused preparation, complete paralysis to 'indirect' stimulation occurs within 30–60 s. The preparation recovers within 1 min, and an experiment may be begun.

Determination of oxygen consumption is possible in this preparation if the organ bath is filled with liquid paraffin and effluent medium is collected from a tap inserted into the bottom of the chamber. Flow-rate determination, required for the calculation of oxygen consumption, is by simple collection of the effluent in the apparatus described. Alternative methods are discussed on p. 103.

*Characteristics of perfusion*

The preparation has been characterized by a succession of workers (Whaler 1960; Hollanders 1968; Rowlands 1969) including the important comparisons with the nerve–muscle preparation (Bülbring 1946) and with incubation techniques. At perfusion pressures of 20–40 mmHg, flow rate is steady (2–3 ml/min per g) for up to 3 h. Muscular activity, recorded as a response to electrical stimulation (q.v.) is also well maintained, and the relevance of this to the tubocurarine test of complete perfusion in this preparation has been discussed above. Metabolic activities are such that the perfused diaphragm produces lactate and consumes oxygen in the presence of 6 mM glucose in the standard medium. The rate of lactate formation, initially 60 $\mu$g (0·7 $\mu$mol)/min, falls to a plateau of 35 $\mu$g (0·4 $\mu$mol)/min per g at 45 min, and is there maintained for the 3 hours of perfusion. The rate of lactate formation is not increased during muscle stimulation. Oxygen uptake, on the other hand, is doubled when the diaphragm is stimulated at a rate of 12/min (Rowlands 1969). The rate of glucose uptake was not determined in these experiments so that it is not possible to construct a balance; nevertheless, it is possible to infer that glucose is oxidized, from experiments in which lactate formation and oxygen uptake were recorded together. In one experiment, only 37 per cent of glucose could be converted to lactate to account for the observed oxygen uptake. Both oxygen uptake and lactate formation may be used as parameters of function, since they are linear with time.

Comparison with the other preparations of diaphragm on this basis indicates that the perfused preparation produces more lactate (35 against 5 µg/min per g) than the incubated diaphragm. The rate of lactate formation in the Bülbring nerve–muscle preparation is intermediate, with a higher initial rate but a lower plateau (25 µg/min per g) than observed in perfusion.

These observations would seem to support the use of perfusion in study of the metabolism of diaphragm. There is a place for more exhaustive tests of function to establish the uniformity of the preparation, and some of the tests discussed in the following section for application to the perfused hind-limb might be applied here.

### Hind-limb preparation

Apart from the diaphragm preparation just described there is no simple preparation of skeletal muscle. Various groups have independently developed a perfusion of the hind-quarters of the rat, in which the major tissue perfused is skeletal muscle (Mortimore et al. 1959, Mahler et al. 1968, Ruderman et al. 1971). The entire vascular bed, which is supplied by the lower abdominal aorta, excluding the skin, pelvic viscera, and tail, is perfused in this preparation. Most of the tissue remaining is muscle, but the perfused tissue includes skin, adipose tissue, and bone, up to 23 per cent of the total (Ruderman et al. 1971). It is perhaps surprising that a preparation of such heterogeneity of tissue should give consistent results in biochemical studies but, particularly in the recent preparation of Ruderman et al., this is the case, and the preparation is in use for the study of metabolic fuels and the effects of muscular work on metabolism (see, for example, Houghton, Ruderman, and Hems 1970).

*Anatomy*

The lower half of the rat—lumbar spine, vertebral muscles, posterior abdominal wall, hip joints, gluteal and thigh regions, in addition to the hind-limbs proper—is supplied with blood from the abdominal aorta. This artery divides on entering the region, to form left and right iliac vessels which enter the thigh on either side. Only minor regions of the anterior abdominal wall and the organs of the pelvis are supplied with blood from the same source, and these can readily be excluded from the circuit. The details will become obvious during a description of the operative preparation.

*Venous drainage.* The venous drainage follows a similar pattern to the arteries; major veins draining into the femoral and iliac veins unite to form the inferior vena cava. This has some smaller lumbar branches which drain tissues with which the present perfusion is not concerned, but these are relatively easily excluded from the circuit, and cannulation of the lower inferior vena cava drains the entire hind-limb preparation.

*Lymphatic drainage.* Lymph vessels accompany the blood vessels, draining the region through a large number of small channels. None of the perfusions to be described has been concerned with them.

*Nerve supply.* The major nerve supply to the hind-quarters is preserved in the present dissection and may continue to influence the preparation in an unpredictable way. Both motor and sensory fibres are contained in the femoral and sciatic nerves. Sympathetic and parasympathetic fibres, which are principally concerned with vaso-motor control, also survive the dissection. The exception are components of the hypogastric plexus. It is a feature of the hind-limb preparation that the rate of perfusion may vary during an experiment (Ruderman, personal communication), and the persistence of vasomotor nerves may explain this.

## Microscopic anatomy

In striated muscle the basic unit is the fibre, a long cylindrical multi-nucleate cell. These are arranged in parallel in groups to form a 'fascicle'. Blood-vessels course in connective tissue septa and ramify into a rich capillary bed, around individual muscle fibres. The capillaries are sufficiently tortuous to accommodate changes in length during contraction and relaxation of the muscle fibres.

Muscle fibres vary in thickness, from 10 to 100 $\mu$m, and may vary in response to use (hypertrophy) or disuse (atrophy). The cross-striations which give the muscle its name are illustrated in standard texts, and their contribution to contraction is currently explained by the sliding filament hypothesis of Hanson and Huxley (1958) (see Bloom and Fawett 1968).

## Physiology of striated muscle

An understanding of the nature of 'work' and of the elastic properties of muscle, together with a discussion of the methods of producing and recording isometric or isotonic contraction, may be obtained from standard physiology textbooks. In biochemical studies of exercising muscle, such as are possible in perfused limb preparations, much depends on the controlled production and recording of muscular work.

## Hind-limb perfusion

Two very similar preparations exist and the present description is based on the method introduced by Ruderman *et al.* (1971) for experiments on respiratory fuels in muscle. The earlier preparation of Mahler, Szabo, and Penhos (1968) is described later for comparison; enough differences exist between the two techniques for a formal experimental comparison to be of value.

The hind quarters of the rat (see above for definition) are perfused through the aorta. A semi-synthetic medium, containing washed aged human red cells (cf. Hems *et al.*, Chapter 3 but with Hb = 7–8 g%) is pumped directly into the aorta after passing through a multibulb oxygenator. A by-pass, with a valve, allows the perfusion pressure to be regulated between 80 and 140 mmHg (or 80–100 mmHg in a preparation with the skin left on; see below). The venous outflow, collected by a cannula in the inferior vena cava, flows back into a reservoir beneath the animal tray. In this preparation, the hind-quarters are not separated from the donor rat.

### *Apparatus*

The apparatus for one perfusion is enclosed in a cabinet of the design described in Appendix 1. Minor modifications to the interior of the cabinet are desirable to accommodate a second pump, a manometer, a reduction valve, and the larger reservoir.

The glassware is essentially that used by Hems *et al.* (q.v. Chapters 1 and 3) for perfusion of the liver. In particular, the reservoir, increased in volume to contain the initial 200 ml perfusion medium, has a centrally placed funnel welded into the side wall (identical with the arrangement described by Hems *et al.* for liver perfusion (1966) but since modified for that purpose (see Chapter 3, Fig. 3.9)).

Medium leaving the preparation may flow either directly into the reservoir, for recirculation, or into the funnel, and thence to waste. This allows the non-recirculating wash-out period (see below) to be achieved without disturbing the animal tray. Two pumps are required; one to supply the oxygenator (a glass, multibulb oxygenator of the design discussed on p. 49) and the second to drive medium under arterial pressure into the aorta. Watson–Marlow pumps are suitable.

The circuit is illustrated in Fig. 5.18, with the direction of flow indicated by arrows.

Alternatively a single pump may be used, with a double clamp track (4·8 mm bore) to feed the oxygenator and supply the perfused tissue through separate channels.

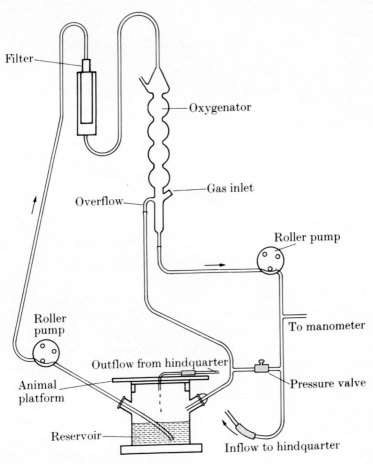

FIG. 5.18. Apparatus for perfusion of isolated rat hind-quarter. The circuit used by Ruderman *et al.* (1971) is illustrated; arrows indicate the direction of flow of medium. For details, see p. 308.

*Filter.* A nylon mesh filter of the type used in Hems liver perfusion, i.e. disposable drip-set filter, is used. (There has been no discussion of the causes of failure in this preparation, but a more effective filter (see Chapter 1, Filters) might be necessary for prolonged perfusions).

*Cannulae.* The arterial cannula is a length of No. 18 polyethylene catheter o.d. 1·2 mm. A disposable catheter (Portex, size) with a Luer fitting to allow of rapid transfer has been used. The venous cannula is the Frankis–Evans stainless-steel cannula already described by Hems *et al.* for use with the perfused liver (portal vein cannula).

## Perfusion medium

The medium of Hems *et al.* is used, modified as follows:

| | |
|---|---|
| bovine serum albumin ($\#V$) | 4·0 g/100 ml, |
| aged, washed human red cells | 7·8 g Hb/100 ml, |
| glucose | 5·5 mM |
| sodium pyruvate | 0·15 mM |

(which makes the lactate/pyruvate ratio 15:1 approx.).

At this haemoglobin concentration, glycolysis by red cells, even when 4 weeks old (see p. 34), becomes significant. The rate of glycolysis should therefore be determined for each batch of cells.

## Operative procedure

Two alternative operations are used: in the one, the perfusion is established without removing the skin; in the other the skin is removed from the lower half of the animal before continuing the dissection. The advantage of removing the skin is twofold: (1) the mass of tissue to be perfused outside muscle is appreciably reduced; (2) direct and rapid sampling of the perfused skeletal muscle becomes possible.

Female rats of 180–230 g are anaesthetized with intra-peritoneal Nembutal (0·1 ml/100 g). With the animal lying on its belly a $\frac{1}{2}$-in strip of loose skin is excised over the lateral aspect of each thigh, and extending to the mid-dorsal region. The two wedges are joined by a transverse mid-dorsal cut through the skin, and the resulting flap is stripped gently caudally. An electric cautery is recommended by the authors to minimize blood loss from small vessels of the skin. A dorsal lumbar fat pad is removed with the skin-flap. Turning the animal to lie on its back, an anterior, mid-abdominal incision through skin is made, and extended to either side at the level of the xiphisternum, completing the skin-flap. This is stripped down to the upper thighs. Tying the superficial epigastric vessel on each side allows the lower flaps of the anterior abdominal wall to be excised. The skin-flap is left in place for the duration of the operation and serves to prevent dessication of the muscle surfaces. At a later stage (just before aortic cannulation) the skin flap is removed by making circumferential incisions at the level of the upper thigh. The dissection proper now begins. The animal is fixed to the operating tray, on its back, with limbs extended, and the abdomen is opened by means of the usual mid-line incision. The incision is extended laterally, after clamping the descending epigastric vessels on either side and the intestine is swept to the animal's left. This exposes the posterior abdominal wall structures and the pelvic contents.

The following ligatures are placed and tied, to allow the uterus, ovaries, and descending colon and rectum to be removed, while restricting the distribution of circulation from the abdominal aorta (see Fig. 5.19):

(1) behind the base of the uterus, to include the uterine artery on each side;

(2) around the ovary and its vessel on each side;

(3) inferior mesenteric artery, close to the lower aorta (see also Chapter 7);

(4) and (5) around the lower and upper ends of the descending colon: the uterus, ovaries, and the enclosed segment of large bowel may then be excised.

Fat, which lies around the bladder, is cleared to expose the hypogastric vessels on either side (they supply the pelvic viscera from above), and these vessels are ligated twice (6) and cut between ligatures.

The pubis is split in the mid-line (with scissors or with cautery), which allows the bladder and remaining pelvic viscera to be removed from above, downwards. As the rectum is continuous with the skin at the anal margin, the block of tissue is released by a circumferential incision about the tail, 1 cm from its base. The prepared skin-flap is removed at this stage (q.v.). A firm ligature round the exposed tail base occludes the caudal artery and excludes the tail from the perfusion. Further ligatures are now emplaced:

(7) a single ligature is passed behind the ilio-lumbar vessels on each side and tied;

(8) and (9) loose ligatures round the inferior vena cava above the and below the renal vessels;

(10) and (11) loose ligatures round the aorta, below the renal vessels and just above the left renal artery (careful separation of the inferior vena cava from the aorta in this region is achieved by opening and closing a curved forceps parallel to the aorta;

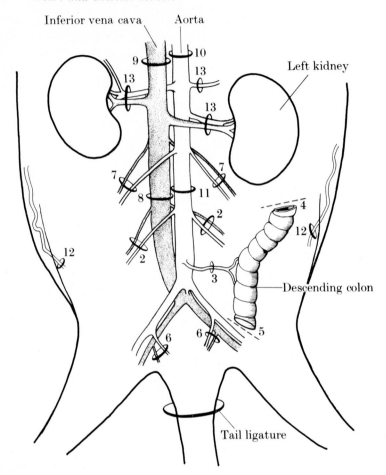

Inferior vena cava     Aorta

Left kidney

Descending colon

Tail ligature

FIG. 5.19.   Operative procedure for perfusion of the isolated hind quarter (preparation of Ruderman *et al.*). The vessels of the posterior abdominal wall are exposed. Ligatures are numbered according to the description on p. 311.

(12) behind the inferior epigastric vessels, as they course from the lateral abdominal wall on to the lower flap of the anterior abdominal wall; these ligatures are tied and the excess flap of abdominal wall removed;

(13) etc: all the major vessels of the upper abdomen are ligated, as a somewhat arbitrary precaution against mixing of perfused and non-perfused circuits.

*Cannulation.* The animal tray is turned through 90° to bring the head to the operator's right. Ligatures (9) and (10) are tied, occluding the upper aorta and vena cava and introducing the 'anoxic phase'. The aorta is incised below the ligature and the cannula, prepared with heparinized saline to the tip, is introduced and tied in place with its end midway between the iliolumbar artery and the aortic bifurcation. Venous cannulation is performed directly through the vessel wall with the Frankis–Evans cannula. Back-flow occurs when the trocar is removed, and the cannula, with its tip at the same level as the arterial cannula, is secured by tying ligature (8). Heparin (0·2 ml = 200 units) is injected through the aortic cannula which is then clamped to prevent bubbles entering the circuit while the preparation is transferred to the perfusion cabinet and attached to the arterial circuit at a perfusion pressure of 100 mmHg approximately.

Operative time in the hands of experts is some 20 min, with an anoxic phase of some 2–3 min. The first 70 ml of medium is discarded through the central funnel of the apparatus (an extension to the outflow cannula allows medium to be collected from the reservoir outflow), and the circuit may then be completed to allow recirculation (see Fig. 5.18). If an initial perfusion pressure of 100 mmHg is used, this may be reduced if the flow rate is too high to measure, or in order to allow an increased arterio-venous oxygen difference to develop.

## Characteristics of perfusion

The preparation has been characterized in a number of ways (see Rudermann *et al.* for detailed account).

*Perfusion flow and pressure.* The minimum flow rate accepted in this preparation is 6 ml/min and, more normally, flow rates of 8–12 ml/min are achieved with perfusion pressure of 80–140 mmHg. In preparations in which the skin is not removed (see below) the pressure required appears to be lower (80–100 mmHg) while the metabolism of the two preparations is essentially similar.

*Appearance of the preparation.* The authors report rapid re-establishment of circulation through the feet and the *in vivo* colour of muscle is restored within 15 min. Persistent cyanosis of the feet or flow rates below 6 ml/min are taken to be gross signs of failure. On electron microscopy of perfused muscle no defect is seen.

*Areas and tissue perfused.* The present authors have also conducted a useful survey of the composition of the tissues perfused in the preparation they describe. Using Evans Blue dye, followed by KOH digestion, it appears that bone (12 per cent), skin (12·3 per cent) and muscle plus fat (77·7 per cent) are perfused. 30 g muscle/180 g body weight of rat represents a reproducible unit of tissue perfused, and this is the basis on which the results of these workers are expressed.

Finally, to compare with the hemi-sected preparation of Szabo *et al.* (1969), the loss of perfusate has been measured by an isotope distribution technique. The equivalent of 1·8 ml (of 200 ml perfusate) may find its way into the systemic circulation, a figure that is considered low enough to make hemi-section unnecessary.

*Metabolism.* Linear rates of glucose uptake (1·2 $\mu$mol/min per 30 g muscle), of lactate production (2·1 $\mu$mol/min per 30 g muscle), and of oxygen consumption (9 $\mu$mol/min per 30 g muscle) are maintained for 30 min, and standard experiments are conducted over this period. Glycerol release is reported to be higher in the first 15 min of perfusion. As tests of function the following additional parameters may be used: (1) the response of the preparation to insulin (glucose uptake increases sixfold; oxygen consumption increases by 40 per cent, and glycerol release is reduced by 50 per cent); (2) release of potassium into the medium; none occurs in the 30-min perfusion period. In more prolonged perfusions (2 h), oxygen uptake, lactate release, and the response to insulin appear to be preserved, but potassium efflux occurs.

*Effect of work.* The use of a stimulator (compare diaphragm preparation) allows controlled muscle contraction to be induced during perfusion of the hind-limb and extends the usefulness of this preparation as a model for muscle *in vivo*. In addition, animals may be trained (e.g. by swimming) before perfusion. Similar metabolic tests have been applied to this preparation.

*Analysis of perfused tissue.* Tissue concentration of high-energy intermediates, suggested as a test of tissue integrity by Schimassek (see Chapter 3, Liver) have been determined for this preparation. After 40-min perfusion, the concentration of creatine phosphate, ATP, and ADP are similar to those found in muscle in the intact animal.

The conclusion is that this preparation, of which nearly 80 per cent is skeletal muscle, provides a useful *in vitro* model for the study of metabolism of resting and working muscle.

## Alternative method: preparation of Mahler, Szabo, and Penhos

This preparation, in fact, a fore-runner of the technique described above, differs from it in the following details.

(1) Perfusion is under constant hydrostatic pressure from a 'head' of 100 cm.

(2) Fresh, heparinized rat blood is used as perfusion medium, with 100 mg % glucose added. It is gassed with 95% $O_2$:5% $CO_2$.

(3) The hind-quarter perfused is isolated from the remainder of the rat by transection just below the level of the kidneys, and the preparation is suspended in the vertical position by means of a thread on either side.

The operative procedure does not differ significantly from that already described; these authors pay greater attention to the lumbar branches of the aorta, with the definite exclusion of the vertebral canal from the perfusion. Characteristics of perfusion are that at the pressure used (100 cm $H_2O$) a flow rate of 5 ml/min is accepted by the authors as 'normal', and satisfactory perfusion is continued for 70 min. Glucose uptake, at a low rate, is linear throughout this time, and stimulation by insulin results in a greatly increased glucose uptake, which is still linear with time.

## Femoral artery perfusion

A preparation that has not found biochemical application as yet is the hind-limb perfusion of Gool and van Ladiges (1969) in which the femoral artery and vein are cannulated and perfused. With fresh citrated rat blood as the medium, perfusions are continued for up to 6 h. However, at the flow-rates recorded (0·75 ml/min approx.) it is unlikely that the volume of muscle supplied is fully oxygenated (cf. Ruderman *et al.*).

## Single hind-limb perfusion: method of Kuna *et al.*

An earlier preparation, which was used to investigate the release of cells from the perfused bone marrow and may have biochemical applications, is the single hind-limb perfusion, devised by Kuna, Gordon, Morse, Lane, and Charriper (1959). In this preparation the aorta is cannulated and the cannula advanced into one or other iliac artery, the opposite vessel being ligated. The non-perfused limb is excised and serves as a 'control' to the perfused leg. Outflow is collected in the inferior vena cava cannula. The preparation, which appears to be stable for 4 h, takes only 5 min to set up.

*Apparatus* (modified in Dornfest *et al.* (1962))

A Sigmamotor pump provides the perfusion pressure, with non-pulsatile flow. (The original method stresses the use of the Dale–Schuster pump, with pulsatile flow.) Temperature control is by means of a coil of Tygon tubing, immersed in a water bath at 29–30°C. No oxygenator is used in non-recirculation experiments, but a column of glass beads with gas flowing in the reverse direction is used and in this paper the use of a 'rotating-plate' oxygenator is discussed.

*Cannulae.* Arterial and venous cannulae of polythene (o.d. = 2 mm) are suitable.

*Perfusion medium*

Heparinized rat blood, obtained by heart puncture, and to which penicillin and streptomycin (see Chapter 1, Antibiotics) are added, both initially, and at intervals through the perfusion, is used and gassed with 95% $O_2$:5% $CO_2$.

*Operative procedure*

Under Nembutal anaesthesia, the abdomen is opened in the mid-line in the usual way, and the intestine is diverted to the animal's left. The great vessels are exposed in the lower abdomen by blunt dissection and loose ligatures are placed individually around the common iliac artery and vein of the limb to be perfused. The vessels of the other limb may be included in a single ligature. If the left leg is to be perfused, the right iliac artery and vein are ligated. The arterial cannula is inserted into the lower aorta, occluding the blood flow temporarily with a pair of curved forceps. The cannula is advanced into the internal iliac artery and ligated in place. Flow begins at once, and the inferior vena cava is cannulated to collect the effluent.

*Characteristics of perfusion*

This preparation has been used principally as a perfused bone, to study the efflux of cells from the bone marrow (Kuna *et al.* 1959, Dornfest *et al.* 1962). In consequence, little is known of its metabolic characteristics. Perfusion with 60–160 mmHg pressure (80–220 cm

$H_2O$) gives flow rates of 0·25–1·5 ml/min (Dornfest *et al.*) at which flow rates glucose consumption was observed at the rate of 1·2 μmol/min.

In the original experiments of Kuna *et al.* a flow rate of 3–6 ml/min was achieved with pressures of 60–80 mmHg (80–110 cm $H_2O$) and glucose consumption was between 0·7 and 1·2 μmol/min. The weight of tissue perfused is approximately 12 g in 180 g rat, about half that perfused in the hind-quarter preparation of Ruderman *et al.* (q.v.). The flow rates used by Kuna *et al.* are probably adequate, while the lower rate of flow suggested by Dornfest *et al.* leads almost certainly to an anoxic perfusion. The rates of glucose uptake are of limited significance since this function is insulin dependent (Ruderman *et al.* 1971, Szabo *et al.* 1969). Further rat blood as a perfusion medium has the disadvantage of containing an unknown insulin concentration.

The parameter of 'release of cells' might be of some general value as a test of function; it appears to occur at twice the rate found in the isolated perfused femur preparation (see Chapter 8) and to be dependent both upon the rate of flow of perfusate and upon the number of cells in the medium (Dornfest 1962).

## REFERENCES

ARNOLD, G. and LOCHNER, W. (1965) Die Temperaturabhängigkeit des Sauerstoffverbrauches stillgestellter, künstlich perfundierter Warmblüterherzen zwischen 34° und 4°C. *Pflügers Arch. ges. Physiol.* **284**, 169.

BASAR, E., RUEDAS, G., SCHWARTZKOPF, H. J., and WEISS, CH. (1968) Untersuchungen des zeitlichen Verhaltens druckabhängiger Änderungen des Strömungswiderstandes in Coronargefäß system des Rattenherzens. *Pflügers Arch. ges. Physiol.* **304**, 189.

BLEEHEN, N. M. and FISHER, R. B. (1954) The action of insulin on the isolated rat heart. *J. Physiol., Lond.* **123**, 260.

BLOOM, W. and FAWCETT, D. W. (1968) *A textbook of histology*, 9th edn, Chapter 11. Saunders, Philadelphia.

BRODIE, T. G. (1903) The perfusion of surviving organs. *J. Physiol., Lond.* **29**, 266.

BROWN, J. W., CRISTIAN, D., and PARADISE, R. L. (1968) Histological effects of procedural and environmental factors on isolated rat heart preparations. *Proc. Soc. exp. Biol. Med.* **129**, 455.

BROWNLEE, G. and STRAUGHAN, D. W. (1957) Motor nerve stimulation and acetyl-choline release in the perfused rat phrenic nerve diaphragm preparation. *J. Physiol., Lond.* **136**, 6p.

BÜLBRING, E. (1946) Observations on the isolated phrenic nerve diaphragm preparation of the rat. *Br. J. Pharmac. Chemother.* **1**, 38.

BURGEN, A. S. V., DICKENS, F., and ZATMAN, L. J. (1949) The action of botulinum toxin on the neuro-muscular junction. *J. Physiol., Lond.* **109**, 10.

CHAIN, E. B., MANSFORD, K. R. L., and OPIE, L. H. (1969) Effects of insulin on the pattern of glucose metabolism in the perfused working and Langendorff heart of normal and insulin-deficient rats. *Biochem. J.* **115**, 537.

CREESE, R. (1954) Measurement of cation fluxes in rat diaphragm. *Proc. R. Soc.* **B142**, 497.

—— HASHISH, S. E., and SCHOLES, W. W. (1958) Potassium movements in contracting diaphragm muscle. *J. Physiol., Lond.* **143**, 307.

DORNFEST, B. S., LOBUE, J., HANDLER, F. S., GORDON, A. S., and QUASTLER, H. (1962) Mechanisms of leukocyte production and release. (i) Factors influencing leukocyte release from isolated, perfused rat femurs. *Acta haemat.* **28**, 42.

ENSER, M. B., KUNZ, F., BORENSZTAJN, J., OPIE, L. H., and ROBINSON, D. S. (1967) Metabolism of triglyceride fatty acids by perfused rat heart. *Biochem. J.* **104**, 306.

FISHER, R. B. and WILLIAMSON, J. R. (1961) The oxygen uptake of the perfused rat heart. *J. Physiol., Lond.* **158**, 86.

GAMBLE, W. J., CONN, P. A., EDALJI-KUMAR, A., PLEUGE, R., and MONROE, R. G. (1970) Myocardial oxygen consumption of blood-perfused, isolated, supported, rat heart. *Am. J. Physiol.* **219**, 604.

GOOL, J. and VAN LADIGES, N. C. (1969) Production of foetal globulin after injury in rat and man. *J. Path. Bact.* **97**, 115.

HOLLANDERS, F. D. (1968) The production of lactic acid by the perfused rat diaphragm. *Comp. Biochem. Physiol.* **26**, 907.

HOUGHTON, C. R. S., RUDERMAN, N. B., and HEMS, R. (1970) Utilisation of glucose and acetoacetate by the exercising perfused rat hind-quarter. *Proc. Eur. Soc. Diabetes* **6**, 645.

HUHMANN, W., NIESEL, W., and GROTE, J. (1967) Untersuchungen über die bedingungen für die Sauerstoffversorgung des Myokards an perfundierten Rattenherzens. *Pflügers Arch. ges. Physiol.* **294**, 250.

JAFFÉ, R. and GARALLÉR, B. (1958) *Pathologie des Laboratoriumstiere 'Kreislauforgane'*, Vol. 1 (eds. Cohrs, Jaffé, and Meesen). Springer-Verlag, Berlin.

JORDAN, J. and LOCHNER, W. (1962) Über den anaeroben und aeroben Stoffwechsel des stillgestellten, künstlich perfundierten Warmblüterherzens. *Pflügers Arch. ges. Physiol.* **275**, 164.

KADAS, T. and OPIE, L. H. (1963) Isolated perfused rat heart adapted for simultaneous measurement of left ventricular contraction, electrocardiogram and metabolism of $^{14}$C labelled substrate. *J. Physiol., Lond.* **167**, 6P.

KESSLER M. and SCHUBOTZ R. (1968) Die $O_2$-versorgung der hämoglobinfrei perfundierten Rattenleber. *Stoffwechsel der isoliert perfundierten Leber* (eds. W. Staib and R. Scholz), p. 12. Springer-Verlag, Berlin.

KRAUPP, O., ADLER-KASTNER, L., NIESSNER, H., and PLANK, B. (1967) The effects of starvation and of acute and chronic alloxan diabetes on myocardial substrate levels and on liver glycogen in the rat *in vivo*. *Eur. J. Biochem.* **2**, 197.

KUNA S., GORDON,, A. S., MORSE, B. S., LANE, F. B., and CHARRIPER, H. A. (1959) Bone marrow function in perfused, isolated hind-legs of rats. *Am. J. Physiol.* **196**, 769.

LANDMARK, K., GLOMSTEIN, A., and ØYE, I. (1969) The effect of thioridazine and promazine on the isolated contracting rat heart. *Acta pharmac. tox.* **27**, 173.

LANGENDORFF, O. (1895) Untersuchungen am überlebenden Saügertierherzen. *Pflügers Arch. ges. Physiol.* **61**, 291.

LEARY, W. P. P. (1970) Aspects of the renin angiotensin system. D.Phil. thesis, University of Oxford.

LEICHTWEISS, H. P., SCHRODER, H., and WEISS, CH. (1967) Die Beziehung zwischen Perfusionsdruck und Perfusionsstromstärke an der mit Paraffinöl perfundierten isolierten Rattenniere. *Pflügers Arch. ges. Physiol.* **293**, 303.

MAHLER, R. J., SZABO, O., and PENHOS, J. C. (1968) Antagonism to insulin action on the perfused hind limb by a reduced insulin B-chain complex. *Diabetes* **17**, 1.

MANSFORD, K. R. L. (1969) Ph.D. Thesis, University of London.

MORGAN, H. E., HENDERSON, M. J., REGEN, D. M., and PARK, C. R. (1961) Regulation of glucose uptake in muscle. I. The effect of insulin and anoxia on glucose transport and phosphorylation in the isolated, perfused heart of normal rats *J. biol. Chem.* **236**, 253.

MORTIMORE, G. E., TIETZE, F., and STETTEN, D. (1959) Metabolism of insulin I[131]. Studies in isolated, perfused rat liver and hind-limb preparations. *Diabetes* **8**, 307.

NEELY, J. R., LIEBERMEISTER, H., BATTERSBY, E. J., and MORGAN, H. E. (1967) Effect of pressure development on oxygen consumption by isolated rat heart. *Am. J. Physiol.* **212**, 804.

—— LIEBERMEISTER, H., and MORGAN, H. E. (1967) Effect of pressure development on membrane transport of glucose in isolated rat heart. *Am. J. Physiol.* **212**, 815.

NEWSHOLME, E. A. and RANDLE, P. J. (1964) Regulation of glucose uptake by muscle. *Biochem. J.* **93**, 641.

OPIE, L. H. (1965) Coronary flow rate and perfusion pressure as determinants of mechanical function and oxidative metabolism of isolated perfused rat heart. *J. Physiol., Lond.* **180**, 529.

—— (1968) Metabolism of the heart in health and disease. *Am. Heart J.* **76**, 685.

ØYE, I. (1965) The action of adrenaline in cardiac muscle. *Acta physiol. scand.* **65**, 251.

—— BUTCHER, R. W., MORGAN, H. E., and SUTHERLAND, E. W. (1964) Epinephrine and cyclic 3′ 5′ AMP levels in working rat hearts. *Fedn Proc. Fedn Am. Socs exp. Biol.* **23**, 562.

RABITZSCH, G. (1968) Koronarperfusion isolierter Warmblütterherzen mit geringen Umlaufsvolumina und Kontinuierlicher Kontractions—und koronarflußregistrierung. *Acta biol. med. germ.* **20**, 33.

RÖSENBLECK, H., HUHMANN, W., GLOY, U., and NIESEL, W. (1967) Anwendung eines Dünnschichtdialysators zur Gasäquilibrierung von Perfusionslösungen bei Dürchströmmungsversuchen an Organen. *Pflügers Arch. ges. Physiol.* **294**, 88.

ROWLANDS, S. D. (1967) Some aspects of the metabolism and mechanical performance of the perfused rat diaphragm. Ph.D. thesis, University of London.

—— (1969) Oxygen consumption, citrate levels and lactate production of the perfused rat diaphragm. *Comp. Biochem. Physiol.* **29**, 1215.

RUDERMAN, N. B., HOUGHTON, C. R. S., and HEMS, R. (1971) Evaluation of the isolated perfused rat hind-quarter for the study of muscle metabolism. *Biochem. J.* **124**, 639.

SZABO, A. J., MAHLER, R. J., and SZABO, O. (1969) Glucose uptake by the isolated perfused rat hind limb during rest and exercise. *Horm. metab. Res.* **1**, 156.

WEISSLER, A. M., KRAGER, E., BABA, N., SCARPELL, D. A., LEIGHTON, R. F., and GALLIMORE, J. K. (1968) Role of anaerobic metabolism in the preservation of functional capacity and structure of anoxic myocardium. *J. clin. Invest.* **47**, 403.

WHALER, B. C. (1960) Some properties of the perfused rat diaphragm. *J. Physiol., Lond.* **150**, 15P.

WIGGERS, C. J. (1921) Studies on the consecutive phases of the cardiac cycle. *Am. J. Physiol.* **56**, 415.

—— (1952) *Circulatory dynamics* (Modern medical monographs No. 4). Grune and Stratton, New York.

WILLIAMSON, J. R. and KREBS, H. A. (1961) Acetoacetate as fuel of respiration in the perfused rat heart. *Biochem. J.* **80**, 540.

WOLLENBERGER, A., RISTAU, O., and SCHOFFA, G. (1960) Eine einfache Technik der extrem Schnellen Abkühlung grosserer Gewebestück. *Pflügers Arch. ges. Physiol.* **270**, 399.

# 6. Endocrine Organs

## PANCREAS

THE pancreas is a highly vascular endocrine organ in which the anatomy of the blood supply lends itself to the method of perfusion. Studies of both exocrine (Khayambashi and Lyman 1969) and endocrine (Grodsky *et al.* 1963, Sussman, Vaughan, and Timmer 1966, Loubatieres, Mariani, Chapal, and Portal 1967) function of the gland have been conducted in the perfused pancreas of the rat. Exocrine function has also been studied in an isolated perfused cat pancreas preparation (Case, Harper, and Scratcherd 1968).

The advantage of the perfused preparation over other techniques used *in vitro*—isolated islet incubation (Lacy and Kostianovsky 1967), slices (Light and Simpson 1956), tissue fragments (Coore and Randle 1964), superfusion, known also by the less satisfactory term 'perifusion' (see Burr *et al.* 1969)—are essentially those of organ perfusion generally. They are summarized by Loubatieres with particular reference to studies of endocrine function.

(1) Superficial and deep islets are equally provided with oxygen, which is constantly replaced. Substrates and effectors arrive at the cell in the physiologically normal way.

(2) The islets remain in anatomical relationship with other cells and tissues of the organ, including blood-vessels and nerves.

(3) Only extrinsic nerve control is lost, and any part which intrinsic nerves play in regulating the gland's function may be preserved.

(4) The maintenance of the tissues' cellular integrity can be confirmed at the end of an experiment by electron microscopy. By extension, individual islets or other fragments of tissue can be studied by standard methods after perfusion.

(5) There is the possibility of a control and an experimental period of study in the same pancreas. Such controls in other *in vitro* systems require multiple incubations.

Perhaps of greater significance is the very particular advantage of perfusion for the study of exocrine function. Cannulation of the segment of duodenum into which the exocrine duct drains allows collection of these secretions. The enzymes produced are kept separate from the cells that produce them, and the well-known destructive effect of pancreatic enzymes is avoided. In addition, interplay between exocrine and endocrine pancreas (Henderson 1969) might be demonstrable in such a preparation. The disadvantages of the technique have not been clearly outlined and will become apparent during the descriptions that follow.

(1) Technical difficulty of the surgical preparation, and the consequent lengthy period required for the preparation are the most obvious drawbacks. It is still probable that the very satisfactory tissue fragment method is more simply applied for biochemical studies (see also Hokin and Hokin 1959).

(2) Heterogeneity of the tissue perfused makes interpretation of results more difficult. Thus, within the pancreas, islets comprise only a small proportion by weight and, as will become apparent, the existing perfusion techniques for the pancreas include a proportion of non-pancreatic tissue, for example, duodenum and stomach.

**Anatomy**

The pancreas lies between the two layers of mesentery that enclose the loop of the duodenum and, in the rat, is diffuse in structure. It is partly covered by both the stomach and the liver when viewed through a ventral incision and is exposed by sweeping the abdominal contents, including the small intestine, to the animal's left. In the rat it is described as 3 cm × 5 cm (Cohrs, Jaffé, and Meesen 1958); it is whitish in colour, dendritic in form, and may be distinguished from fat in the mesentery which is slightly yellow. The main pancreatic duct and the bile duct, which traverses the pancreas for half of its course to the duodenum, are seen when the containing loop of bowel is held up to the light. The pancreatic ducts are multiple (Green 1935) and their entry into the intestine is variable (Cohrs *et al.* 1958). Doerr and Becker (1958) illustrate the main duct as emptying into the bile duct rather than into the duodenum. Sussman's technique of perfusion takes this into account by cannulation of the distal bile duct to relieve 'back pressure' on the pancreatic duct. In the guinea pig and the rabbit (Doerr and Becker 1958) the main pancreatic duct enters the duodenum itself. The general rule is made that in carnivores the ducts empty near or into the common bile duct, while in herbivores they empty more distally. This may be of some importance in devising perfusion studies in which exocrine function such as protein synthesis is to be investigated.

*Arterial supply*

All blood to the pancreas comes from the aorta via the coeliac axis and the superior mesenteric artery, and these are the important vessels in perfusion studies. The vessels directly concerned are the pancreatico-duodenal arteries and the gastroduodenal branch of the hepatic artery. The tail of the pancreas receives branches from the splenic artery, but this itself arises from the coeliac axis (Fig. 6.1). Perfusion therefore

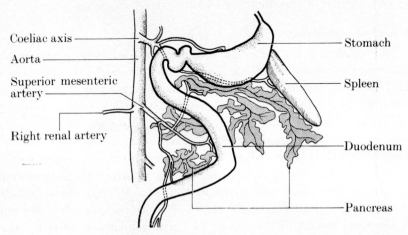

FIG. 6.1.   Anatomy of pancreatic perfusion techniques. The superior mesenteric artery supplies the duodenum and most of the pancreas (Sussman technique). The organs supplied by the coeliac axis—the stomach, spleen, and pancreas—are included in the preparation of Grodsky. The liver is excluded from the circuit by the ligature which includes the hepatic artery and portal vein (see Fig. 6.2).

usually resolves itself into cannulation of either the coeliac axis (Grodsky *et al.* 1963) or of the segment of aorta from which arise the coeliac and the superior mesenteric arteries (Sussman *et al.* 1966). It should be noted, therefore, that in intestinal perfusion techniques (see Chapter 7) the perfusion of the superior mesenteric artery of necessity includes part of the pancreas in the circulation.

*Venous drainage*

Blood leaves the pancreas through numerous small vessels, all of which drain into the portal vein and pass through the liver, before entering the general circulation. Necessarily, not all blood entering the portal vein is derived from the pancreas, a fact that must be considered in devising *in situ* perfusion methods.

### Lymphatic drainage

Although much studied in larger animals including man, the very extensive lymphatic drainage of the pancreas has not been studied in the rat. It probably ends in the superior mesenteric lymph duct (Bollman, Cain, and Grindlay 1948) and pancreatic lymph could be collected by cannulation of this duct (see Chapter 7).

### Microscopic anatomy

The pancreas is a gland, with both internal (endocrine) and external (exocrine) secretory functions. Its microscopic anatomy reflects this dual function and is described in detail in physiology and anatomy texts. There is no substantial difference in structure of the pancreas of the rat and that of other mammals. Most of the tissue is arranged in acini which perform the exocrine function, draining secretions into the duct system already described. The endocrine function is performed by the islets of Langerhans with insulin and glucagon as their main product in the rat. The islets are distributed evenly through the pancreatic tissue, with more islets near the spleen. They comprise about 1 per cent of the total weight of the organ. The arrangements of islets in other species is described by Henderson (1969) and by Doerr and Becker (1958).

### Methods of perfusion

Two methods are described; the first is a gastro-duodenal pancreatic perfusion introduced by Anderson and Long (1947) and brought to a high degree of perfection by Grodsky *et al.* (1963). In this preparation the entire coeliac bed, which includes considerable tissue mass outside the pancreas, is perfused. Nevertheless, useful indicator studies have been performed and the preparation has applications within the limits set by its anatomical and biochemical complexity. The preparation has been characterized by its oxygen consumption, its flow rate, which is linear for about 2 hours, and its insulin production. Other parameters of function will be discussed in a later section and together make this a well-documented preparation.

Technically, its relative ease of preparation gives it advantages over the second method, the pancreatic-duodenal preparation of Sussman *et al.* (1966) and of Loubatieres *et al.* (1967). This is known as an isolated pancreas preparation. However, although its authors suggest that it may be modified without difficulty to become a truly isolated preparation, if the original description of Sussman *et al.* (1966) is followed, the pancreas itself and the adjacent length, some 3–4 cm of duodenum into which the pancreatic duct drains, are perfused.

The segment of aorta from which the coeliac and superior mesenteric arteries arise is cannulated, ligating their unwanted branches. Oxygenated medium is pumped into this arterial circuit and collected from a venous cannula in the portal vein. The duodenum may be cannulated to collect the exocrine secretion. Flow rate, oxygen consumption, and insulin production have been determined for this preparation and allow some objective comparisons to be made between the two. Since insulin production is a unique function of the pancreas, this may be used to isolate the effects of perfusion conditions upon the pancreas from those upon the other tissues present in the two preparations (see calculations below).

A major disadvantage of the pancreatic-duodenal preparation is the operative time required. Sussman suggests that it may take 75 min and although later workers have reduced this time to about 30 min, the objections to such a prolonged operation cannot be overlooked (see Chapter 1).

*Recirculation circuit*

In the general introduction (p. 61) the advantages of recirculation in perfusion studies are presented. The complexity of the available pancreas preparations, however, offers an argument against the use of a recirculating circuit. If some agent that influences insulin production were released from the duodenum, for example, pancreozymin, secretin, or glucagon, it would appear in the portal circulation beyond the pancreatic outflow. No means is recorded whereby this could influence insulin production by a 'retrograde' path but it could be carried in the medium to reach the arterial circuit and hence the pancreatic islets, in a recirculating-perfusion circuit. A 'once-through' method overcomes this objection.

As yet, there have been no studies reported on glucagon or glucagon-like activity produced in the perfused pancreas; clearly the duodenum with its known activity in this respect will influence the results and it will not be possible to exclude them by the simple device of 'once-through' perfusion. The development and characterization of a formally isolated perfused pancreas preparation is therefore urgently required.

**Method I. Perfusion of coeliac axis and the associated organs. Method of Grodsky** *et al.* **and Khayambashi and Lyman**

This method is based upon a perfused pancreas preparation first reported in detail by Anderson and Long (1947). Together with the stomach, spleen, and duodenum as a unit the pancreas is removed from the rat through a ventral abdominal incision, and transferred to a warmed support. The coeliac axis is cannulated for the inflow and the portal vein for the outflow. A peristaltic pump (Khayambashi and Lyman used a Sigmamotor; see Chapter 1, Pumps) supplies medium (usually half and half rat blood and a semisynthetic medium) to the coeliac axis at a pressure of 40–100 mmHg (50–135 cm $H_2O$) and the flow is adjusted to 10 ml/min. A simple film oxygenator and water-jacketing for temperature control completes the apparatus.

*Apparatus*

The original apparatus of Anderson and Long (1947) has been modified by Grodsky. Continuous dialysis of the medium during perfusion is omitted.

*Pumps*. All workers have used pulsatile flow although no critical study has been made of the importance of this in pancreas. Anderson and Long used a piston and valve pump with the inherent disadvantages of leaking, haemolysis, and difficulties in cleaning. Khayambashi, and probably also therefore Grodsky, from whom the modified method is derived, use a Sigmamotor finger pump (see Chapter 1) to provide a constant flow of about 10 ml/min.

*Perfusion pressure*. Curry, Bennett, and Grodsky (1968) suggest that a mean pressure of 60–70 mmHg (80–95 cm $H_2O$) will give the required flow rate of 10 ml/min. Khayambashi and Lyman, using a medium that contained plasma, recommend pressures of 40–100 mmHg (55–135 cm $H_2O$, mean = 95 cm $H_2O$) for the same flow rates.

While no figures are available for the coeliac axis pressures *in vivo* in the rat, it is probably safe to assume that a constant hydrostatic pressure of 95 cm $H_2O$ would be equally effective in perfusing this organ system.

*Temperature control*. Temperature control is by water-warmed jacketing of the organ support and the other main components of the apparatus at a temperature of 37°C (Khayambashi and Lyman).

*Oxygenator.* Oxygenation is achieved by means of either a simple bubbler (Khayambashi and Lyman) or an elementary falling film oxygenator (Grodsky *et al.*). The latter consists of a sloping 18-in glass tube down which medium flows as a stream, with gas (95% oxygen and 5% $CO_2$, with a bicarbonate buffer system) flowing in the opposite direction. Neither method is particularly efficient (see Chapter 1) and could well be replaced by the multibulb oxygenator or rotating disc apparatus as discussed in Chapter 1.

*Perfusion medium*

Some of the characteristics of perfusion medium required to support insulin synthesis, or release from the perfused pancreas have been analysed by Grodsky *et al.*

*Medium* I (reported by Grodsky *et al.* (1963)). Haemolysis creates a problem in the estimate of insulin production by the perfused pancreas preparation, since even without perfusion insulin is destroyed at a rapid rate by haemolysed media (Grodsky *et al.* 1963). Since this information is relevant to all studies in which insulin production is used directly or indirectly as a test of function, it is summarized here. (Note that there is now some doubt concerning the destruction of insulin by haemolysed red cells and the elaborate precautions outlined below may be unnecessary. Many workers prefer to use a 'once-through' perfusion system so that newly released insulin is in contact with the medium for only a very few seconds before being collected for storage.) Each medium tested by Grodsky *et al.* consists of a 1:1 mixture of fresh heparinized rat blood, prepared in the standard way, with the component named, made up in 0·9% saline; The percentage of insulin destroyed in the time stated is shown in brackets:

Saline only (92%, 3 h); 3% albumin (65%, 4 h); 6% Dextran (92%, 4 h); White's solution (96%, 3 h); 3·5% polyvinylpyrrolidine (18%, 4 h): 1% gelatin (35%, 4 h); 2% gelatin (11%, 4 h).

Obviously the authors chose the last named for this series of experiments. However, without comment, later papers from the same group use the media described below.

*Medium* II. 4% Dextran (Cutter Laboratories) in Krebs–Henseleit medium, prepared as Umbreit (see Media) and gassed with 95% $O_2$, 5% $CO_2$. This is the standard medium mentioned in later publications from Grodsky's laboratory (for example, in that of Curry, Bennett, and

Grodsky (1968)). No information about insulin breakdown is given however. Media of this basic design have been quite satisfactory in other perfusion systems. Alternatively, 4% human albumin is used in place of Dextran with no apparent difference in insulin production, medium flow, rate or duration of perfusion. In both of the preceding media, the pH is reported to be 7·24 to 7·30. This can only be so if bicarbonate has been lost during the preparation of the medium, something that is known to occur with bovine serum albumin (see Hems *et al.* (1966)). To ensure reproducible and constant pH and bicarbonate concentration the precautions suggested in Chapter 1 (Media) should be followed. Note that the figure of 29 mM bicarbonate, given by Curry *et al.* (1968) must be in error; it does not correspond to the figures given by Umbreit (see Chapter 1 for reference), nor is the pH achieved that which would be expected from the Henderson–Hasselbach equation.

*Medium* III. Khayambashi and Lyman used a medium containing fresh rat plasma (for preparation see Chapter 1, Media), diluted 1:1 with Ringer's saline (see Media) or with 0·9% saline. No details of the gas mixture or of the final pH are given but it is to be expected that if 5% $CO_2$ were used a final pH of 7·15 ($HCO_3 = 12·5$ mM) if 0·9% NaCl, and 7·3 ($HCO_3 = 18$ mM) would result if Ringer's saline were used to make up the plasma. Addition of a calculated amount of bicarbonate in each case could provide a pH of 7·4.

Comparison of Media II and III is only possible to a limited extent, since the two groups of workers have investigated different aspects of pancreatic function. Flow rate (10 ml/min at mean pressure of 95 cm of $H_2O$) and the duration of the experiments (1–2 h) is the same. No figure for oxygen consumption is given by Khayambashi and Lyman (1968), beyond stating its constancy. However, since the groups are connected, it may be assumed that any gross discrepancy would have been documented.

*Medium* IV. Anderson and Long (1947) perfused the pancreas for 3–4 h with heparinized whole rat blood. Apart from having lower flow rates than those obtained with Media I–III above (2–2·7 ml/min at 150 cm $H_2O$ mean pressure), the preparation appears to have behaved similarly. Haemolysis was given as the reason for cessation of flow. Insulin was released, but these results, using a bio-assay technique, cannot be compared with later ones, obtained by radio-immuno-assay.

*Operative procedure*

Rats 250 g in weight and starved overnight (Grodsky *et al.*) or 300–350 g and starved for 17 h (Khayambashi and Lyman) are used, and anaesthetized with intraperitoneal Nembutal in the usual way. The object of the dissection is to remove in one block the stomach, spleen, pancreas, and duodenum and to attach cannulae to the coeliac axis and to the portal vein. With the animal on its back and with its limbs extended, the usual mid-line incision through skin from pubis to manubrium sterni and through the linea alba is made. Lateral incisions are made after clamping the descending superior epigastric arteries about midway between the xiphisternum and the pubis. A saline pack is applied to the exposed abdominal contents to prevent drying. The superior mesenteric artery is located (see Fig. 6.2), tied, and cut between two ligatures. This allows the entire intestine below the duodenum to be separated and removed from the rat without blood loss, which simplifies the subsequent dissection. The oesophagus is ligated as high within the abdomen as possible and cut above the ligature. This frees the upper end of the stomach and of the block of tissues to be removed. Loose ligatures are now placed around the 'gastro-hepatic ligament', i.e. the collection of connective tissue containing the portal vein, bile duct, and hepatic artery, which runs towards the hilum of the liver. An additional loose ligature around the portal vein, just behind the previous one, allows of rapid cannulation at a later stage in the operation.

The aorta is exposed in the upper abdomen, between the crura of the diaphragm, necessitating careful dissection above and to the left of the liver. Loose ligatures (2) are placed well above the origin of the coeliac axis and another well below the coeliac axis, near the recently tied stump of the superior mesenteric artery (Fig. 6.2).

The ligature round the 'gastro-hepatic ligament' is tied and the vessels cut between the ligature and the liver. Now the block of tissue can be released from the abdomen by cutting the aorta between the upper ligatures and below the lower one. With the organs on the tissue tray the aorta is opened longitudinally to expose the origin of the coeliac axis on the anterior wall. The arterial cannula, well filled with warm medium, is now inserted into the exposed orifice and tied firmly in place with the prepared ligature. (This procedure obviously involves a period of anoxia and it may be advisable to reduce this to a minimum by cannulating the coeliac axis while the aorta is still *in situ*. Using bulldog clips to occlude the aorta above and below, it is possible to do this with only a few seconds of anoxia.)

The portal vein is incised in the vicinity of the ligature and flow commences. Once effective cannulation and perfusion have been confirmed, the flow is arrested for the short time it requires to cannulate the portal vein and flow begins again immediately. In the original

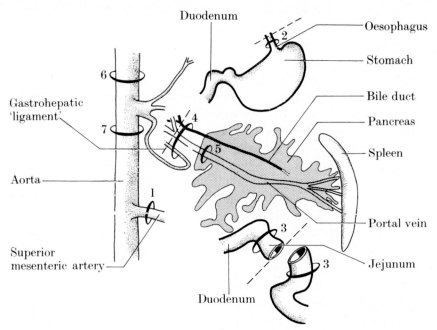

Fig. 6.2.   Operative procedure for pancreatic perfusion technique of Grodsky *et al.* (1963). Ligatures are placed in numerical order. Three incisions are required (shown as broken lines) to free the block of tissues for perfusion. Note that, for clarity, the duodenum is not shown.

account, Grodsky gives the following figures: total anoxic time during preparation less than 5 min; flow rate 7–9 ml/min; inflow pressure (generated by the pump) 100 mmHg approximately.

*Criteria of function of the gastro-duodenal pancreatic perfusion preparation*

Several parameters of physiological and biochemical function are reported by Grodsky *et al.* (1963).

(1) *Histology at the end of perfusion.* Granules are preserved to light microscopy after 2 h perfusion, and diminished in number roughly in proportion to the measured release of insulin into the medium.

(2) *Oxygen consumption.* No figures are available for the *in vivo* oxygen consumption of the pancreas and neighbouring organs. However, Grodsky *et al.* (1963) report arterio-venous oxygen differences of

between 150 and 280 mmHg, at flow rates of 10 ml/min. The oxygen consumption was stable for 2 h in the absence of added substrates, and the preparation may be assumed to be stable for this length of time. The rate of oxygen uptake by the Grodsky preparation may be calculated from the data to be approximately 4·5 $\mu$mol/min per g dry weight. This is low when compared with that of the perfused liver (7·5 $\mu$mol/min per g dry weight, see Chapter 3) or perfused kidney (23 $\mu$mol/min per g dry weight), but in the absence of *in vivo* figures no further comment can be made.

At low flow rates an approximately linear increase in oxygen consumption with increasing flow rate up to 10 ml/min is described. Grodsky attributes this to inadequate oxygenation at the lower flow rates. From the figures given, and assuming the weight of the pancreas to be 1·3 g wet weight (see Doerr and Becker), i.e. 0·3 g dry weight, a flow rate of 1·2 ml/min should be adequate to provide the oxygen requirements of the perfused pancreas of the rat. (This figure is calculated from the oxygen content of the perfusion medium, i.e. 0·024 ml $O_2$ per ml.) By analogy with kidney, another and perhaps more interesting explanation of the linear relationship of oxygen uptake and flow rate may be that an active transport process (e.g. of sodium) which is oxygen-requiring and flow-dependent (cf. Kramer and Deetjen, Chapter 4) functions in the pancreas. Case *et al.* (1968) in the perfused cat pancreas preparation have documented the secretion of $Na^+$ into the exocrine ducts, although the energetics of this process are obscure. The contribution of the non-pancreatic tissues of this preparation to oxygen consumption must not be overlooked (see p. 337 for comparison with Sussman technique).

(3) *pH*. Inlet and outflow pH were the same, indicating that the buffer capacity of the medium was adequate. Since only in the first 10 min of perfusion was recirculation used, and the long term experiments were 'once-through', a high buffer capacity is not important. If full oxygenation is ensured $H^+$ production might be expected to be low.

(4) *Ionic composition of the medium*. Calcium ion concentration was varied and was without effect on the oxygen consumption over the range 0·68–15·3 mEq/l. On the other hand, insulin release in response to glucose was affected by this variation. It is inhibited at low calcium concentrations and may therefore be a more sensitive parameter of pancreatic function than is oxygen consumption itself.

(5) *Insulin output*. This function of the pancreas is carefully divided by Grodsky and other workers in this field into insulin release, as

induced by glucose or by tolbutamide, and insulin synthesis, which has been demonstrated to occur in response to glucose in the perfusion medium but not to tolbutamide. Under standard conditions insulin production is linear and the response to glucose stimulus can be repeated at various times during the course of a simple perfusion.

(6) *Inhibitors of protein synthesis.* As in other organ perfusion systems (e.g. liver), puromycin added to the perfusion medium inhibits the synthesis of new protein  In this case, the synthesis of insulin determined by the incorporation of $^{14}C$ valine, is inhibited.

(7) *Synthesis and/or release of enzymes.* The formation of amylase (and therefore presumably also lipase), has been shown to be linear for 40 min in this preparation (Khayambashi and Lyman) but then falls, despite continuing linearity of the flow rate. This may be a reflection of the discharge of existing enzyme, or of the inability of the isolated perfused organ to synthesize new enzymes. A response of the perfused pancreas to the composition of the medium was demonstrated in this work. Thus, serum prepared from rats fed on soya beans and trypsin-inhibitor, stimulated amylase secretion significantly in the isolated perfused pancreas of a normally fed rat. The total amylase produced was considerably above that formed in the similar pancreas, perfused with serum from normally fed animals. This lends further weight to the suggestion that failure of amylase production is not an indicator of failure of the perfusion system but rather is the stimulus to its formation and release absent in the isolated perfused organ.

(8) *Tissue sampling, before or during perfusion.* The sampling of tissue to demonstrate effects that may be reflected in the medium is clearly of import in assessment of the perfused organ. It has not yet been attempted. Numerous quick-sampling methods *in vivo* and in cultured islets have been published (Matschinsky, Rutherford, and Ellerman (1968)) and it would be of interest to use these techniques in pancreas in the same way as they have been used in the study of liver metabolism. The dendritic form of the rat pancreas lends itself to sampling by tying off a single arm of the pancreas with continuation of perfusion of the remaining tissue.

## Method II. Pancreatic-duodenal preparation. Method of Sussman *et al.*

This method is described in considerable detail by Sussman *et al.* (1966) and a similar method is described by Loubatieres (1967). The technique has not been characterized in such detail as the preceding method. Nevertheless, a comparison of oxygen consumption and insulin production can be made, which enables the two preparations to be assessed.

The pancreas is isolated from all but its adjacent loop of the duodenum and medium is pumped into the segment of aorta from which arises the coeliac axis and the superior mesenteric artery. Originally a special atraumatic roller pump and an apparatus similar to that of Miller *et al.* (1951) was used. More recently an apparatus has become commercially available (Ambec Inc.) which employs the same pump, a rotating film oxygenator, and a large enclosed water-bath for temperature control. In contrast to the preparation of Grodsky *et al.* the medium incorporates rat erythrocytes.

*Apparatus*

Sussman *et al.* describe their original perfusion apparatus as one based on that of Miller (1951). A warmed cabinet is used and this does not differ substantially from the perfusion cabinet described in the appendix. An extra pump is required to deliver medium directly to the pancreas. The first pump supplies a multibulb oxygenator in the usual way.

*Pump.* In their original description Sussman *et al.* used a conventional pulsatile pump providing 2·0–2·5 ml/min flow at pressures of 20–70 mmHg (25–95 cm $H_2O$). Loubatieres describe their pump as 'pulsatile', with mean pressures between 40–55 mmHg (55–70 cm $H_2$)) applied directly to the arterial side of the perfusion. It is additionally described as automatically regulated, presumably being able to adjust flow to the pressure required. In a later paper Sussman and Vaughan (1967) devote considerable attention to the pump. They conclude by using a pump (described in Chapter 1) which provides constant flow at any pressure and reduces haemolysis by means of a large diameter roller mechanism allowing of low speeds. This is the pump that has been incorporated into the commercial (Ambec) apparatus.

*Cannulae.* The arterial cannula is a paediatric plastic cannula (1·5 mm external diameter) with an attachment for the measurement of intraluminal pressure. The venous cannula is polythene, o.d. 2 mm. As duodenal cannula the cannulae described on p. 371 (Kavin *et al.*) are suitable.

The details of the perfusion chamber are shown in Fig. 6.3. Perfusate is pumped from the reservoir (1), through silicone rubber tubing (2), to a glass lung for aeration (3). The perfusate is then pumped (4), through the flow meter (5) and courses through the filter (from blood transfusion packs) (6) to the enclosed perfusion chamber. The perfusate enters the tissues through the arterial cannula and pressure is measured in mmHg using an aneroid manometer. The arterial blood may be sampled through a stopcock (8) and venous blood through stopcock (11).

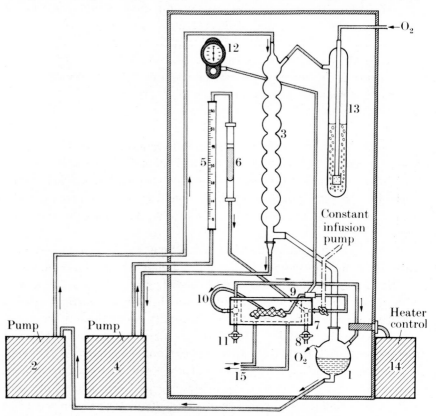

FIG. 6.3.   Perfusion apparatus for pancreas–duodenum preparation of Sussman, Vaughan, and Timmer (1966). The apparatus is based upon the design of Miller *et al.* (see Fig. 3.6 for comparison). The numbers refer to the text (p. 333).

Oxygen is directed through the lung, after passing through a humidifier (13). The perfused tissue is immersed in Krebs–Ringer bicarbonate containing bovine albumin at 37°C and circulated through warmed glass coils (15).

*Perfusion medium*

*Medium* I. Fresh rat blood (heparin is added during collection) is diluted 1:5 with Krebs bicarbonate buffer, prepared as Umbreit (see

Chapter 1, Media). Crystalline bovine albumin 2·5 g % is added and the haemoglobin content of this mixture is approximately 3 g %.

*Medium* II (Loubatieres *et al.* 1967). These authors recommend a red-cell-free medium consisting of 2 g/litre bovine serum albumin in a bicarbonate saline prepared as by Umbreit (see Chapter 1). For unexplained reasons the authors gassed with $93\%$ $O_2$ $7\%$ $CO_2$ but claim a pH of 7·4. (This is incorrect and $5\%$ $CO_2$ in $95\%$ $O_2$ is preferable.)

### Operative procedure

The rat, 350 g weight, is starved overnight and this facilitates the operation by mobilizing omental fat. Anaesthesia is with Nembutal and it is recommended that the rat receives a supply of oxygen from a simple funnel over the snout throughout the long operative procedure. With the rat on its back and the limbs extended, the operator sits to the animal's right.

The fur is shaved or wetted with alcohol. A mid-line incision through skin and linea alba and through the skin over the anterior thorax is required. Lateral incisions are made as described for the liver operation with clamps to occlude the descending epigastric artery either side of the mid-line. The intestine is moved to the animal's left and covered with wet saline packs and the aorta with its superior mesenteric branch is clearly visible. No attempt is made at this stage to locate the pancreas, the handling of which is kept to a minimum. The descending colon is easily identified, as it remains attached to the lower posterior abdominal wall by a short mesentery. This fine layer of connective tissue is cut along the length of the colon in a plane in which no blood vessels are found, thereby mobilizing the lower gut. It will later be removed. By blunt dissection, the pancreas is separated from the overlying colon. The jejunum is ligated and severed just behind its pancreatic attachment; this ligature is left long for orientation afterwards. The distal jejunum has a copious blood supply and its vessels are tied for the last half inch of the distal segment to facilitate its later removal. The superior mesenteric vein (running into the portal vein) is seen over the surface of the pancreas. It is tied just beyond the pancreas, with double ligatures, and cut between them. The pancreas has attachments to the descending colon which are ligated and cut. At this stage the superior mesenteric artery and the coeliac axis are identified and are preserved throughout the following dissection. The attachment of pancreas to the spleen is dealt with as follows. The fine membrane covering the spleen and closely adherent to it is carefully picked off with sharp forceps. The main splenic vessels, entering towards the upper pole are tied twice and cut between the ligatures. Four separate vessels course through the mesentery to cross between the pancreatic tail and the spleen and if tied together these pedicles can 'bunch' the vessels in the tail of the pancreas. It is better to tie each of them individually. One marker thread is left long and denotes the tail of the pancreas when it is isolated.

The stomach is pulled downwards (caudally) and the major left gastric artery, which runs into the upper and medial aspect of the stomach near the oesophagus, is tied. One ligature encloses both the vessel and the oesophagus (see Fig. 6.1) while the second and third occlude the vessel and oesophagus separately higher up. Both the vessel and the oesophagus are cut between ligatures. Lifting the stomach to the right, exposes vascular connections to the posterior wall. These are tied (a single distal tie is sufficient) and cut. Finally, the pylorus is tied twice and cut between the ligatures, releasing the stomach which can now be removed.

Heparin is given at this stage (0·1 ml intravenously) either into the femoral vein or inferior vena cava (see Chapter 4 for technique).

The right renal pedicle—artery, vein, and ureter—is cleared with blunt dissection and a ligature passed behind the artery and vein with curved forceps. Double ligatures allow the vessels to be cut and the kidney is removed. By careful dissection, the aorta is exposed at this level, to clear it from the inferior vena cava between the left renal artery and the superior mesenteric artery. A ligature passed behind the aorta at this level can now be tied. Three loose ligatures are passed round the portal vein as it leaves the pancreas for the liver. The uppermost ligature should include all the structures of the portal tract—vein, hepatic artery, and bile duct—while the other two include only the vein.

Cannulation is delayed until the aorta has been cleared along its length, preserving the coeliac and superior mesenteric branches. This entails passing ligatures round and tying particularly the lumbar arteries. The technique applied to the ligature of aortic branches in preparation of the perfused colon (p. 388) could be applied usefully here. The vessel must be freed from the inferior vena cava from the level of the diaphragm to the ligature below the left renal artery. Loose ligatures are positioned around the aorta, just below the diaphragm, avoiding the origin of the coeliac artery.

*Venous cannulation.* A retrograde cannulation of the portal vein may now be performed, tying the uppermost ligature first and making an incision in the anterior wall of the vein. The cannula is filled with heparinized saline before insertion, but in any case flow-back along the cannula begins immediately. When flow is assured, aortic cannulation is performed. The rat is turned round with its head towards the operator, and a mid-line incision through the thorax is made, cutting down the diaphragm to the aorta. From the moment of entering the chest the anoxic phase begins and speed is essential. A small retractor is useful to spread apart the walls of the chest, exposing the now collapsed lungs, and, along the left border of the vertebral column, the thoracic aorta is seen as it lies behind the oesophagus. It is separated from this structure with curved forceps and as soon as it is sufficiently free to be grasped in a pair of forceps a small incision is made in its left lateral wall. The cannula is inserted and advanced until its tip just passes the diaphragm. The abdominal aortic ligature is tightened, holding the cannula in place, and flow begins. The final step is to complete the isolation of pancreatic circulation by tying the ligature which has already been prepared around the lower inferior cava. Now the perfusing pancreas must be transferred to the organ chamber. The inferior vena cava is cut below the last tied ligature, and it should be possible by grasping the two cannulae and the loop of duodenum to lift the pancreas clear of the rat. The organ is orientated in the tray by means of the identifying ligatures placed during the preparation, and successful uniform perfusion is obtained if this step is carefully observed.

Flow should be about 2 ml/min with a perfusion pressure of about 40 mmHg. If the pressure is over 80 mmHg either at the start of the perfusion or during an experiment it is usual to reject the preparation. Sussman *et al.* suggest that 75 min are required for the total operation with some 5 min of anoxia. The present preparation is somewhat quicker but still takes 30 min in the hands of experts. The period of partial anoxia (open chest, circulation continued) is from 3 to 5 min.

*Characteristics of perfusion*

At the pressures chosen by Sussman (27–95 cm $H_2O$) flow rates of 2·0–2·5 ml medium/min are obtained. Both pressure and flow are lower than those used by Grodsky *et al.* in the saline-perfused preparation or by Khayambashi and Lyman in the plasma perfusion of the same preparation. This may be due to the presence of red cells in the present preparation, since Anderson and Long, with blood as the perfusion medium, obtained similarly lower flow rates of 2·0–2·7 ml/ min. As will become apparent, these flow rates are in each case appropriate to the required oxygen supply. (The same cannot be said of the flow rates of 3–6 ml/min used in the much larger cat pancreas preparation of Case, Harper, and Scratcherd 1968.) In the present preparation the figures suggest that flow is linear for at least 2 h, and may even increase in the second hour. As already mentioned, a rise in the perfusion pressure required to maintain this flow rate is taken as a sign of failure. The perfusion pressure remains steady, around 45 mmHg (60 cm $H_2O$) for the 2-h period. Arterio-venous oxygen difference is constant at about 120 mmHg (21 per cent = 0·005 ml $O_2$) for 2 h, and since in the experiments reported flow rate is some 20 per cent higher in the second hour than in the first, the oxygen consumption of the pancreas has increased.

Sussman suggests the following criteria of viability in order of usefulness.

(a) The response of the perfused organ to a glucose load; the pancreas which is satisfactory discharges insulin into the medium, although no quantitative data has been given for the normal limits.

(b) Arterio-venous oxygen differences, already discussed, are a more generally applicable criterion of function and are quantitative. A rough estimate of oxygen consumption can be made from the figures obtained with oxygen electrodes in the Sussman experiments. The electrode detects oxygen in solution and the haemoglobin bound oxygen must be ignored (see p. 120). Assuming an oxygen solubility of 0·024 ml/ml medium and an arterio-venous difference of 124 mmHg, a flow rate of 2·0–2·5 ml/min of medium suggests an oxygen consumption per gramme dry weight of pancreas of 1·24 $\mu$mol/min. This is low when compared with the figure of 4·5 $\mu$mol/min per g dry weight obtained by Grodsky *et al.* at a flow rate of 10 ml/min but, remembering Grodsky's observation of the flow dependence of oxygen consumption by the pancreas (p. 331), the Sussman figure for 'isolated' pancreas falls on the curve constructed from Grodsky's data. This observation is particularly relevant when the question of extra-pancreatic tissue being perfused is considered. From appropriate perfusion data in the rat, the oxygen consumption of small intestine and of spleen can be calculated, (see Chapter 7) and is sufficiently close to that of the pancreas for the two very different preparations to have a uniform oxygen uptake when expressed on a weight basis.

(c) Increasing vascular resistance is a sign of failure and an invariable event in the Sussman preparation after about 2–3 h. Grodsky claims no longer for the alternative preparation, but Anderson and Long perfused for from 3 to 4 h with the same preparation, using rat blood medium. In the cat Harper perfuses with a simple saline medium for 6 h without change in perfusion pressure so that the explanation for failure of the rat preparation after only 2 h is not immediately available. Certainly, interstitial oedema is a recurring feature of the available methods of pancreatic perfusion in the rat and could account for a rise in pressure.

(d) As the least reliable sign, Sussman offers the physical appearance of the perfused organ.

Apart from measurements of oxygen uptake by the gland, metabolic parameters of function are entirely lacking. Where different sugars are used to stimulate insulin output, Sussman points out that there are some differences between the perfused pancreas and the pancreas slice, and further there is in the case of mannose a difference, albeit irregular, between the two perfusions. More discerning metabolic tests are required before this preparation is established as a reliable method of biochemical experimentation in the pancreas.

### THYROID GLAND

Biochemical studies concerned with the thyroid gland have been centred upon the effects of the hormone on its target organs (for example, Tata 1967). In addition, there is an extensive literature about the gland itself (see Wolff (1964) for review), particularly on the uptake and release of iodine and iodinated compounds by the gland. Standard biochemical techniques, including slices (Lerner and Chaikoff 1945, Freinkel and Ingbar 1955), homogenates (Rall, Robbins, and Lewallen 1964), and subcellular fractions have been studied, but perfusion has had no general application.

Two preparations exist that may have broader applications to biochemistry. The isolated perfused thyroid gland of the rabbit (Williams 1966) has been used to study the electrophysiological properties of the follicular membrane, but its author found it difficult to apply the same technique to the rat or guinea-pig. In the rat, the technique of 'single-pass *in situ* perfusion of the pre-labelled thyroid gland' (Inoue, Grimm, and Greer 1967) is the only currently practicable preparation. These two preparations will be described.

## Anatomy

The thyroid gland is a bilobed organ which lies in the neck, the two lobes joined by a narrow isthmus that crosses in front of the trachea.

The blood supply of the gland is therefore derived separately from both sides of the body, and in the dog this has been used to provide two 'identical' perfusions from the same organ (Folkman, Cole, and Zimmerman 1966). There is some cross circulation between the two lobes (see below) and ideally the gland would be treated as a single organ. The weight of the gland is expressed as a proportion of body weight by the formulation

$$100. \sqrt[3]{\left(\frac{\text{thyroid weight}}{\text{body length}}\right)}.$$

*Arterial supply*

(1) *Rabbit.* The main arterial supply is via the inferior thyroid artery, the only branch of the common carotid artery. When this segment of the carotid is cannulated and perfused with Evans Blue, the whole of the ipsi-lateral thyroid lobe, the inner surface of the larynx, some of the trachea, and a short segment of the oesophagus are all perfused. In addition, the contra-lateral lobe becomes faintly coloured.

(2) *Rat.* Two main arteries supply the thyroid in the rat; the inferior thyroid artery (see above) arises as a branch of the profunda cervicalis which is in turn derived, via the costocervical trunk, from the subclavian artery (Green 1935). The superior thyroid artery arises from the external carotid artery at the level of the larynx, and the two thyroid vessels have extensive anastomoses. A simple perfusion, such as that developed for the rabbit via the inferior thyroid artery is therefore not available in the rat, and the method to be described cannulates the aorta, leaving patent only the innominate artery.

*Venous drainage*

(1) *Rabbit.* As rat.

(2) *Rat.* No single vein drains the gland. The superior and middle thyroid veins empty into the internal jugular vein, while the inferior thyroid vein, which is described as 'relatively small' (Green 1935), drains into the superior vena cava, either directly or via the profunda cervicis vein. Although no information is available concerning the relative importance of these three vessels in draining blood from the thyroid, the method of 'perfusion' described by Inoue, Grimm, and Greer (1967), collects medium that flows from the cut end of the inferior thyroid vein. (Note that the parathyroid glands derive a blood supply from and drain into the same vessels as the thyroid gland; it is difficult

in a small animal such as the rat or rabbit to reliably exclude them from a perfusion circuit.)

*Nerve supply*

A copious network of sympathetic nerves from the superior cervical ganglion supplies the thyroid gland and remains intact in the *in situ* preparation.

*Lymphatic drainage*

A network of lymph vessels surrounds the follicles of the thyroid gland and emerges as lymph trunks which follow the veins (Cohrs *et al.* 1958).

*Microscopic anatomy*

The gland consists of numerous follicles, each with a lining epithelium, the shape of whose cells is determined by the state of functional activity of the gland as a whole. The follicles store thyroglobulin 'colloid' which is synthesized and secreted by the epithelial lining cells, and a rich capillary network surrounds the follicles, supplying and removing iodine and hormones. The follicular cells are of uniform appearance, but in addition there are 'basket' cells (C cells, medullary cells), long known to occur outside the basement membrane of the follicles (see Bloom and Fawcett 1968), but more recently considered to be of greater importance since they have been recognized as the source of thyrocalcitonin, a hormone controlling calcium metabolism (Foster *et al.* 1964).

*Physiology* (see Tata 1967 for review)

The thyroid gland has the almost unique property of concentrating iodide against a gradient. In addition, the gland synthesizes the hormones $T_3$ and $T_4$ by means of now well-established enzymic processes (Trotter 1962). Hormone not immediately discharged from the gland is stored as 'colloid' (thyroglobulin) in the follicles of the gland. Thyroid-stimulating hormone (TSH) regulates the synthesis and release of hormone from the gland *in vivo*.

**Experimental preparations**

Apart from *in vivo* experiments, using inhibitors [131]I, [125]I, thyroidectomy and other adjuncts are supplemented for biochemical experiment by a number of standard *in vitro* techniques.

(1) *Slices.* Lerner and Chaikoff (1945), Freinkel and Ingbar (1955), and many others have made use of the standard tissue-slice technique to study oxygen consumption, hormone synthesis, and iodide concentration.

(2) *Incubated gland.* Single lobes of rat thyroid gland continue to concentrate $^{131}$I from the medium when incubated under standard conditions. However, Shimoda and Greer (1964) demonstrate that only the outer follicles continue to concentrate under these conditions. Histology confirms that after prolonged incubation (3 h) there is degenerative change in the central follicles of the gland, and suggest that thyroglobulin release, demonstrable in such preparations, is probably only due to such degeneration of central follicles (Grimm and Greer 1966). Such observations suggest that a perfused preparation may offer considerable improvement.

(3) *Perfused thyroid gland.*

## Methods of perfusion

I. Isolated perfused thyroid gland (rabbit): Williams (1966, 1969); Folkman *et al.* (1966).

II. *In situ* perfused thyroid gland (rat): Inoue *et al.* (1967).

## Method I. Isolated perfused thyroid gland of rabbit

One lobe of the thyroid gland of a rabbit, together with its attachment to the trachea, is perfused through the segment of internal carotid artery from which the inferior thyroid vessel arises. Tyrode buffer, with or without haemoglobin in solution, is used as perfusion medium in a non-recirculating system, and the viability of the preparation confirmed by the existence of the expected membrane potential, and by its ability to incorporate $^{32}$P into phospholipid (Williams 1966).

### Apparatus

The apparatus used is not described, nor are details of perfusion pressure, flow rate, or duration of the experiments given. However, the recirculation apparatus suggested by Folkman *et al.* for perfusion of thyroid gland in the dog or rabbit, and referred to in the sections on thymus gland (Chapter 8) and kidney (Chapter 4) could be used. A peristaltic pump delivers medium at a pressure of 120–180 mmHg to the arterial cannula. A membrane oxygenator (see Chapter 1, Oxygenators) is used, and flow rates of 2 ml/min per g achieved with Medium III. The total volume of medium in this apparatus is 25 ml.

*Cannula.* PE 160 polythene tubing (o.d. 1·5 mm) is recommended by Williams (1966) or, alternatively, a glass cannula, o.d. 1 mm at the tip can be inserted via the carotid artery into the inferior thyroid vessel itself.

## Perfusion medium

*Medium* I. Tyrode solution, including 5·6 mM glucose, and gassed with 95% $O_2$:5% $CO_2$ gives perfusions that are satisfactory for 45 min (Williams 1966).

*Medium* II. Tyrode solution, to which haemoglobin, prepared according to the method of D'Silva and Neil (1954), is added to give a final concentration such as to contain 5 vols. % $O_2$ when equilibrated with air (=3·5 g Hb), allows perfusion to be continued for 2–4 h (Williams 1969).

*Medium* III (Folkman *et al.* 1966). Human haemoglobin (prepared as by D'Silva and Neil 1954), 1 g %, in Eagle's medium (see Chapter 1, Media), plus glucose (400 mg%), penicillin, neomycin, and mycostatin, gives a medium with the following final ion concentrations: $Na^+$ 150 mE/l; $K^+$ 7·1 Cl' 153; pH 7·4–7·5, and allows the very protracted perfusions of thyroid gland. Perfusions for 2–3 weeks may apparently be conducted with survival of the tissue, as judged by visual criteria.

*Medium* IV (Gimbroni *et al.* 1969). Successful perfusion of dog thyroid gland in the apparatus of Folkman *et al.* (q.v.) have been conducted with the following media (see also Chapter 1, Media).

(a) Undiluted, heparinized (2·5 mg/100 ml) autologous plasma prepared 2 h previously by plasmapheresis is rendered either rich or poor in platelets, by centrifugation. In either case, it is free of cells. With (a) platelet-poor plasma, successful perfusion is obtained in which histological damage to the follicles is seen after 3 h.

(b) Platelet-rich plasma is prepared in exactly the same way but at lower speeds of centrifugation (not given in the original publication) so that a high platelet count remains. This medium allows 3 h of perfusion without oedema formation and with histological normality of the follicles (Gimbroni *et al.* 1969).

The trachea, with the thyroid gland attached, is excised and cut in half longitudinally, and the perfusing half-thyroid gland pinned securely to a support within the organ chamber. (The chamber is described as being illuminated from below, which greatly assists the manipulations required for microelectrode measurements.) It is immersed in Tyrode solution, which contains 5·6 mM glucose (for composition see Chapter 1, Media), and perfused in a 'once-through' system. A stopcock on the inflow line allows two media to be interchanged rapidly in the experiments of Williams (1966).

### Characteristics of perfusion

This study involved the electrophysiology of the thyroid epithelial cell and established that the organ perfused with Medium II maintains a negative potential of 43·4 mV (*in vivo* figure = 50 mV) for up to 4 h. Modifying the medium by increasing $K^+$ concentration had the expected effect of reducing the membrane potential. This is the only test of viability recorded in such a preparation, apart from the histological data already discussed and obtained by Folkman *et al.*, and by Gimbroni *et al.* It is stated by Williams (1966) who does not record the figures, that viability was also assessed by the ability of the perfused thyroid gland to incorporate $^{32}P$ into phospholipid. Such tests need to be extended before this preparation is fully characterized, but its potential in biochemical experimentation is clear.

### Method II. *In situ* perfusion of rat thyroid gland (method of Inoue *et al.*)

The method used *in situ* is somewhat primitive for biochemical studies. The arch of the aorta, with all but the two common carotid branches tied, is perfused under gravity, and medium is collected with a pipette as it emerges from the cut end of the inferior thyroid vein. Tissue other than the thyroid is perfused, and effluent medium is in contact with the 'gutter' of neck muscles as it emerges from the gland. However, the authors consider that cannulation of the inferior thyroid vein (Matsuda and Greer 1965) impeded flow through the gland. The effectiveness of the method lies in the prior labelling of the gland with radio-iodine, so that specific studies on the thyroid gland are possible.

### Apparatus

A flask of oxygenated medium provides a hydrostatic pressure of 50–90 cm $H_2O$. No pump or oxygenator is used, but a filter of Dacron or glass-wool (see Filters) is used, and rough temperature control is provided by a light bulb.

*Cannula.* PE 205 polythene tubing (o.d. $= 2 \cdot 0$ mm) is used for the aortic cannulation.

### Perfusion medium

Citrated ($0 \cdot 5$–$1 \cdot 0\%$ citrate) bovine blood, which has been stored at $4 °C$ for 2–3 days is oxygenated by rotating in a flask in air immediately before use.

### Thyroid labelling technique

Adult rats of 300 g wt. are maintained for 1 week on a low iodine diet (approx. 30 $\mu$g $^{127}$I/kg). Forty to fifty $\mu$Ci carrier-free $^{125}$I is injected intra-peritoneally, 24 h before perfusion, and 200 $\mu$Ci $^{131}$I given intra-venously exactly 10 min before exsanguinating the animal.

### Operative procedure

The rat is anaesthetized with ether, the abdomen opened, and the animal killed by exsanguination from the abdominal aorta. The chest is opened by vertical incisions through the ribs on either side of the sternum, removing the sternum (cf. lung perfusion, Chapter 9). The dissection may be accomplished under direct vision, although its authors recommend $\times 10$ magnification with a dissecting microscope. Ligatures are placed about the descending aorta (1), the left subclavian artery (2), and the right subclavian artery (3) and all are tied. In addition, the pulmonary vessels are ligated on each side at the hilum of the lung (5). A loose ligature (6) is placed around the ascending aorta. The left ventricle is incised and the cannula, filled to the tip with medium, is inserted through this incision into the aorta. When the tip lies just below the origin of the innominate artery, ligature (6) is tied, holding it in place. Perfusion begins at once.

*Venous cannulation.* The outflow may be cannulated by inserting a cannula through the inferior vena cava into the right atrium, to drain both superior venae cavae. Alternatively, the inferior thyroid vein may be cannulated directly with the reservations already discussed. In most experiments conducted with this preparation the venous outflow from the thyroid is allowed to escape by cutting the inferior thyroid vein.

### Characteristics of perfusion

This is a perfusion of the common carotid distribution and only the previous labelling of the thyroid gland with $^{125}$I, $^{131}$I makes it at all specific to this gland. Nevertheless, histological changes, analysis of I-labelled thyroid products in the effluent, and the effects of TSH added to the perfusion medium, on the discharge of hormones, have all been established with this preparation. Perfusion continues for 2–3 h, with steady flow-rate (3 ml/min) and more or less linear secretion of $T_3$ and $T_4$ during this time.

On the basis of histological appearances after perfusion, this preparation is a considerable improvement on the simple incubation of the gland.

## ADRENAL GLAND

The standard preparation of isolated perfused bovine adrenal gland is that devised by Hechter *et al.* (1953). In the first publications from that laboratory, three alternative methods were discussed: an arterial perfusion; a venous perfusion (reverse flow); and a venous perfusion with laceration of the capsule and adrenal cortex to increase flow rate. The last-mentioned has found general applications (Schneider, Smith, and Winkler 1967; Banks 1965; Kirschner, Sage, and Smith 1967). Other methods of perfusion exist which may have applications to biochemistry.

### Methods of perfusion

    I. Bovine adrenal gland. Method of Hechter *et al.* (1953).
    II. Cat adrenal gland. Method of Douglas and Rubin (1961).
    III. Rat adrenal gland(s). Method of Cession-Fossion (1964, 1967).

### Method I. Bovine adrenal gland. Method of Hechter *et al.*

(A) *Arterial perfusion of bovine adrenal gland via the relevant segment of aorta with an outflow cannula in a main adrenal vein*

A number of defects in this preparation have made it of little general relevance, thus:

    (a) Arterial perfusion required a considerable degree of surgical skill and could not be standardized due to anatomical variation.
    (b) Many experiments failed owing to leaks in the aorta.
    (c) Infusion of dye after perfusion showed patchy distribution, with up to 5 per cent of the gland not supplied with medium.
    (d) Flow rate was variable and, finally, the preparation included extra-adrenal tissues which can only be removed with damage to the vessels of the gland.

## (B) *Venous perfusion*

Reverse perfusion was considered feasible on anatomical grounds. A special precaution is necessary, however, if adequate flow rates are to be obtained. Numerous lacerations are made through the capsule and the adrenal cortex to a depth of 3 mm or so. When this is done, flow rate increases from 0·11 ml/min per g to 2·42 ml/min per g (Hechter *et al.* 1953). This compares favourably with the rates of flow obtained in the best arterial preparations (q.v.), and by the criterion of 'steroid conversion' (11-hydroxylation), venous-perfusion appears to be more satisfactory.

## *Perfusion medium*

*Medium* I. Hechter *et al.* (1953) used citrated bovine blood, which was filtered and then passed twice through the isolated perfused liver of a rat. This was required 'to remove vaso-active substances'.

*Medium* II. More recent investigators have substituted Tyrode solution (11·6 mM $NaHCO_3$) (Schneider *et al.*) or Locke's solution (6·3 mM $NaHCO_3$) gassed with 95% $O_2$:5% $CO_2$ as perfusion medium (see Chapter 1 for criticism). Using this medium, flow rates of 10–15 ml/min are obtained or, approximately, 1 ml/min per g gland. Note that the information on oxygen requirement of the adrenal gland provided by Banks (1965) gives the adrenal medullary oxygen consumption as 0·04 ml $O_2$/min per g wet weight (1·8 $\mu$mol $O_2$/min per g). This is for strips of adrenal medulla incubated in Tyrode buffer, at 37°. It is twice the oxygen content of the medium flowing through the gland in unit time in the perfusion methods described above, using Medium II. Hechter's original preparation, on the other hand, with Medium I, probably provided adequate oxygen to comply with the requirement estimated from Banks's figures: the oxygen content of the medium traversing the gland in unit time with Medium I is approximately 0·4 ml/min (=18 $\mu$mol $O_2$). Incidentally, Banks records that noradrenaline release and its response to carbachol is not dependent upon oxygen, which makes it an insensitive parameter of function.

## *Apparatus and operative procedure*

A water-jacketed, gravity-fed perfusion apparatus is illustrated by Hechter *et al.* (1953) and details of the simple operative procedure required for reverse-perfusion are given.

*Characteristics of perfusion:* (*see above*)

For biochemical experiments, it is important to establish conditions of uniform fully-oxygenated perfusion. These appear not to obtain with the existing, haemoglobin-free perfusion media for which it may be necessary to substitute a medium resembling that of Schimassek (see p. 20).

## Method II. Arterial perfusion of the adrenal gland of the cat

Douglas and Rubin (1961) describe in great detail, the preparation of an *in situ* arterial perfusion of the cat adrenal gland. The notable features are a constant perfusion pressure, provided by a cylinder of compressed air, and the use of Locke's solution without red cells as the perfusion medium. These conditions allow perfusion flow rates of 4 ml/min to be maintained for 1–2 h. Comparison is made with the physiological blood-flow through the cat adrenal, said to be 1–3 ml/min, but of course the oxygen supplied dissolved in Locke's solution is only 10 per cent of that in whole blood.

## Method III. Arterial perfusion of adrenal gland of the rat

Cession-Fossion (1964, 1967) describes an *in situ* preparation in which both adrenal glands of the rat are perfused through the aorta, with effluent collected from the abdominal inferior vena cava. The complexity of the adrenal circulation in the rat is illustrated and perfusion is not restricted to adrenal tissue. Unfortunately, little technical information is given; the perfusion medium is Tyrode solution, gassed with $95\% \ O_2 : 5\% \ CO_2$, but the flow rate is not recorded.

Noradrenaline is released in to the effluent and responses to various stimuli (for example, changes in the ionic composition of the perfusion medium) occur. Perfusions of up to 30 min duration are reported, but most experiments appear to have been shorter. Thus, although this preparation may serve as a guide to future techniques, it must at present be considered inadequate for biochemical requirements on two grounds:

(i) perfusion is not restricted to adrenal tissue,

(ii) the weight of tissue perfused is, in the rat, so little, that even micro-techniques may not be adequate to following its metabolism.

## OVARY

Biochemical studies have been conducted in a perfused ovary preparation from the cow (Romanoff and Pincus 1962). This preparation will be briefly described, since there is no published account of a preparation in a smaller animal. The anatomical problems of such a preparation in the rat are apparent.

**Perfusion of isolated bovine ovary** (Romanoff and Pincus (1962))

The ovary, removed with the attached broad ligament, is perfused through the ovarian artery, with a medium of citrated bovine blood.

*Apparatus*

The apparatus, consisting of a simple revolving oxygenator-reservoir, a glass organ-chamber, and a pump providing pulsatile pressure (48–96 'beats'/min) is contained in an incubator at 40°C. The pump is described as having a special and characteristic slow build up of systolic pulse which is achieved by having a piston stroke: piston diameter greater than 4:1. A more rapid build up of pressure is said to result in increased resistance. The details of the pump are not given in this publication.

*Perfusion medium*

*Medium* I. Chilled citrated glucose saline (600 ml ACD/gal of saline) is perfused first, until the ovary is cold and free of blood. The perfusion medium proper (II) is then introduced.

*Medium* II. Citrated whole bovine blood containing (per gal) $10^3$ units penicillin, I g streptomycin, and 50 mg Terramycin, is filtered before use, through cheese-cloth. The medium is gassed for 10 min with 100% oxygen before use.

*Operative procedure*

The authors describe the operation as follows. The bovine ovary is obtained from the slaughter-house and should include sufficient of the broad ligament to enable the ovarian artery to be cannulated before it branches. After cannulation, perfusion is begun, first with Medium I to wash out blood, and then with Medium II. When the flow is established (details of pressure and flow rate are not given in this publication), the organ is held up to the light to enable branches of the ovarian artery that are not destined for the ovary to be tied. The veins should not be ligated. The excess of broad-ligament is removed and the preparation is mounted in the organ chamber.

*Characteristics of perfusion*

Little information is given, but the metabolic activity of the ovary is confirmed by the incorporation of $^{14}C$ acetate label into some secretory product of the gland. Counts continue to appear at a rate that is linear over the course of 3–4 h perfusion, and the rate increases in response to horse pituitary gonadotrophin added to the medium.

## TESTIS

Biochemical studies in this tissue have so far been conducted in tissue slices (Elliott, Greig, and Benoy (1937)) and there is one report of successful perfusion, in the rabbit (Vandemark and Ewing 1963). Perfusing with defibrinated homologous blood, the organ was maintained for some 7 h. Studies of glucose uptake and of testosterone formation and release have been undertaken, with some comparison with tissue-slice experiments.

*Anatomy*

The testis is supplied by a single artery, the internal spermatic artery, which arises high on the aorta (at the level of the renal vessels) and has a long course through the abdomen. It supplies other structures, the most important of which in the present context is (in the rat) the epididymal fat pad. The details of the blood supply are discussed in Chapter 8, p. 400. Venous drainage of the testis is by means of multiple veins (pampiniform plexus) and does not lend itself to direct cannulation. A single spermatic vein is formed on each side and drains into the inferior vena cava.

**Perfusion of rabbit testis** (Vandemark and Ewing (1963))

In this preparation, the isolated testis is perfused under direct pulsatile pressure through a cannula in the internal spermatic artery itself. The venous side is not cannulated but the medium is collected and recirculated through a simple bubble oxygenator. Early experiments (1963) were of this type, with rabbit blood as medium; later (1966) the system was modified to a 'once-through' circuit.

*Apparatus*

A simple rotary pump (Model PA 23, New Brunswick) is used to provide a pulsatile flow, through a glass-wool filter, to the spermatic artery. The testis is contained in a glass chamber, and venous effluent drips from the cut veins into a simple oxygenator. This is basically a falling film device. From the oxygenator-reservoir, medium passes to a dialysing compartment (later abandoned) whence it is collected by the pump.

The temperature of the medium is 36·5°C, but there are no special provisions for thermoregulation in the apparatus described.

*Cannula.* A length of PP 10 polythene tubing is used for cannulation of the spermatic artery.

## Perfusion medium

Rabbit blood (from four animals) is defibrinated with glass beads, filtered twice, and stored overnight at 4°C before use. In addition, penicillin (100 000 units) and dihydrostreptomycin (0·165 g per 150 ml blood) are added, and the glucose content is made up to 1500 $\mu$g/ml (8·3 mM). The medium is gassed with 95% $O_2$:5% $CO_2$.

## Operative procedure

Operative details are scanty in the publications of Ewing and others. Essentially, a rabbit is anaesthetized, the scrotal skin is shaved, and one testis is removed. Then, under a dissecting microscope, the cut end of the spermatic artery is located and cannulated with a length of PP10 polythene tubing, previously filled with Locke's solution or in later experiments, with 0·25 M sucrose.

## Characteristics of perfusion

Pressure in the early reports (1963) was 50–55 mmHg, with flows of 4·5–5·9 ml/g per h. In later experiments, the perfusion pressure was routinely 100 mmHg, which was apparently without harm to the preparation, and resulted in identical flow rates (Ewing and Eik-Nes 1966). The tests of 'survival' include glucose uptake, which is to some extent dependent upon medium concentration (Vandemark and Ewing 1963), but remained constant for 10 h if the supply of substrate was sustained.

Flow was constant for at least the same time. Testosterone production was constant at a low rate, and $^{14}$C acetate and $^{14}$C cholesterol both transferred label to this product. Gonadotrophin added to the perfusion medium resulted in the expected increase in testosterone production, and this higher rate of formation was also linear with time. Histology after 7 h of apparently successful perfusion showed, however, that many tubules were degenerate and therefore presumably not well maintained with oxygen.

A serious fault in this preparation is revealed if the oxygen consumption of slices of testis is compared with the oxygen delivered to the perfused testis in unit time. While the slice consumes approximately 5 $\mu$l $O_2$/mg dry wt. per h ($\simeq 1\cdot 0$ $\mu$mol $O_2$/min per g wet; assuming a wet/dry ratio of 4), the medium provides only $0\cdot 5$ $\mu$mol/min per g at the rate of flow indicated. It must be concluded that these perfusions are conducted under anoxic conditions, and the histological picture may thereby be explained. A minimum flow rate of 1 ml/min per g is required and an alternative preparation is required. A suggested modification of the operative preparation of the perfused epididymal fat pad (see Chapter 8) in the rat allows the testis to be perfused with an adequate flow rate, but no formal studies have been undertaken.

## PROSTATE GLAND

A perfusion of the ventral prostate gland of the rat with homologous heparinized blood (Farnsworth, Brown, and Lawrence 1963) is based upon a similar preparation for the dog (Hudson, Butler, Brendler, and Scott 1950) and has been used to investigate the fate of testosterone added to the medium. The operation, which depends on cannulation of the abdominal aorta and the inferior vena cava, tying off all but the inferior vesical branch of the common iliac vessels and the superior vesical branch of the common iliac artery, takes up to 2 h. However, the anoxic period is 1–2 min only, and satisfactory perfusion is said to continue for 2–$2\frac{1}{2}$ h.

### Apparatus, perfusion medium, and operative procedure

The simplest sort of once-through perfusion apparatus, with direct pumping of medium to the arterial cannula, is used. For this, and for the operative details of the isolation of the blood supply to the prostate gland the reader is referred to the original paper of Farnsworth *et al.* (1963). The gland is left *in situ*; indeed it is difficult to see how it could be anatomically isolated for the present purpose.

### Characteristics of perfusion

Flow rate of medium is given as $0\cdot 08$ ml/min and is stable for 2–$2\frac{1}{2}$ h as reported. In addition, a dye injected into the inflow delineates the tissue perfused by this route. The whole prostate gland, and no other significant area of tissue, is supplied. At the present time, this preparation is of academic interest; very little is known of the biochemistry of the prostate gland and the basic information might be more conveniently obtained in standard incubation techniques.

## REFERENCES

ANDERSON, E. and LONG, J. A. (1947) The effect of hyperglycaemia on insulin secretion as determined with the isolated rat pancreas in a perfusion apparatus. *Endocrinology* **40**, 92.

BANKS, P. (1965) Effects of stimulation by carbachol on the metabolism of bovine adrenal medulla. *Biochem. J.* **97**, 555.

BLOOM, W. and FAWCETT, D. W. (1968) *A textbook of histology*, 9th edn.

BOLLMAN, J. L., CAIN, J. C., an dGRINDLAY, J. H. (1948) Techniques for the collection of lymph from the liver, small intestine and thoracic duct of the rat. *J. Lab. clin. Med.* **33**, 1349.

BURR, I. M., STAUFFACHER, W., BALANT, L., RENOLD, A. E., and GRODSKY, G. (1969) Dynamic aspects of proinsulin release from perifused rat pancreas. *Lancet* (ii) 882.

CASE, R. M., HARPER, A. A., and SCRATCHERD, T. (1968) Water and electrolyte secretion by the perfused pancreas of the cat. *J. Physiol., Lond.* **196**, 133.

CESSION-FOSSION, A. (1964) Perfusion *in vitro* des glandes surrénales du rat. *Rev. belge Path. Méd. exp.* **30**, 349.

—— (1967) Action des amines sympathicomimétiques à action indirecte sur la médullo-surrénale du rat perfusée *in vitro*. *Archs int. Physiol. Biochim.* **75**, 303.

COHRS, P., JAFFÉ, R., and MEESEN, H. (1958) Pathologie der Laboratoriumstiere, Vol. 1. Springer-Verlag, Berlin.

COORE, H. G. and RANDLE, P. J. (1964) Regulation of insulin secretion studied with pieces of rabbit pancreas incubated *in vitro*. *Biochem. J.* **93**, 66.

CURRY, D. L., BENNETT, L. L., and GRODSKY, G. M. (1968) Dynamics of insulin secretion by the perfused rat pancreas. *Endocrinology* **83**, 572.

DOERR, W. and BECKER, V. (1958) 'Bauchspeicheldrüse' (pancreas). in *Pathologie der Laboratoriumstiere*, Vol. 1, p. 130. Springer-Verlag, Berlin.

DOUGLAS, W. W. and RUBIN, R. P. (1961) The role of calcium in the secretory response of the adrenal medulla and acetylcholine. *J. Physiol., Lond.* **159**, 40.

D'SILVA, J. L. and NEIL, M. W. (1954) The potassium, water and glycogen con tents of the perfused rat liver. *J. Physiol., Lond.* **124**, 515.

ELLIOTT, K. A. C., GREIG, M. E., and BENOY, M. P. (1937) The metabolism of lactic and pyruvic acids in normal and tumour tissues. III. Rat liver, brain and testis. *Biochem. J.* **31**, 1003.

EWING, L. L. and EIK-NES, K. B. (1966) On the formation of testosterone by the perfused rabbit testis. *Can. J. Biochem.* **44**, 1327.

FARNSWORTH, W. E., BROWN, R. J., and LAWRENCE, M. H. (1963) Perfusion of the isolated rat prostate *in situ*. *Endocrinology* **73**, 498.

FOLKMAN, J., COLE, P., and ZIMMERMAN, S. (1966) Tumour behaviour in isolated perfused organs: *in vitro* growth and metastases of biopsy material in rabbit thyroid and canine intestinal segment. *Ann. surg.* **164**, 491.

FOSTER, G. V., BAGHDIANTZ, A., KUMAR, M. A., SLACK, E., SOLIMAN, H. A., and MACINTYRE, I. (1964) Thyroid origin of calcitonin. *Nature, Lond.* **202**, 1303.

FREINKEL, N. and INGBAR, S. H. (1955) Effect of metabolic inhibitors upon iodide transport in sheep thyroid slices. *J. clin. Endocr. Metab.* **15**, 598.

GIMBRONI, M. A., ASTER, R. H., COTRAN, R. S., CORKERY, J., JANDL, J. H., and FOLKMAN, J. (1969) Preservation of vascular integrity in organs perfused *in vitro* with a platelet-rich medium. *Nature, Lond.* **222**, 33.

GREEN, E. C. (1935) *Anatomy of the rat.* Trans. Am. phil. Soc. Hafner, New York (3rd printing, 1963).

GRIMM, Y. and GREER, M.A. (1966) Radioautographic studies on the site of radio-iodine metabolism of rat thyroid lobes incubated *in vitro. Endocrinology* **79**, 469.

GRODSKY, G. M., BATTS, A. A., BENNETT, L. L., VCELLA, C., McWILLIAMS, N. B., and SMITH, D. F. (1963) Effects of carbohydrates on secretion of insulin from isolated rat pancreas. *Am. J. Physiol.* **205**, 638.

HECHTER, O., JACOBSEN, R. P., SCHENKER, V., LEVY, H., JEANLOZ, R. W., MARSHALL, C. W., and PINCUS, G. (1953) Chemical transformation of steroids by adrenal perfusion: perfusion methods. *Endocrinology* **52**, 679.

HEMS, R., ROSS, B. D., BERRY, M. N., and KREBS, H. A. (1966) Gluconeogenesis in the perfused rat liver. *Biochem. J.* **101**, 284.

HENDERSON, J. R. (1969) Why are the islets of Langerhans? *Lancet* (i), 469.

HOKIN, L. E. and HOKIN, M. R. (1959) Studies of pancreatic tissue *in vitro. Gastroenterology* **36**, 368.

HUDSON, P. B., BUTLER, W. S., BRENDLER, H., and SCOTT, W. W. (1950) Vascular perfusion of the prostate gland. *J. Urol.* **63**, 319.

INOUE, K., GRIMM, Y., and GREER, M. A. (1967) Quantitative studies on the iodinated components secreted by the rat thyroid gland as determined by *in situ* perfusion. *Endocrinology* **81**, 946.

KHAYAMBASHI, H. and LYMAN, R. L. (1969) Secretion of rat pancreas perfused with plasma from rats fed soybean trypsin inhibitor. *Am. J. Physiol.* **217**, 646.

KIRSCHNER, N., SAGE, H. J., and SMITH, W. J. (1967) Mechanism of secretion from the adrenal medulla. *Mol. Pharmacol.* **3**, 254.

KRAMER, K. and DEETJEN, P. (1964) Oxygen consumption and sodium reabsorption in the mammalian kidney. In *Oxygen in the animal organism* (eds F. Dickens and E. Neil), I.M.B. Symposium Series, Vol. 31, p. 411.

LACY, P. E. and KOSTIANOVSKY, M. (1967) Method for the isolation of intact islets of Langerhans from the rat pancreas. *Diabetes* **16**, 35.

LERNER, S. R. and CHAIKOFF, I. L. (1945) The influence of goitrogenic compounds on respiration of thyroid tissue. *Endocrinology* **37**, 362.

LIGHT, A. and SIMPSON, M. S. (1956) Studies on the biosynthesis of insulin. 1. The paper chromatographic isolation of $^{14}$C-labelled insulin from calf pancreas slices. *Biochem. biophys. Acta* **20**, 251.

LOUBATIERES, A., MARIANI, M. M., CHAPAL, J., and PORTAL, A. (1967) Action frenatrice de faibles doses d'adrenaline et de noradrenaline sur l'insulino-secretion etudiée sur le pancreas isolé et perfusé du rat. *C. r. Seánc. Soc. Biol.* **161**, 2578.

MATSCHINSKY, F. M., RUTHERFORD, C. R., and ELLERMAN, J. E. (1968) Accumulation of citrate in pancreatic islets of hyperglycaemic mice. *Biochim. biophys. Res. Commun.* **33**, 855.

MATSUDA, K. and GREER, M. A. (1965) Nature of thyroid secretion in the rat, and the manner in which it is altered by thyrotropin. *Endocrinology* **76**, 1012.

RALL, J. E., ROBBINS, J., LEWALLEN, C. G. (1964) The thyroid. In *The hormones* V (eds. G. Pincus, K. V. Thimann, and E. B. Astwood), Chapter 3. Academic Press, New York.

ROMANOFF, E. B. and PINCUS, G. (1962) Studies of the isolated perfused ovary; methods and examples of application. *Endocrinology* **71**, 752.

perfusion is established the animal must be killed; opening the chest or severing the carotid artery is the method chosen here.

### Characteristics of perfusion

The preparation as described is viable for 5 h. But to achieve the results that are discussed below, the two drugs shown in the section on Media must be present in the perfusion fluid. Noradrenaline is infused continually, and a glucocorticoid is given as a single dose. The alternatives, pentobarbitone or propanalol, are offered in another paper by the same authors (see also p. 392). Without these additions the failure of the perfusion is ascribed by the authors to excessive 'secretion' of fluid into the lumen, hypermotility of the gut (a common fault of intestinal perfusion methods; see Dubois *et al.* (1968)), low vascular resistance, and progressive epithelial necrosis. All these undesired effects are reversed by the drugs named and the authors publish qualitative data based on the criteria of failure listed above (Table 7.1).

(1) Twelve to fifteen g of small bowel and caecum (wet weight) are perfused in the preparation described. No details of wet/dry ratio are given, so that extrapolation of quantitative results is difficult. The ratio given (4·0 Kavin *et al.*, see below) is for unperfused tissue. The possibility that pancreas too, which receives almost one-third of its blood supply from the superior mesenteric artery (see Chapter 6), might be perfused has not been mentioned by the authors. The venous effluent from this tissue is certainly excluded by the method of cannulation of the superior mesenteric vein, but in future studies with *in situ* preparations it may be advisable to formally exclude the circulation to the pancreatic-duodenal branch of the superior mesenteric artery.

(2) Oxygen consumption is said to be constant for 5 h, but figures are not presented. The physical parameters measured also remain constant, viz. blood flow (10–20 ml/min), arterial pressure (*c.* 95 mmHg; 130 cm $H_2O$), and lymph flow (1–2 ml/h).

(3) In addition, chylomicrons appear in the lymph collected about 2 h after an infusion of fat into the lumen. The dead space of the lymphatic cannula accounts for only 5–6 min delay. Under somewhat different conditions of perfusion (propanalol, $10^{-5}$M in the perfusate in place of glucocorticoid and noradrenaline), $^3$H-leucine added to the perfusing medium results in the labelling of $D > 1·006$ and $D$ 1·006–121 lipoprotein in lymph while the plasma lipoprotein remains virtually unlabelled. The preparation therefore demonstrates transoprt and

## TABLE 7.1

*The physical response of the perfused intestine to drugs and hormones*

(from Windmueller *et al.*)

| Addition | Dose (perfusate conc. M) | hyper-secretion | hyper-motility | hyper-aemia | low vascular resistance | Epithelial necrosis 5 h |
|---|---|---|---|---|---|---|
| None | – | + | + | + | + | not examined |
| Dexamethazone (DEX) | $6 \times 10^{-7}$ | + | + | – | + | not examined |
| Propanalol+DEX | $1 \times 10^{-5}$ | – | – | – | + | + |
| Pentobarbital+DEX | $2 \times 10^{-4}$ | – | – | – | + | + |
| Ethyl-ether+DEX | anaesthetic | – | – | – | + | + |
| Atropine+DEX | $9 \times 10^{-5}$ | – | – | – | + | + |
| Noradrenaline+DEX | $4-8 \times 10^{-8}$ | – | – | – | – | – |

(+ positive; – negative response)

For details of perfusion method see above and for doses of drugs see, in addition, Windmueller *et al.* (1970).

metabolic conversion in two directions: lumen → lymph and vasculature → lymph. No demonstration of metabolites of lumen components
appearing in the perfusate has been reported, but the necessity of
replacing glucose in the perfusion medium every hour to maintain a
concentration of 1–3 mg/ml indicates that there is glucose uptake and
possible utilization in this preparation.

**Method IA. Small bowel perfused *in situ*, with irrigation of the lumen.
Method of Kavin, Levin, and Stanley.**

The second method to be described in this group of *in situ* perfusions
of small bowel, with irrigation of the lumen, is older than that of
Windmueller *et al.* but no other publications using the method have
appeared. It is technically more difficult than the last method described,
requires a more elaborate apparatus, and, since the lymph duct is not
cannulated, it is limited in its general application. However, an
important difference, which may make this preparation of value is the
achievement of apparently satisfactory conditions of perfusion, confirmed by several independent tests of viability, without the addition of
drugs to the perfusion medium. Nor are any drugs, apart from heparin,
injected into the donor rat.

Direct, pulsatile perfusion of the small intestine is achieved through
the abdominal aorta, all branches of which, apart from the superior
mesenteric artery, are tied off. Outflow of medium from four parallel
loops of bowel, irrigated continuously, is via the portal vein. Again,
branches other than the superior mesenteric vein are tied off. A simple,
semi-synthetic medium containing washed bovine erythrocytes and
low-molecular weight Dextran in a bicarbonate buffered saline is used
and is oxygenated by means of a thin-film oxygenator. Perfusions are
continued for 1–2 h. The lumen of the 4 loops is irrigated continuously.

*Apparatus*

The apparatus is illustrated in Fig. 7.4. A box cabinet, 75 cm × 60 cm
× 50 cm, houses the animal and the apparatus at 37°C. The cabinet
described in the Appendix would be suitable. An additional feature
of the cabinet described here is the maintenance of humidity by the
discontinuous injection of a jet of steam.

The oxygenator is a rather complex falling-film device, described on
p. 57. Any film-oxygenator could be used, and the multibulb glass
oxygenator has the advantages already discussed in Chapter 1.

25

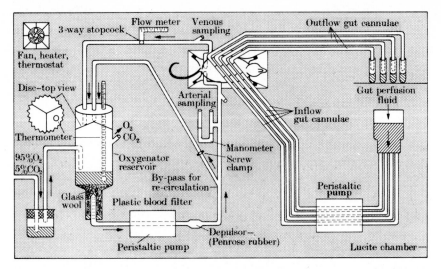

FIG. 7.4.    Apparatus for perfusion *in situ* of small-intestine loops: method of Kavin *et al.* (1967). For details of method see p. 369.

*Filter.* Beneath the reservoir of the oxygenator is a filter of packed glass-wool, through which medium flows to the peristaltic pump. Glass wool is thought by the authors to offer special protection against vaso-motor components of the medium, and may possibly explain why no drugs are required in the medium. This has not been tested. Gas (95% $O_2$, 5% $CO_2$) is blown over the surface of the medium in the reservoir (or through the 'lung').

*Pump.* A Harvard peristaltic pump, with the wave form described in Chapter 1 is used, but a 'Penrose' rubber chamber acts as a de-compressor between the pump and the arterial cannula. There is consequently no information concerning the wave form reaching the intestine. A manometer and overflow circuit are provided along the arterial line.

The venous outflow, with a flow meter in its course, returns to the oxygenator reservoir.

*Lumen circuit.* As will be described below, four separate loops of intestine are perfused iso-peristaltically and in parallel, i.e. the anatomi-cal direction of the loops is the same as the direction of flow through the lumen. A four-channel peristaltic pump (the Harvard finger pump would be suitable) supplies medium. The rate is not given in this publication; 1 ml/min per loop would be suitable (cf. Dubois *et al.* (1968)).

*Cannulae.* The arterial (aorta) is of polythene Sterilon, 18 gauge (1·4 mm o.d.). The venous cannula is of the same material and size. For the intestine 4 pairs of inflow and outflow cannulae are required and those of 19 mm, 3 mm o.d. stainless steel are recommended.

## Medium

(1) *Perfusion medium.* Red cells from fresh heparinized or citrate bovine blood are washed three times with sterile Ringer's solution. Eighty millilitres of cells are suspended in 160 ml of the medium, which contains Rheomacrodex at 22 g/litre and is otherwise similar to Krebs–Henseleit saline.

(2) *Lumen fluid.* No details are given, but possible solutions are described in Table 1.7.

## Operative procedure

Rats of 250–400 g weight are starved overnight. Ether anaesthesia is used. Fig. 7.5 shows the position of ligatures which are emplaced after the abdomen has been opened in the usual way (see p. 88). Before placing ligatures, however, the length of intestine to be perfused

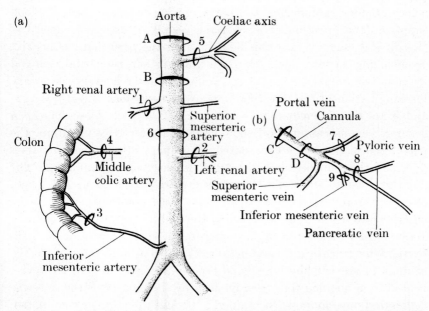

FIG. 7.5. Operative procedure for intestinal perfusion (method of Kavin *et al.* (1967); see p. 372 for details). (a) Ligatures required for arterial cannulation. (b) Ligatures required for venous cannulation.

is defined. Cannulae, as described, are inserted into the jejunum at the duodeno-jejunal flexure (see Fig. 7.1(a)) and distally, about 2 cm before the ileo-caecal junction. The gut is irrigated gently in the reverse direction (i.e. colon → jejunum) with Ringer's solution at 37°C and then divided into four roughly equal lengths. (There is some advantage in making this precise, since the four loops are then comparable.) These should be marked with ligatures at the time, to ensure that all are mounted in the same peristaltic direction. The remaining three sets of inflow/outflow cannulae are tied in place. The authors emphasize that the intestine must be preserved at the site of cannulation and this is best achieved by tying only a very thin rim of bowel to the cannulae, and by not occluding the adjacent mesentery.

With the intestine lying towards the animal's left, the right renal pedicle is tied (1). Now the intestine is displaced to the animal's right and the remaining ligatures (2)–(5) illustrated in Fig. 7.5 are placed and tied. This isolates a segment of aorta, from the diaphragm to just beyond the origin of the superior mesenteric artery, with only the latter patent, and its supply is now restricted to the small bowel. Ligatures A and B, around the aorta are loosely tied, the one below the diaphragm (A) and the other (B) just above the origins of the superior mesenteric artery. Before cannulating the aorta, which ends the normal blood supply to the intestine, it is advisable to place the venous ligatures (Kavin *et al.* suggest that this be done after the perfusion has begun). By swinging the intestine back to the animal's left, the portal vein and its main tributaries are seen. Fig. 7.5 illustrates the ligatures necessary for cannulation of the intestinal venous outflow alone. Ligatures 7, 8, and 9 are tied immediately, while C and D are left loose. Since, in general, venous cannulation should precede arterial cannulation, the portal vein can now be cannulated. The upper (distal) ligature, C, is tied, and the vein becomes distended, which assists cannulation. The polythene cannula described is plugged loosely with tissue paper and inserted through a small incision, made with pointed scissors in the anterior wall of the vein, just behind the tied ligature (C). The cannula is tied in place when the tip lies opposite the superior mesenteric vein, by tightening ligature D, and ligature C is tied around the back of the cannula to prevent kinking. Blood flows into the cannula, and is prevented from flooding the operative field by the plug. Now the intestine is diverted once more to the animal's left and the aorta is cannulated. The cannula is filled to the tip with medium before its insertion. Ligature A is tied, which introduces the anoxic phase, and the cannula is inserted from above, downwards, through an incision made just below ligature A. It is tied in place when the tip lies near the orifice of the superior mesenteric artery. Ligatures B are tightened, to close the distal aorta, and ligature A is used to steady the cannula. To reduce the period of anoxia, the cannula in the aorta should be attached to the perfusion apparatus immediately, or even at the time of cannulation, so that flow can begin at once. The animal is killed at this stage by opening the chest.

On the arterial circuit, a pressure of 90–105 mmHg (=120–140 cm $H_2O$) is applied which in the hands of Kavin *et al.* gives initial flow rates of 1–2 ml/min (20–35 ml/min per 100 g wet tissue). Higher rates of flow occur after 20 min perfusion and continue for 1–2 h (i.e. 3·5–5·0 ml/min; cf. 8·5 ml/min given by Windmueller). The venous pressure is kept at −2 cm $H_2O$ without which the flow rates described cannot be achieved in this preparation.

*Characteristics of perfusion*

The preparation appears to be viable for 1–2 h, with flow rate, perfusion pressure, glucose absorption, and glucose uptake, medium pH and oxygen consumption as parameters of function. Data from composited experiments (Table 1, Kavin *et al.* (1967)) indicate that oxygen consumption, medium pH, and blood flow are uniform over the time 12–114 min. Linearity of these functions in any one experiment is more difficult to ascertain from the information given. The rate of oxygen consumption, *c.* 0·09 ml/min per g dry wt. (4 $\mu$mol/min per g dry wt.) compares with that of isolated 'slices' of intestine, incubated in bicarbonate buffer at the same temperature and pH as those used here (see Long (1961)). At the flow rates given, the early phase of perfusion may be in danger from anoxia; at 20 ml/min per 100 g wet tissue (q.v.) the oxygen provided is 5·8 $\mu$mol/min per g dry wt.; oxygen uptake is 4 $\mu$mol/min per g dry weight. This is the worst prediction, with maximum $O_2$ uptake and minimum flow. Nevertheless, it may become of importance where substrates provoking increased oxygen uptake are added to the medium. The rate of glucose consumption (0·03 mg/100 mg dry wt. = 0·3 mg/g dry wt. = 1·7 $\mu$mol/min per g dry wt.) would, if completely oxidized under optimal conditions, require approximately 10 $\mu$mol $O_2$. This is more than the observed oxygen consumption under the present conditions, and it must be assumed that the major part of the glucose removed from the medium is not completely oxidized. In fact 60 per cent must be otherwise metabolized, for example, to lactate.

(Calculated on the basis of 6 oxygen atoms per mol of lactate oxidized to $CO_2$ and water $= 6$ mol oxygen per mol glucose, since glucose $\rightarrow 2$ lactate.)

By comparison with the method of Windmueller *et al.* (1970), there is a very much lower flow rate, despite the use of a similar inflow pressure. The haematocrit of the medium used by Kavin (33 per cent) is not very different from that of the rat blood used by Windmueller, and the difference is presumably to be attributed to the drugs—noradrenaline, glucocorticoid, or propanalol—used by the latter authors. Quantitative data on comparable facets of metabolic activity are not yet available for the Windmueller preparation, but a comparison of *in situ* versus isolated intestine perfusion may be made with the preparation of Dubois, Vaughan, and Roy (see below).

## Method IB. *In situ* perfusion without irrigation of the lumen. Method of Hestrin-Lerner and Shapiro

This very simple method of perfusion was introduced by its authors to study the then hypothetical utilization of glucose during its uptake by intestine. It is a possible alternative method to the more complex methods described in Method IA, but is not to be recommended without further characterization of the preparation.

### Method

A medium of washed bovine erythrocytes suspended in dialysed bovine serum, with added glucose, and at a temperature of 40°C, is infused under hydrostatic pressure of 80–100 cm $H_2O$, into the aorta. The aorta is tied off above and below the origin of the superior mesenteric artery, and the portal vein is cannulated. A closed loop of intestine is filled with a known volume of experimental medium and its content analysed by sampling with a needle from time to time. The vascular perfusion is 'once-through'.

### Apparatus

The apparatus required is very simple. An elevated reservoir with facilities for bubbling gas (see p. 44) is supported at the required height of 80–100 cm above the rat. A filter, such as the standard blood-transfusion filter described, is set below the reservoir and medium flows via a polythene tube, with roller clip or clamp, to the arterial cannula. The cannula is not described, but that of Kavin *et al.* (q.v.) would be suitable. A similar portal vein cannula may be used. Medium flows freely from the portal vein cannula and is collected into a receiver. Obviously, little modification is required to make this a recirculation method.

## Medium

(1) Bovine erythrocytes, from fresh, citrated or heparinized blood are washed three times in twice their volume of Ringer's solution. Bovine serum is dialysed for 48 h against Tyrode solution (Krebs–Henseleit medium should be used if it is intended to gas the medium with 95% $O_2$, 5% $CO_2$, see Media) of which at least three changes are desirable. Equal parts of cells and dialysed serum are mixed, and glucose added to a final concentration of 100 mg %. The medium is gassed with air.

(2) A standard saline medium (e.g. Ringer's) with varying concentrations of glucose is recommended for these experiments.

## Operative procedure

The operation is essentially similar to that described by Kavin *et al.* However, a much less elaborate preparation of the aorta is undertaken. No branches are tied, and, in particular, the right renal artery is left patent. A ligature is passed round the aorta above the origin of the superior mesenteric artery, and another below it. The aorta is occluded with forceps above the upper ligature and cannulated through a small incision. The ligatures are tied, the upper one around the cannula and the lower one beyond the tip of the cannula as it lies opposite the superior mesenteric artery. The aorta is thereby occluded and medium flows into the mesenteric (and probably also the right renal) beds. Out-flowing medium is collected by cannulating the portal vein as described by Kavin *et al.* (q.v.) but no precaution is taken to isolate the outflow from the superior mesenteric vein. It is assumed that at the flow rates used (4–6 ml/min) all the effluent passes through the intestinal vasculature. For an isolated preparation, the operation of Kavin *et al.* is required. Finally, a loop of intestine is isolated by tying ligatures at either end. The sac thus formed is injected with medium of known glucose content. Portal effluent samples are collected and analysed for glucose, so that a comparison of glucose (or any other component) in the outflow and inflow should give a semi-quantitative indication of transport from the lumen.

*Characteristics of perfusion*

The results published are scanty, and show no active glucose transport. This may mean that this approach is too simple, but there may be some application in this approach to metabolic studies of the perfused intestine.

## Method IB. *In situ* perfusion without irrigation of the lumen. Method of Gerber and Remy-Defraigne

This preparation appears likely to have a general application.

### Method

A roller pump delivers a medium of fresh heparinized rat blood, diluted by 20 per cent with Ringer's solution to the segment of aorta from which the superior mesenteric artery arises. Outflow is via the portal vein, and the medium is oxygenated by a rotating-disc oxygenator, such as has also been used in liver perfusion studies (see Oxygenators). Medium is recirculated and the intestine is left *in situ*, with the lumen either irrigated or not.

### Apparatus

The apparatus is a small-volume perfusion system (40 ml approx.) adapted from Miller's liver perfusion system, housed in a cabinet at 37°C.

A roller pump (type not specified) delivers medium directly to the arterial cannula in the aorta at a pressure of 70 mmHg (95 cm $H_2O$). Venous effluent from the portal vein cannula flows to the rotating-disc oxygenator whence it is collected by the pump. The details of the oxygenator are given on p. 54. A manometer tube (PE 100) is interposed in the arterial line which in turn is polythene PE 260.

*Cannulae.* The aortic cannula is specially designed to deliver medium from the side rather than the end. An 18-gauge needle with the point filed down and sealed, has an orifice $2 \times 4$ mm drilled in the side, 5 mm from the tip. When the cannula is tied in place, this orifice should lie opposite the superior mesenteric artery, while ties above and below occlude the aorta. A venous cannula of polythene tubing of 1·4 mm o.d. as described by Kavin, is suitable.

### Medium

This consists of 30–35 ml fresh heparinized rat blood plus 10 ml Ringer's solution (see Media) plus 150 mg glucose (=approximately

21 mM). The haematocrit is 33 per cent and the medium is gassed with 95% $O_2$:5% $CO_2$ (a pH of 7·1–7·2 is expected).

## Operative procedure

Under ether anaesthesia, a mid-line abdominal incision with lateral extension (after clamping the descending epigastric vessel on each side) is made in the usual way. The rectum and descending colon are excised between ligatures, leaving only that part of small and large intestine which is supplied by the superior mesenteric vessels. The intestine is displaced to the animal's left, and the connective tissue overlying the aorta and inferior vena cava in the mid-lumbar region, i.e. between the kidneys, is cleared. This exposes the aorta, free from the inferior vena cava over a length of some 5 mm above and below the origin of the superior mesenteric artery. Comparing Fig. 7.5, ligatures A and B are loosely tied, while 3,3′ and 4,4′, round the renal vessels and coeliac axis are tied and the vessels cut beyond them. The portal vein is exposed and two loose ligatures passed round it ((C) and (D)). The gastro-duodenal (7) and the splenic (8) veins are tied off.

Aortic cannulation is from below; a bulldog clip occludes the aorta just below the renal artery and the cannula is inserted through a cut in the vessel below this. Medium should be flowing through the cannula at the time, and the tip is pushed beyond the bulldog clip, as far as the superior mesenteric artery orifice, before ligatures A and B are tied. Anoxia is therefore almost entirely avoided.

As soon as possible the venous cannula is inserted. The portal vein is held near the liver in curved forceps and an incision made in its anterior wall. Through this, the cannula is inserted and tied with two ligatures, the uppermost of which, occludes the portal vein and steadies the cannula. Recirculation is begun at once with a flow of 4–6 ml/min.

The intestinal lumen is not usually cannulated in this *in situ* preparation. The operation time is 10–15 min.

## Characteristics of perfusion

Perfusion is said to continue for from 1 to 3 h and histology of the mucosa, by light-microscopy, is described as normal after this time. The biochemical studies undertaken are useful indicators of the stability of the preparation. Thymidine uptake from and dihydrothymine appearance in the medium were linear for at least 1 h. Both are proportional to the initial concentration of thymidine in the medium and this too may be taken as a good test of the preparation.

No information on transport can be obtained under the experimental conditions described.

This preparation is the only perfusion as yet in which use has been made of the method to analyse the composition of the gut itself. Metabolic transformation of thymidine and incorporation of its $^3$H label into DNA has been demonstrated and points the way to further biochemical applications.

This perfusion system appears to be satisfactory without the addition of drugs to the medium, which differs from that used by Windmueller *et al.* only in its dilution from a haematocrit of 45 to 35 approximately.

## Method IIA. Small intestine isolated, with irrigation of the lumen. Method of Lee and Duncan

In this method an upper jejunal segment of intestine is excised and suspended in an oil bath. In the authors' description, cannulation of the superior mesenteric artery and the portal vein are accomplished afterwards and with some delay and the arterial circuit is perfused without a pump. Gas pressure from a cylinder of 95% $O_2$:5% $CO_2$ supplements the inflow under gravity, to achieve a pressure of 70 mmHg (95 cm $H_2O$) and no additional oxygenator is used. The intestinal lumen is circulated with a modified Fisher–Parsons apparatus at a rate of 20 ml/min.

*Medium*

(1) Perfusion medium of whole heparinized rat blood is used, with addition of glucose (to a final conc. 12 mM) and pentobarbitone 5 mg/100 ml. The latter is used to reduce gut motility (see p. 368).

(2) The lumen fluid used is Krebs–Henseleit bicarbonate medium +27 mM glucose, circulated at a rate of 20 ml/min.

*Operation*

The operative technique described by the authors has little to recommend it, entailing as it does, a prolonged ischaemic anoxic phase. The operative procedure of Dubois *et al.* (see below) can be used.

*Characteristics of perfusion*

Perfusion rate is constant for about 1 h in this preparation, but it then falls. Lymph-flow and evidence of water absorption are even more short-lived and decrease after about 30 min. In its present form, therefore, this preparation is of limited value.

**Method IIA. Small intestine isolated, with irrigation of the lumen. Method of Dubois, Vaughan, and Roy**

The second method in this group, appears to offer a better possibility of maintaining the isolated perfused intestine for periods sufficient to allow biochemical studies to be undertaken. It is based on the established technique of perfusion of the pancreas, devised by Sussman *et al.* (1966), and uses the commercially available apparatus (Ambec) described in detail in Chapter 1 and in the appendix.

A loop of 20 cm of small intestine is perfused in isolation through the aorta and superior mesenteric artery, with outflow from the portal vein, using a 'once-through' method. The loop is contained in the organ compartment of the Ambec apparatus and the lumen is irrigated by means of an additional pump (Sigmamotor).

*Apparatus*

A detailed description of this apparatus is contained in Chapter 1 and further information on its use for pancreatic perfusion is included in Chapter 6.

A non-peristaltic roller pump feeds the arterial cannula, collecting medium from an oxygenator of revolving mesh design. The whole apparatus is mounted on or within a warmed water-bath at 37°C.

*Cannulae*. The aortic cannula is 2·5 mm diam. with a side-arm to allow pressure measurements. The one illustrated is of polythene, with metal fitments and side-arm, but no details are published. The venous cannula is of polythene; P.E. 160 diam. 1·5 mm is suitable. Intestinal cannulae are not described  but those of Kavin *et al.* (p. 371) or of Windmueller *et al.* (p. 362) are suitable.

*Medium*

(1) *Perfusion medium*. Washed bovine red cells from citrated blood are added to Krebs–Henseleit buffer to an haematocrit of 20 per cent (the details of ionic composition of the medium published here are, as those of Kavin *et al.*, obscure. It has been assumed that standard concentrations were in fact used, and these are found in Chapter 1, Media). Further additions to the medium are bovine serum, albumin (2·5 g %), Dextran, M.W. 78 000 (2·5 g %), and glucose 100 mg % (5·5 mM) and a non-ionic polyol surfactant, pluronic F-68, (6 mg %) is added to prevent frothing. (From the description given, it appears that these additions and concentrations refer to the medium before addition of

blood cells. The concentrations should therefore be reduced by approximately 20 per cent and become albumin 2·0 g  Dextran 2·0 g  and glucose 80 mg (4·4 mM) per 100 ml. The surfactant is apparently added later, and the concentration therefore remains unchanged: 6 mg %). The pH of the medium is brought to 7·4 with N orthophosphoric acid and the medium gassed with 95% $O_2$:5% $CO_2$. (See p. 24 for criticism.)

(2) *Lumen fluid.* Isotonic saline (0·9% NaCl) is infused at the rate of 1·0 ml/min with a Sigmamotor peristaltic pump.

*Operative procedure*

Intra-peritoneal pentobarbitone anaesthetic is used, in rats of 250–350 g weight, starved overnight. The standard skin incision in the mid-line is extended to the upper thorax. The abdominal wall is incised in the mid-line, with lateral extensions  after clamping the descending superior epigastric vessels. The jejunum is found as it lies to the animal's right beneath the liver and the mesentery cut down to the duodeno-jejunal flexure. Vascular connections between the colon and the pancreas are severed between ligatures  exposing the jejunum more fully.

The first and second jejunal arteries are identified—they travel in the transparent mesentery from the superior mesenteric artery to the intestine, where they are seen as they form vascular arcades. Double ligatures (3/0 silk) are placed around the jejunum between these two vessels, and further ligatures are placed 20 cm distally along the jejunum, thereby isolating the loop to be perfused.

The aorta is carefully cleared, up to the diaphragm, and ligatures placed around the following vessels:

(1) superior mesenteric artery, beyond its jejunal branches;
(2) coeliac artery, renal vessels, middle colic, right colic and inferior mesenteric vessels;
(3) branches of the portal vein, apart from the superior mesenteric.

All these ligatures are tied. Loose ligatures are placed in addition round:

(4) the aorta, just below the diaphragm, just above the superior mesenteric artery, and below the same artery;
(5) the portal vein, between the liver and the site of entry of the superior mesenteric vein.

The intestinal loop is now separated from the remaining intestine by severing it between the two sets of ligatures, and the remaining intestine can be removed. This helps the subsequent dissection. Intestinal cannulae are inserted and tied in place, with the minimum of damage to the wall of the jejunum. Heparin (100 units) is given into the femoral vein in the normal way and then the arterial cannula is prepared by filling it to the tip with perfusion medium or with saline (see below).

The chest is opened in the mid-line and retracted with a suitable instrument, taking care not to damage the heart or great vessels. The lungs collapse and can be displaced to the animal's right (turn the animal around to do this, so that the head is to the operator's right). The aorta is easily located behind the oesophagus, and can be cannulated as described for the pancreas (p. 337) by grasping the vessel above, and incising it at about 1 cm above the diaphragm. The cannula through which saline is already flowing is inserted until the point lies just proximal to the superior mesenteric artery and tied in place with the ligatures (4) already described. Venous cannulation is performed, inserting the polythene cannula through an incision in the anterior wall of the portal vein and then tying the ligatures. Finally, the loop of isolated bowel and its attached cannulas must be transferred to the apparatus and laid in the warmed tray. Flow of medium must be interrupted briefly to achieve this. The authors recommend that saline be infused for the first 1–3 min, until all the rat's own blood has been washed out, when the standard medium is introduced. The period of anoxia during preparation, apart from the introductory saline perfusion, may be up to 4 min, but this is a considerable improvement on the method suggested by Lee and Duncan (q.v.). Perfusion is begun at a rate of 2·0–3·0 ml/min, with an arterial pressure between 30 and 60 mmHg (40–80 cm $H_2O$) and is reported to remain stable at this level for 3–5 h. As already mentioned, the experiments here reported did not use recirculation perfusion. Lumen irrigation, again on a 'once-through' basis, is begun at 1·0 ml/min.

## Characteristics of perfusion

Since the preparation was introduced with the aim of studying intestinal absorption and transport (Dubois *et al.* 1968), the emphasis in the results presented is on the uptake of glucose, water, and 3-0 methyl glucose. The tests of viability applied are the perfusion rate and pressure and the histology at the end of perfusion. No alteration in flow rate of perfusion within the 3- to 5-hour experimental period is reported, but the actual reason for the ending of the perfusion is not recorded. Pressure does not change. Histologically, however, there is evidence of mucosal damage at the tips of the villi. The final dry weight of the perfused loop appears to vary widely, from 390 to 1050 g. The problem of oedema is not mentioned by the authors, although it is a common problem in this field (see Windmueller). Excessive peristalsis is taken as a qualitative sign of failure of the perfusion. No additions to the medium are made to reduce it, although, from the report of Lee and Duncan, it is probable that pentobarbitone anaesthetic reduces motility. It does not appear to have been a major problem to these authors, and a comparison with the Windmueller *in situ* technique is clearly relevant.

Active transport of glucose against a concentration gradient and glucose utilization are demonstrated. Quantitative information is unfortunately too scanty to allow of accurate comparison with other methods. Oxygen consumption, however, at 0·04–0·08 ml/min per g dry weight, compares closely with that observed by Kavin *et al.* in their preparation *in situ*. This allows the general conclusion that this fundamental parameter is unaffected by the minor differences in medium between these two preparations, viz. bovine versus rat erythrocytes, albumin and Dextran versus Rheomacrodex alone, and pentobarbitone versus ether anaesthesia, and an *in situ* versus an isolated preparation. The pH, ionic composition of the medium, and flow rate were the same. Kavin *et al.* used higher perfusion pressures to obtain similar flow rates, however. This in itself is of significance in excluding an effect of such a mechanical difference on a metabolic parameter such as oxygen consumption. The importance of establishing these details in a formal comparative experiment must, however, not be overlooked. Finally, Dubois *et al.* point out that experimental results from different perfusions are better compared on a dry-weight basis. This convention has already been adopted in the perfused kidney (Nishiitsutsuji–Uwo *et al.*) where oedema is a problem, and it is probably applicable to the intestine for the same reason.

## Method IIB. Small intestine isolated, without irrigation of the lumen. Method of Forth

This is a 'once-through' perfusion system for an isolated segment of small intestine in which the lumen is not cannulated but experimental fluids are introduced as single doses. A semi-synthetic medium is passed into the superior mesenteric artery under hydrostatic pressure, with gas pressure to supplement this. No pump and no other oxygenator is used. It is therefore a simple preparation. Unfortunately, until now it has been used only to study iron and cobalt transport, and there is no comparable data with any other published method of perfusion.

*Apparatus*

The apparatus is illustrated diagrammatically in Chapter 1 (see Fig. 1.2). Gas pressure from a cylinder of 95% $O_2$:5% $CO_2$ forces medium from a reservoir, through a heating coil, which is immersed in a water-bath, into the superior mesenteric artery cannula at a pressure of 60 mmHg (80 cm $H_2O$). Outflowing medium is collected in a fraction-collector.

*Cannulae.* The cannulae are illustrated here, the arterial cannula being Polythene, 1 mm o.d., and the venous cannula glass, o.d. 3 mm. No intestinal cannulae are required for this preparation.

*Medium*

(1) To wash through the perfusion-bed, 20 ml 0·9% NaCl, containing heparin 4·0 mg %, papaverine 10 mg %, and promethazine 4 mg % is used. (2) Unfortunately the published details of the perfusion medium proper are a little vague; electrolytes, amino acids, and glucose are said to be present at the same concentrations as *in serum*. The closest published description would seem to be that of Schnitger (see p. 22). Heparin 4 mg % and papaverine 10 mg % are also present. To this medium are added the following: washed fresh human red cells, to a p.v.c. 30 per cent; polyvinylpyrollidine (PVP, M.W. 25 000) 5·6 g %; bovine serum albumin, 3·4 g %. The final heparin and papaverine concentrations become 2·7 mg % and 6·7 mg % respectively. When gassed with 95% $O_2$:5% $CO_2$, the medium achieved pH 7·0 (this corresponds to a bicarbonate concentration of 9 mM, i.e. the basic medium, before dilution by the additions mentioned, contained 12 mM bicarbonate (Tyrode solution, see Chapter 1).

*Operative procedure*

The operative procedure is said to be that of Gerber and Remy-Defraigne (1966), but obviously differs from that described on p. 377 since an isolated loop of intestine is prepared.

Under ether anaesthesia the abdomen is opened in the mid-line, with
lateral extension in the usual way. The intestine is displaced to the
animal's left and the superior mesenteric artery is identified after clearing the middle section of aorta of connective tissue. Ligatures are placed
around the coeliac artery, the right renal artery and vein, the left renal
artery and the aorta, immediately below the origin of the superior
mesenteric artery. Loose ligatures, two in number, are positioned
around the superior mesenteric artery itself, with attention to the

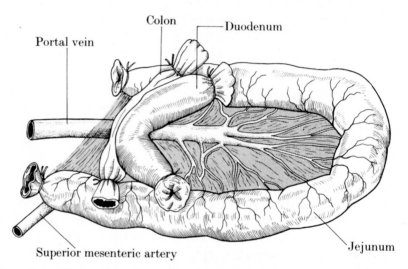

FIG. 7.6.   Preparation of an isolated loop of small intestine for perfusion by
the method of Forth (1968). The operative preparation is described on p. 383
and the hydrostatic pressure perfusion apparatus used by Forth is illustrated in
Fig. 1.2.

mesenteric lymph duct that runs alongside (Fig. 7.2). It may be
impossible to avoid damage to this structure.

   The portal vein is similarly prepared by placing two loose ligatures
round it and tying the splenic and pancreatico-duodenal tributaries.
Before cannulating the vessels, the loop of jejunum is prepared. The
loop is 20 cm long, beginning at the duodeno-jejunal flexure (as
Dubois *et al.*). To avoid unnecessary dissection, Forth suggests the
following procedure. The loop is tied at either end and cut between
ligatures. The loop of duodenum and of colon which is included between
the 'arms' of the jejunal loop thus isolated are each tied at the end, and
cut. Their vascular connections are tied, but no attempt is made to
remove them. The preparation that results is illustrated in Fig. 7.6.

At this stage, the loop of intestine is free apart from its vascular connections, the portal vein, and the superior mesenteric artery. The ligatures prepared are tied. The portal vein is cannulated and the cannula tied in place. Then the superior mesenteric artery is cannulated directly, the cannula being tied in place with the loose ligatures prepared. Flow begins at once, and the period of anoxia is limited to the half minute required for cannulation.

*Characteristics of perfusion*

In order to achieve net water absorption in this preparation, the medium must be supplemented with the anti-histamine promethazine (N(2′-dimethylaminopropyl)phenothiazine). Flow rate is 6·5 ml/min at a pressure of 60 mmHg (80 cm $H_2O$). Perfusions were continued for only 30 min. Reabsorption of water and of glucose, studied by injecting 1·0 ml 0·9% NaCl with 100 mg % glucose into the lumen of the closed loop, are reported to be similar to the values obtained with closed loops isolated, but with an intact circulation *in vivo* (Forth 1968).

In iron and cobalt absorption studies, these ions appeared in the portal effluent at a nearly linear rate for the short experimental period recorded (9–14 min). The rate of flow of medium had no influence on the rate of absorption of iron from the lumen, but neither was oxygen required for this process, so that little significance attaches to the observation.

The method, though simple, may have limitations in its biochemical application as it appears to be viable for so short a time. Perhaps with a more physiological pH, as suggested by Kavin *et al.* for intestine or Hems *et al.* for liver, survival times may be prolonged into the useful range.

**Method III. Perfused large intestine, *in situ* preparation with irrigation of the lumen. Method of Windmueller *et al.***

This perfusion technique, already described in Method IA as a preparation of small intestinal perfusion, includes that part of the large intestine that is supplied by the superior mesenteric artery (see Fig. 7.1(b)), that is to say, the upper half to two-thirds of the colon, as far as the middle of the descending colon. Its authors suggest that the preparation can be limited to perfusion of the small intestine by simply tying gut and vasculature at the appropriate level. It may thereby be possible, by comparing the results of initially perfusing the whole preparation and then the restricted preparation, to obtain information on the metabolism of the large intestinal component. Although a very indirect approach, it is justified by the relative simplicity of the original preparation. Alternatively, the technically more difficult but eminently successful preparation of Powis may be used (see below).

## Method IV. The isolated perfused colon, with irrigation of the lumen. Method of Powis (Parsons and Powis)

In this preparation perfusion fluid is passed through the superior and inferior mesenteric arteries via the aorta, without interruption of flow during their preparation. The superior mesenteric vein is cannulated for the outflow, and recirculation of medium is accomplished by means of a roller pump. A semi-synthetic medium, similar to that developed for the liver (see p. 20) and containing ox erythrocytes, is oxygenated with a simple rotating reservoir-type oxygenator (see Chapter 1). A second circuit, with a pump but without an oxygenator, is provided for the lumen of the bowel, and Krebs–Henseleit medium is provided for this circuit. No attempt has been made to cannulate the lymphatic outflow, which is much less significant than that in the small intestine (q.v.), but studies of the effect of nerve stimulation have been undertaken.

### Apparatus

The apparatus has been adapted from that described for the perfused liver by Powis (1970), based on that of Blakeley, Brown, and Ferry (1963) for the perfusion of spleen. It consists of two circuits, one for the lumen and the other for the vascular perfusion.

(1) *Lumen circuit.* Two jacketed glass reservoirs are connected to a peristaltic pump (either Harvard, model 600-000 or Sigmamotor 'kinetic clamp' pump). The two reservoirs are interchangeable, with a three-way tap before the pump, and beyond the pump there is another tap that allows the entire contents of the lumen to be emptied. The outflow returns to the reservoir or, more usually, is allowed to run to waste. Dead space of this circuit is 3·5 ml.

(2) *Vascular circuit.* To allow the medium to be rapidly changed during an experiment, the vascular circuit is duplicated—two reservoir-oxygenators, two pumps, and two filters joining to pass medium through a single heat-exchanging coil, manometer, and bubble trap, before entering the arterial cannula. Blood is drawn from the reservoir by a non-pulsatile roller pump (Varioperpex LKB 12000) and filtered through a compressed stack of Fipoca filter elements (see p. 76). The

medium next passes through a heat-exchanger, which is a coil of nylon tubing (Portex, nylon 6, $1.9 \times 2.76$ mm) and then through a 1-ml glass bubble-trap. Both of these are immersed in the liquid paraffin in the perfusion reservoir. The perfusion pressure is measured as the height of a column of saline, through a T-tube inserted between the heat-exchanger and the bubble trap. The venous outflow is set at a level of the mid-point of the colon, into a graduated tube with a tap in the base. This serves as a flow meter, and empties into the reservoir or, in 'once-through' studies, to waste. Dead space of the vascular circuit is $2.5$ ml beyond the pump.

*Organ support.* A steel plate, $9 \times 22$ cm, stands upright in a reservoir (jacketed with water at $37°C$) containing $1.25$ litres liquid paraffin. Two spring clips and two adjustable Perspex clips are mounted on the plate, the last-named on two pot magnets, allowing them to be mounted independently and to hold the inlet and outlet cannulae. The first-named clips hold the colon preparation in a vertical position, upper end down. Thus, the lumen circulation must enter through a tube sealed into the base of the reservoir. All other cannulae pass over the rim of the reservoir.

The vertical position of the colon allows gas bubbles to be dispelled from the lumen of the colon. This is especially important in the $CO_2/HCO_3$ transport studies undertaken by Powis (1970) because stationary gas bubbles would upset concentration gradients. It is not, however, critical to vascular perfusion.

*Tubing.* Portex Nylon 6 ($1.9/2.76$ mm for the vascular and $2.44/3.24$ mm for the luminal) tubing is used throughout on the basis of its minimal oxygen/$CO_2$ permeability (quoting Powis). Tygon vinyl tubing (S 50-HL) for joints, and Silescol silicone-rubber tubing for the Vario-perpex pump, were used in addition.

*Cannulae.* The arterial cannula is of Portex polyethylene tubing PP190 $1.19/1.7$ mm) drawn out in a flame. The venous cannula is of square-ended tubing of the same diameter. The lumen cannula consists of $3 \times 5$ mm glass tubing, with $60°$ bevel and a constriction beyond the tip.

*Medium*

(1) Krebs–Henseleit buffer, plus $0.2\%$ glucose and 'antibiotics'. Umbreit *et al.* is followed, and $Ca^{++}$ always reduced to $1.25$ mM. Antibiotics are benzyl penicillin and streptomycin sulphate.

(2) Krebs–Henseleit buffer containing 3 g % bovine serum albumin (Armour), and 7% p.c.v. erythrocytes (ox cells, washed three times in saline). Again $Ca^{++}$ is added at half the normal concentration.

(3) The lumen fluid is Krebs–Henseleit buffer or bicarbonate saline (Na 145 mM, Cl 120 mM $HCO_3$ 25 mM). See Media.

*Operative procedure*

Rats, 250-g males, starved for 18 h, are anaesthetized with Nembutal (supplemented with ether as required). An anti-histamine, promethazine hydrochloride (2 mg/100 g body weight) is given intravenously before beginning the operation. None is added to the perfusion medium and this appears to be adequate to reduce motility and permit uniform perfusion. This together with the choice of medium are the critical features leading to success in this preparation. The use of promethazine in turn, necessitates a tracheotomy (see Chapter 9).

The abdomen is opened in the mid-line, with lateral incisions in the usual way, and the gut is displaced to the animal's right. The operative technique is designed to free the aorta from the posterior abdominal wall, from its termination at the iliac and caudal arteries to its point of entry into the abdomen through the diaphragm. This must be achieved without occluding the inferior mesenteric artery, which arises very low down, just before the division of the aorta, and the superior mesenteric artery, which has already been described (see Fig. 7.1(b)) as it arises opposite the right renal artery. In addition, the superior mesenteric vein must be located and cannulated. In the process, the entire gut, from pylorus to descending colon, is freed from its attachments to the abdominal organs and from the posterior abdominal wall. The small bowel is excluded from the present perfusion and the colon is then free to be transferred to the organ bath.

(i) The aorta is cleared and separated from the inferior vena cava, in its lower half. Blunt dissection or even separation by firm pressure with both index fingers is preferable to the use of sharp instruments. Most animals are lost at this early stage by tearing the friable inferior vena cava. It is not safe to pass a forcep behind the right iliac artery (q.v.) without first carefully separating its right lateral border from the underlying vein.

(ii) Ligature 1 is passed behind both iliac arteries, then its other end is passed behind the aorta higher up. This has the effect of tying off the caudal artery lying on the posterior aspect of the aorta at its termination and inaccessible to normal methods of ligation. This technique of passing a ligature *twice* behind the aorta is a useful method of tying all the side branches of the aorta in an ascending order, without necessarily visualizing the vessel to be tied.

(iii) Ligatures, two to each vessel or group of vessels, enabling a cut to be made between, are now applied as described to the branches of the aorta in ascending order, using the first ligature, that around the left iliac artery, and left long on purpose, as a means of lifting the aorta ventrally as it is freed.

(iv) The left renal artery and vein may be ligated and cut together. The right renal artery is ligated separately.

(v) Now the lower colon is ligated below the point of reflection of the inferior mesenteric artery. The lumen is opened to allow the insertion of a glass cannula. The upper limit of large intestine to be perfused is defined by a ligature below the caecum, and the upper cannula is inserted at this level. Warmed Krebs–Henseleit saline is used to irrigate the intestine from below, upwards. The direction is important, since irrigation in the direction of peristalsis tends to cause constriction of the gut.

The upper cannulation excludes the caecum from the preparation and all the vessels to the caecum and to the small intestine can now be tied. This includes all branches of the superior mesenteric artery to the small intestine, as far as the duodeno-jejunal junction. Using single ties, the vessels and intervening mesentery can be cut and the intestine thus freed can be removed. It is convenient to start this at the caecum and work upwards.

(vi) The duodenum is freed from the overlying mesentery by tying the vessels concerned in double ligatures and cutting between them.

(vii) Heparin is given intravenously, 250 units, either into the femoral vein or, more conveniently, in this preparation, into the dorsal vein of the penis.

(viii) Arterial cannulation is achieved through the left internal iliac artery, around which loose ligatures are positioned. The cannula is filled with medium; a bulldog clamp controls the internal iliac artery at the bifurcation and an incision is made to receive the end of the cannula. A few ml of medium are injected from a syringe to wash through the cannula.

Cannulation of the superior mesenteric vein (portal vein) is performed, and the cannula tied in place with the prepared ligatures. Blood flows freely from the end of the cannula.

*Mounting the intestine.* The intestine and its four cannulae (two luminal and two vascular) is lifted from the abdomen of the rat and placed vertically on the metal support described above. The lower end of intestine is uppermost, so that reverse irrigation of the lumen dispels gas. The gut and its support are lowered into the paraffin bath and lumen irrigation begun.

The long arterial extension is shortened, with a brief interruption of flow; care must be taken to avoid bubbles. When flow is resumed, the venous outflow is led from the cannula to the reservoir and recirculation can begin.

## Characteristics of perfusion

Perfusions of 4–5 h duration have been achieved by Powis. Perfusion pressure and flow rate are well maintained for this time. Powis also points out the relationship between intra-luminal pressure and the pressure required to perfuse the vascular bed. At 7 cm $H_2O$ luminal pressure, peristalsis is minimal, and the perfusion pressure is 32 cm $H_2O$. Maintenance of circulating volume requires a minimum of 3 g % albumin in the perfusion medium, and with this addition histology of the colon wall appears to be normal, both on light-and electron-microscopy, after 5 h of perfusion.

*Metabolic activity.* (a) Glucose uptake (542 $\mu$mol/h per g dry wt.) and lactate production (382 $\mu$mol/h per g dry wt.) occur in the colon perfused with standard medium. The rate of glycolysis appears to be higher if non-haemoglobin media are used, suggesting that these preparations are anoxic, despite adequate flow rates. (b) Oxygen uptake (9 $\mu$mol/h per g dry wt.) is constant for 4 h of perfusion. (c) As regards ion transport by the colonic mucosa, the central study undertaken by Powis was of bicarbonate and other ion transport from lumen to medium and vice versa. Total ion transport is lower in this perfused preparation than that observed *in vivo*, and the prior injection of promethazine (as described in technique, above) only partly restores this to normal. The active transport of ions appears to be a useful test of viability in this preparation, and it is dependent upon many variable factors—the composition of the perfusion medium, the lumen medium, perfusion pressure, and flow rate. Active transport of $Na^+$ and $HCO_3^-$ has been demonstrated and continues at the same rate for the 4 h of perfusion. Bicarbonate secretion is so rapid that significant changes in the bicarbonate content of the medium occur in recirculation experiments. Since pH too may vary as a result, 'once-through' perfusion is recommended (see discussion in Chapter 1).

## GENERAL COMMENTS ON INTESTINAL PERFUSION TECHNIQUE

Some of the necessary comparisons between the various preparations have already been presented. No major differences have been found between *in situ* and isolated preparations, nor does pulsatile blood-flow alter the character of the perfusion.

Survival time in both instances is similar, and oxygen consumption, flow rate, and perfusion pressure do not differ. However, none of these tests is sufficiently critical and more stringent, perhaps biochemical, parameters of intestinal function need to be established.

Glucose consumption, thymidine uptake, lipoprotein synthesis (or at least, incorporation of labelled H) have been studied by workers in the field and could be adapted to quantitative tests of function. Absorption of substances from the intestinal lumen by active transport is an important test of mucosal function but may be too crude for the present purpose. Unless some metabolic transformation occurs, for example, glucose → lactate or fat → triglyceride or → lipoprotein, absorption *per se* may be a mucosal function that continues after the remainder of the tissue has ceased to be viable. Thus, in everted sacs (Wiseman 1961) glucose uptake continues to occur, together with transport onto the serosal surface, although only the oxygenation of the mucosal layer itself is likely to be adequate.

Iron absorption, cobalt absorption, and transport of other substances may be qualitatively of interest, but it is difficult to devise a quantitative test on this basis (Forth 1968). Furthermore, the observation that $O_2$ is not required for uptake of some ions takes it from the realm of those energy-dependent processes, which are likely to be good tests of cellular integrity (see Chapter 1).

*Composition of the medium. The use of drugs*

The special problem of intestinal perfusion appears to be the ability of peristalsis in the muscular layers to so occlude the blood-vessels that perfusion pressure rises and flow may cease altogether. Extra peristalsis which is seen in these preparations may in turn be a consequence of oxygen lack, although this is at present only surmise (Dubois *et al.*

1968). A variety of methods has been applied to overcome this problem and most authors have resorted to the addition of drugs to the medium. Table 7.2 shows the range of drugs and other modifications tested; unfortunately, each author usually reports only one type of experiment, and since other differences between perfusion media and conditions of perfusion are present, the table can only act as a guide to the drug to be applied in each circumstance.

TABLE 7.2

*Agents that may be used to reduce gut motility and/or improve perfusion of intestine*

| Author | Drug | Other manoeuvre |
|---|---|---|
| Parsons and Pritchard (frog) | None | Low $Ca^{2+}$ and $Mg^{2+}$ in perfusate |
| Dubois *et al.* | None | Pentobarbitone anaesthesia |
| Lee and Duncan | Pentobarbitone (5 mg%) | |
| Windmueller *et al.* | (i) Pentobarbitone $2 \times 10^{-4}$M | |
| | (ii) Propanalol $10^{-5}$M | |
| | (iii) Glucocorticoid $6 \times 10^{-7}$M Plus Noradrenaline $4-8 \times 10^{-8}$M | |
| | (iv) None | Perfusate recycled through second ether-anaesthetized rat |
| Kavin *et al.* | None | Glass wool filter |
| Gerber and Remy-Defraigne | None | |
| Forth | (i) Papaverine 10 mg% Plus Promethazine 4 mg% | |
| | (ii) Papaverine 6·7 mg% | |
| Powis | (i) None | Pre-operative promethazine 2 mg/ 100 g body weight, i.v. |
| | (ii) Promethazine 20 mg% | |

There are objections to this approach, not the least of which is that each of the agents added to the perfusion medium has metabolic effects in its own right and leads away from the aim of perfusion, that of providing a defined *in vitro* system. Powis (1970) suggests the injection of an agent (promethazine) into the animal before perfusion; but this too, probably persists in the intestinal wall throughout perfusion. The fact that some authors, notably Gerber and Remy-Defraigne, and Kavin *et al.* have achieved satisfactory preparations without the use of drugs encourages one to look harder for the causes of failure in other preparations, and to aim for an isolated perfused intestine preparation under fully defined conditions. At present, the preparations of Windmueller *et al.* for small intestine, and of Powis for large intestine appear to be those most suited to biochemical experiment.

## REFERENCES

BABKIN, B. P. (1950) *Secretory mechanism of the digestive glands.* Hoeber, New York.

BLAKELEY, A. G. H., BROWN, G. L., and FERRY, C. B. (1963) Pharmacological experiments on the release of sympathetic transmitter. *J. Physiol., Lond.* **167**, 505.

BOLLMAN, J. L., CAIN, J. C., and GRINDLAY, J. H. (1948) Techniques for the collection of lymph from the liver, small intestine or thoracic duct of the rat. *J. Lab. clin. Med.* **33**, 1349.

BOYD, C. A. R., PARSONS, D. S., and THOMAS, A. V. (1968) The presence of $K^+$ dependent phosphatase in intestinal epithelial cell brush borders isolated by a new method. *Biochem. biophys. Acta* **150**, 723.

DICKENS, F. and WEIL-MALHERBE, H. (1941) Metabolism of normal and tumour tissue. 19. The metabolism of intestinal mucous membrane. *Biochem. J.* **35**, 7.

DUBOIS, R. S., VAUGHAN, G. D., and ROY, C. C. (1968) Isolated rat small intestine with intact circulation. In *Organ perfusion and preservation* (ed. J. C. Norman), p. 863. Appleton Century Crofts, New York.

EXTON, J. H. and PARK, C. R. (1967) Control of gluconeogenesis in liver. General features of gluconeogenesis in the perfused livers of rats. *J. biol. Chem.* **242**, 2622.

FISHER, R. B. and PARSONS, D. S. (1949) A preparation of surviving rat small-intestine for the study of absorption. *J. Physiol., Lond.* **110**, 36.

FORTH, W. (1968) Eisen und Kobalt-Resorption am perfundierten Dunndarm-segment. 3 *Konf. der Gesellschaft für Biologische Chemie* (eds W. Staib and R. Scholz), p. 242. Springer-Verlag, Berlin.

GERBER, G. B. and REMY-DEFRAIGNE, J. (1966) DNA metabolism in perfused organs (II. Incorporation in DNA, and catabolism of thymidine at different levels of substrate by normal and X-irradiated liver and intestine.) *Archs int. Physiol. Biochem.* **74**, 785.

GHOSH, M. N. and SCHILD, H. O. (1958) Continuous recording of acid gastric secretion in the rat. *Br. J. Pharmacol.* **13**, 54.

HEMS, R., ROSS, B. D., BERRY, M. N., and KREBS, H. A. (1966) Gluconeogenesis in the perfused rat liver. *Biochem. J.* **101**, 284.

HESTRIN-LERNER, S. and SHAPIRO, B. (1954) Absorption of glucose from the intestine: II *in vivo* and perfusion studies. *Biochim. biophys. Acta* **13**, 54.

KAVIN, H., LEVIN, N. W., and STANLEY, M. M. (1967) Isolated perfused, rat small bowel—technique, studies of viability, glucose absorption. *J. appl. Physiol.* **22**, 604.

KUCZYNSKI, N. (1890) Beiträge zur Histologie der Brunnerschen Drüsen. *Pam. Tow. lek. warsz.* **86**, 32.

LEE, J. S. and DUNCAN, K. M. (1968) Lymphatic and venous transport of water from rat jejunum: a vascular perfusion study. *Gastroenterology* **54**, 559.

LONG, C. (ed.) (1961) *Biochemists handbook*, p. 798. Spon. London.

MAXIMOW, A. A. and BLOOM, W. (1952) *Textbook of histology*, 6th edn, Chapters 23 and 24. Saunders, Philadelphia.

NISHIITSUTSUJI–UWO, J. M., ROSS, B. D., and KREBS, H. A. (1967) Metabolic activities of the isolated, perfused rat kidney. *Biochem. J.* **103**, 852.

PARSONS, D. S. (1968) Methods for the investigation of intestinal absorption. *Handbook of physiology*, Vol. 3, p. 1177. American Physiological Society.

—— and POWIS, G. (1971) Some properties of a preparation of rat colon perfused *in vitro* through the vascular bed. *J. Physiol., Lond.* **217**, 641.

—— and PRITCHARD, J. S. (1968) A preparation of perfused small intestine for the study of absorption in amphibia. *J. Physiol., Lond.* **198**, 405.

POWIS, G. (1970) Observations of metabolism and transport in organs perfused *in vitro*. D.Phil. Thesis, University of Oxford.

SCHOLZ, R. (1968) *Stoffwechsel der isoliert perfundierten Leber* (eds. W. Staib and R. Scholz). Springer-Verlag, Berlin.

SCHULTE, F. (1958) Speiseröhre, Magen, Darm and Bauchfell. *Pathologie der Laboratoriumstiere*, Vol. 1, p. 99. Springer-Verlag, Berlin.

SUSSMAN, K. E., VAUGHAN, G. D., and TIMMER, R. F. (1966) An *in vitro* method for studying insulin secretion in the perfused isolated rat pancreas. *Metabolism* **15**, 466.

WINDMUELLER, H. G. and SPAETH, A. E. (1969) Vascular perfusion of the isolated rat small-intestine: fluid flux and lymph lipoprotein biosynthesis. *Fedn Proc. Fedn Am. Socs. exp. Biol.* **28**, 323.

—— —— and GANOTE, C. E. (1970) Vascular perfusion of the isolated rat gut: norepinephrine and glucocorticoid requirement. *Am. J. Physiol.* **218**, 197.

WISEMAN, G. (1961) Sac of everted intestine technic for study of intestinal absorption *in vitro*. *Meth. med. Res.* **9**, 287.

# 8. Other Organs and Tissues

APART from preparations of perfused adipose tissue and brain (Andjus, Suhara, and Sloviter 1967; Robert and Scow 1963; Ho and Meng 1964), which were developed specifically to study biochemical problems, most of the methods to be discussed briefly in this section are unsuited in their present form to such an end. Immunological studies have been performed in perfusions of spleen, thymus, and bone; preparations of placenta and of salivary gland have been developed but are still in an experimental phase. The present purpose is to describe the preparations in outline, so that their merits may be assessed by workers interested in their possible biochemical applications. In an anatomical sense, perfusion of an organ or tissue may present no problems: even very small vessels may be perfused if their parent trunk can be cannulated, and all other branches occluded. But the collection of 100 per cent of the perfusate may be difficult, and the maintenance and assessment of normal function requires detailed study in each individual case.

## ADIPOSE TISSUE

On the basis of studies in human embryo, work which has been confirmed in many other species, Wassermann (1965 for review) introduced the concept of fat as an organ. Thus, fat cells are specialized, are always in a specific anatomical relationship to blood-vessels, and adipose tissue can always be recognized as such, even when depleted of fat. Characteristic of this location is the existence of 'fat pads', for example, epididymal, parametrial, lumbar, mesenteric, etc. The concept of organ perfusion has been extended to biochemical studies in adipose tissue (Robert and Scow 1963, Ho and Meng 1964).

### Anatomy

It is generally agreed that adipose tissue does not exist in an avascular form, and its anatomical distribution is therefore determined by blood supply. A number of localized 'organs' exist, of which the epididymal and myometrial fat pads are examples. They each have a distinct blood supply (see Methods for detail), innervation from the autonomic nervous system, and lymphatic drainage. The microscopic anatomy is discussed in detail by Rodbell and Scow (1965) and it is perhaps sufficient in the present context to refer to the very close relationship between the fat cell wall and the capillary endothelium. Chylomicra from the vessel lumen are taken up across the two membranes, and it is probable that the action of lipoprotein lipase, an enzyme that regulates fatty acid release into the blood stream from adipose tissue is located near this interface. In any study of fat tissue the vascular component is of importance, and this has promoted an interest in vascular perfusion of adipose tissue in biochemical investigation.

Alternative experimental methods in biochemical studies of adipose tissue are discussed by Renold and Cahill (1965): (a) *in vivo* techniques; (b) *in vitro* techniques; homogenates and extracts have been widely used.

Isolated intact adipose tissue cells can be prepared by the technique introduced by Rodbell (1964). The epididymal fat pad of rat or mouse also lends itself to incubation experiments, since it is a very fine sheet of tissue in which oxygen supply by diffusion presents no problem.

### Methods of perfusion

Method I. Parametrial fat body. Method of Robert and Scow (1963). Method II. Epididymal fat pad. Method of Ho and Meng (1964). Both methods have remained esoteric and no other authors have published results with these preparations.

### Method I. Perfusion of parametrial fat body in the rat

The authors of this method illustrate the parametrial fat body (Fig. 8.1), a pad of fat, 0·6–1·2 g weight, lying alongside the horn of the uterus on each side. Its blood supply is also illustrated, as it arises from the common iliac vessels. The preparation is perfused by means of direct-pulsatile flow using defibrinated rat blood, which has been diluted by 10 per cent with Tyrode solution containing 4 g % dialysed albumin. The apparatus is illustrated and the operative preparation discussed in detail in an article by Scow (1965). A 'once-through' technique is usually employed.

*Apparatus*

The apparatus of Long and Lyons (1954), introduced for liver perfusion (see p. 189) was used, with modifications, since sterile conditions available with the original apparatus are not required here. A valve pump (see p. 64), a rotating-flask oxygenator, a filter of 'broadcloth',

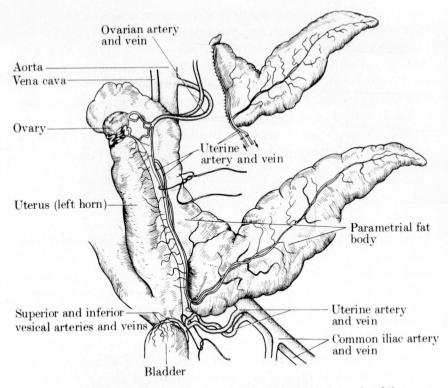

Ovarian artery
and vein

Aorta

Vena cava

Ovary

Uterine
artery and vein

Uterus (left horn)

Parametrial fat
body

Superior and inferior
vesical arteries and veins

Uterine artery
and vein

Common iliac artery
and vein

Bladder

FIG. 8.1. Perfusion of adipose tissue. Anatomy and blood supply of the parametrial fat body: method of Robert and Scow (1963). The fat body is isolated
from the adjacent uterus and from its ovarian blood supply and perfusion is
established through the uterine artery and vein. For details see p. 398.

and a water bath through which a coil of polythene tubing carrying
perfusate to the fat pad passes, constitute the apparatus.

*Perfusion medium*

*Medium* I. Defibrinated rat blood is filtered and mixed in the proportion 1:10 with the albumin mixture described below. Defibrination
is accomplished by stirring blood in a siliconized glass vessel with a
bundle of polyethylene rods, rotated at 65 rev/min for 45 min. The clot
is removed and the blood filtered through 'broadcloth'. Tyrode solution, in which Fraction V bovine serum albumin (see p. 27) is dissolved
to 30 g %, is dialysed for 18 h against two or three changes of Tyrode
solution. This is then diluted to 4 per cent albumin with further Tyrode
solution, penicillin (333 units/ml) and streptomycin (0·022 mg/ml) are
added and the resulting solution is used to dilute the defibrinated blood

to an haematocrit 4·5 per cent. The final glucose concentration of this medium is adjusted to 0·05 g % (2·8 mM approx.).

*Medium* II. Bovine albumin in Tyrode solution, either 4 or 8%, without red cells, has been used as an alternative medium in studies on the accumulation of oedema in perfused fat pad.

*Bathing medium.* The isolated fat pad, apart from its vascular perfusion, is bathed in medium of which the glucose content is said to be critical. The supernatant 'serum' of Medium I is suitable. It is needless to point out that this preparation is now a combined 'perfusion–incubation'.

## Operative procedure

Female rats (150–220 g wt.) are donors of fat, while both male and female donors are used to perfuse. The animal is anaesthetized with ether, and the left uterine horn exposed. Fig. 8.1 shows the vessels and ligatures that are required to isolate and cannulate the uterine artery and vein. A dissecting microscope is required, as the authors recommend that these vessels be severed at their origins from the common iliac vessels. Diathermy is required to ensure that small vessels between the fat pad and the adjacent uterus are occluded and the organ with its cannulae is transferred to the organ chamber. The operation is said to take from 10 to 15 min, and involves 1 min of anoxia during the cannulation.

## Characteristics of perfusion

At a perfusion pressure of 85–95 mmHg, a flow rate of approximately 0·2 ml/min is obtained (Medium I) and maintained for up to 2 h. In the absence of red cells (Medium II), a higher flow rate is obtained— 0·39 ml/min.

The unsolved problem of this perfusion (as of the epididymal fat pad preparation, see below) is oedema formation (Scow 1965). With Medium I, 46 per cent increase in weight occurs within the early stages of perfusion. Medium II results in either 600 per cent increase (with 4 per cent albumin) or some 300 per cent increase (with 8 per cent albumin). None of these preparations shows linearity of fatty acid release, a parameter that might be taken as a test of function. Nevertheless, this preparation has been used to demonstrate an effect of heparin (lipoprotein lipase release) and of adrenaline. The results will be discussed in conjunction with those obtained with the epididymal fat preparation below.

## Method II. Perfusion of isolated epididymal fat pad of rat

The epididymal fat pad is supplied by the internal spermatic artery as is the testis. As described on p. 400, the spermatic artery arises high on the aorta, or even as a branch of the renal artery on the same side. The venous drainage of the area is to the inferior vena cava, and by suitably placed ligatures the epididymis and testis can be excluded from the circulation, leaving an isolated perfused preparation of adipose tissue (Ho and Meng 1964). A technically difficult preparation, the fat pad is contained in a perfusion apparatus like that used by Morgan *et al.* for heart perfusion (Morgan *et al.* 1961; see Chapter 5). A semi-synthetic perfusion medium, 5% bovine albumin, in Krebs–Henseleit buffer, gassed with 95% $O_2$, 5% $CO_2$, is pumped directly into the relevant segment of aorta, and recirculated after passing through a sintered-glass filter. Flow rates of 0·4–5 ml/min are maintained for a 50-min experimental period.

### *Apparatus*

Fig. 8.2 shows the apparatus used, and its similarity to that of Morgan is evident (cf. Fig. 5.4). Pulsation from the pump is damped by the passage of medium through a bubble trap; oxygenation is less carefully considered in this modified apparatus than in the original design, and the gas delivery tube is simply allowed to dip beneath the surface of the medium that accumulates on the surface of the filter. Temperature control is by means of water-jacketing. Some preparations have both inflow and outflow cannulae (see below), but in the absence of an out-flow cannula, effluent medium is allowed to bathe the outer surface of the perfused fat pad, in a manner rather similar to that of the 'perfusion-incubation' of Robert and Scow (q.v.).

### *Perfusion medium*

Five per cent bovine albumin in Krebs–Henseleit buffer gassed with 95% $O_2$:5% $CO_2$. (No adjustment was made for bicarbonate losses—see p. 39.)

### *Operative procedure*

The donor rat (male, 150–200 g) is anaesthetized with intra-peritoneal Nembutal (q.v.). Heparin 5 mg/kg, was given intravenously in early experiments but was later abandoned, apparently without effect on the perfusion or on the results. In the operative procedure suggested, a mid-line abdominal incision is made, through skin and the linea alba,

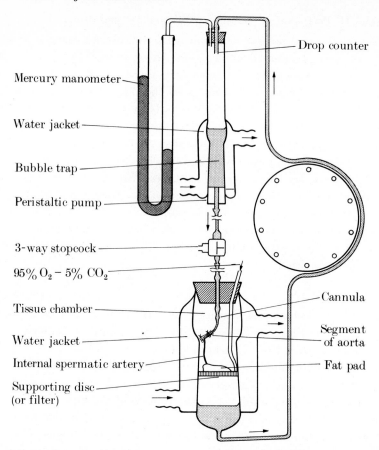

Mercury manometer

Water jacket

Bubble trap

Peristaltic pump

3-way stopcock

95% $O_2$ – 5% $CO_2$

Tissue chamber

Water jacket

Internal spermatic artery

Supporting disc
(or filter)

Drop counter

Cannula

Segment
of aorta

Fat pad

FIG. 8.2.  Perfusion of adipose tissue. Diagram of the apparatus required for perfusion of the epididymal fat pad: method of Ho and Meng (1964). The diagram shows the recirculation of medium during perfusion. A small amount of medium collects above the filter disc and the gas inlet dips into this fluid. With a non-recirculation perfusion the filter disc is replaced by a coarsely perforated Teflon plate.

extending the incision laterally between clamps as previously described. Without disturbing the testis or epididymal fat pad, the ligatures illustrated in Fig. 8.3 are placed. To perfuse the left pad, after tying the right spermatic artery at its origin (Lig. (1)), Ligatures (2) and (3) are placed around the aorta, above and below the origin of the left spermatic artery, with Ligature (3) below the left renal artery. The authors advise displacing the left kidney to the animal's right for this, since the aorta is thereby exposed. Alternatively, the left renal vein may be divided between ligatures and the caval end retracted, with the same effect. The segment of aorta enclosed by the two ligatures is cannulated from above, and Ligatures (2) and (3) tied, to retain the cannula and to occlude the distal vessel. Perfusion begins at once, and this can be seen in the testis and epididymal fat pad if they are now lifted from the scrotum. The block of tissue including the spermatic artery and vein, and the segments of aorta and inferior vena cava to

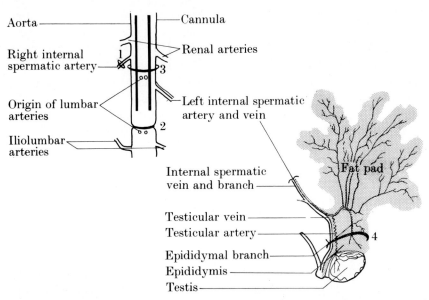

FIG. 8.3. Perfusion of adipose tissue. Operative procedure for perfusion of rat epididymal fat pad (Ho and Meng 1964). The diagram shows the anatomical relationships in the cannulation of the aorta for perfusion. The order of ligations is indicated by numbers 1–4.

which they relate, plus the epididymis, testis, and fat pad are removed to the organ chamber. Ligature (4) can now be placed to exclude the testis and epididymis from the circuit. (For studies on oxygen consumption, the relevant segment of the inferior vena cava is cannulated to collect spermatic vein-effluent, and ideally, this is done before cannulation of the aorta.)

*Alternative operative procedure* (see also p. 352)

A slightly modified operative procedure is suggested below, which allows of simpler aortic cannulation. It might in addition be adapted to perfusion of the rat testis. Apparatus and cannulae similar to those used for kidney perfusion (Nishiitsutsuji–Uwo *et al.* 1967), may be used.

The abdomen is opened in the mid-line with lateral extensions between clamps, as already described (p. 88). The intestine is diverted to the animal's left, and the connective tissue that overlies the aorta and inferior vena cava at the level of the renal vessels and below, is separated, most effectively with the fingers. The left renal vein is examined to exclude the case of anomalous drainage of the spermatic vein into this vessel, and it is then cleared and ligated twice. By cutting the renal vein between these ligatures, the medial end may be retracted to expose the aorta and the origins of the spermatic arteries. The inferior vena cava and aorta are separated by careful use of curved forceps, with particular attention to the right spermatic artery which crosses the inferior vena cava, usually about 1 cm below the renal vein. Ligatures, an upper and a lower, are placed loosely about the inferior vena cava and aorta, and at this stage it is not necessary to define which epididymal fat pad is to be perfused; the aorta, inferior vena cava, spermatic vessels, fat pad, and testis are mobilized to allow rapid removal after cannulation of the great vessels.

*Venous cannulation.* The inferior vena cava is cannulated as for kidney perfusion (p. 234), with the exception that the lower ligature is below the right renal vein.

*Arterial cannulation.* The aortic cannula is most easily inserted from below; the lower ligature is tied, and the aorta is occluded with a bulldog clip, an incision is made in the anterior surface of the aorta and the cannula, through which medium is already flowing, is inserted so that the tip lies at the level of the spermatic vessels. Perfusion begins at once, and is limited to the spermatic vessels by tying the upper aortic ligature. The unwanted (contra-lateral) spermatic vessels are now ligated, and cut below the ligature, the perfusing block of tissues is excised and transferred with care to the organ chamber, and finally the ligature (Lig. 4 in Fig. 8.3) is sited to exclude the testis and epididymis from the circulation. (Alternatively, the fat pad may be excluded, and the preparation serves as a perfusion of testis and epididymis.)

*Characteristics of perfusion* (Ho and Meng preparation)

A constant flow of medium, at 0·4–0·5 ml/min is maintained for 60 min, with some adjustment to the perfusion pressure. The pressure is reported to be about 100 mmHg initially, but falls within 5 min to the steady value of approximately 50 mmHg.

Arterial oxygen tension was below saturation, but at the flow rates used effluent medium was not desaturated of oxygen. The authors give no indication of the linearity of oxygen consumption with time, but report a basal oxygen consumption of $2 \cdot 8 \, \mu\text{mol/min}$ per g, rising to $4 \cdot 2 \, \mu\text{mol/min}$ per g in the presence of adrenaline. The approximate figures for incubated epididymal fat (incubated in the presence of albumin and glucose) are $1 \cdot 2 \, \mu\text{mol/min}$ per g, rising to $1 \cdot 5 \, \mu\text{mol/min}$ per g with adrenaline. Maximal rates of oxygen consumption in the incubation experiments were higher (up to $3 \cdot 5 \, \mu\text{mol/min}$ per g) but do not reach the levels observed in perfusion.

The release of fatty acids follows a linear time course, for the 50-min experimental period, and the addition of adrenaline to the perfusate produces the expected increase in the rate. It remains linear, and the response itself has been used by these authors to test the viability of the preparation. Oedema in this preparation, unlike that of Robert and Scow (q.v.) is not a serious problem. An increase in weight of 15 per cent occurs early in the perfusion, but it does not increase further. The best results of Robert and Scow showed a weight gain of some 46 per cent, while, using a semi-synthetic perfusion medium, such as that used here, the increase was several hundred per cent.

No explanation is immediately available for this difference, but it may derive from the difference in operative technique: the present preparation uses cannulation of a large vessel (the aorta) with a minimum anoxic period. In the preparation of the parametrial fat body by the accepted technique, the terminal artery, the uterine vessel, is cannulated, which is more difficult and time-consuming. Anoxia, and perfusion problems generally, are greater in the latter preparation and may account for the unsatisfactory result of oedema formation. Since this probably has consequences for the biochemical studies (for example, lipoprotein lipase release appears to behave differently in this perfusion from the expectation (Salaman and Robinson 1966)), the epididymal fat pad preparation must be preferred at present.

## BRAIN

Biochemical studies in the brain have been conducted *in vivo* and *in vitro*, in preparations at various levels of organization—tissue slices, homogenates, differential cell fractions, and sub-cellular fractions. Perfusion has found little place in these studies since the technique presents great difficulties. In addition, the brain is heterogeneous, and the metabolic contribution of the various component cell and tissue types would present difficulties in interpretation. One aspect of brain metabolism in particular, namely the blood-brain barrier remains particularly obscure and has been the subject of studies with perfused brain preparations (Geiger and Magnes 1947, Andjus *et al.* 1967, and Thompson, Robertson, and Bauer 1968).

Two preparations relevant to the rat will be described. The simpler is a 'perfused rat head' preparation (Thompson *et al.* 1968), in which the carotid distribution is perfused with no attempt to limit the circulation to the brain. Somewhat more elaborate, and therefore technically more difficult, is the preparation of Andjus *et al.* (1967) which attempts to exclude the muscle of head and neck from the circuit. This preparation is based upon a more elaborate technique developed in the cat by Geiger and Magnes (1947). References to the more frequent use of larger animals (monkey and dog) are included in these papers.

*Uses of brain perfusion*

In addition to the metabolic applications to be discussed below, the technique of perfusion has been used to aid histological fixation (Torack 1969 and Palay, McGee-Russell, Gordon, and Grillo 1962). For this purpose  the arch of the aorta is perfused with a 'balanced salt' solution followed by a fixative. Noteworthy is the use of artificial ventilation to reduce the period of anoxia in this preparation (contrast Thompson below).

**Anatomy**

The macroscopic anatomy of the brain of various species of small mammals is described by Gerhard, Meesen, and Veith (1958); there are differences in the degree to which the various regions (olfactory cortex, hemispheres, cerebellum, etc.) are developed. Microscopic anatomy is described in standard histological texts (Bloom and Fawcett 1968), and is marked by heterogeneity of cell type within the same area of the brain and between different areas.

*Arterial supply.* In the rat, an internal carotid vessel on each side (arising from the arch of the aorta and from the innominate artery respectively), and a vertebral vessel on each side (a branch of the costocervical branch of the subclavian vessel, and therefore derived from the arch of the aorta) provide all the blood to the brain and to the skull. The extra-cranial structures are largely supplied by the external carotid artery and the pterygopalatine artery (see Green 1963) and tying these (Andjus *et al.* 1967) limits the perfused area to the cranium and intracranial structures. In the cat (Geiger and Magnes 1947) the internal carotid artery is less important than is the internal maxillary artery. The special features of cat cerebral vasculature that lend themselves to perfusion are discussed by the authors quoted above.

*Venous drainage.* Blood from cerebral veins enters sinuses that lie on the surface of the brain and all blood leaves the cranium via the internal jugular veins. Cannulation of the superior vena cava (or the right atrium) collects this outflow (cf. Thompson *et al.* 1968).

*Lymphatic drainage.* Rusznyák, Foldi, and Szabo (1967) discuss the lymphatic drainage of the brain and associated structures.

*Cerebrospinal fluid.* Fluid formed by ultra-filtration in vascular plexuses within the ventricles, circulates both in the ventricular system and over the surface of the brain, in the sub-arachnoid space. A separate preparation for 'perfusion' of the ventricles is described (Scremin 1968) and this approach may be supplementary to any more formal technique of brain perfusion.

## Methods of perfusion

Method I. Rat head perfusion. Method of Thompson, Robertson, and Bauer (1968).

Method II. 'Isolated' perfused rat brain preparation (Andjus *et al.* 1967).

Method III. Isolated perfused brain of the cat (Geiger and Magnes 1947).

## Method I. Rat head perfusion. Method of Thompson *et al.*

In an apparatus reminiscent of that of Miller *et al.* (1951), for perfusion of the liver (see p. 148), the aortic arch of the rat is perfused, and the inferior vena cava is cannulated for the outflow. Recirculation of a perfusion medium that contains rat blood maintains the preparation for up to 3 h (see p. 407).

The main advantage of this preparation over the more direct brain perfusions is its relative simplicity of operative preparation. Its limitations in metabolic studies as a heterogeneous tissue are apparent.

### Apparatus

Based on an apparatus designed for the perfusion of rat liver (Thompson *et al.* 1958), a pump provides pulsatile flow at a pressure of 120/90.

An oxygenator of the Hooker type (see p. 57), a Dacron cloth filter, and a heating coil which is immersed in a water bath at 38°C, complete the circuit.

*Cannulae.* Polythene cannulae are used, PE 240 (o.d. 2·4 mm) for the aorta and PE 205 (o.d. 2·0 mm) for the inferior vena cava.

*Perfusion medium.* Fresh (heparinized) rat blood is collected from animals under pentobarbitone anaesthesia and diluted to an haematocrit of 12–16 per cent with a buffer containing 5 per cent bovine serum albumin, Fraction V (Pentex), and glucose, terramycin, and 'salts'. Precise details are not given but it is based on the medium used by Brauer (pp. 20 and 184).

### Operative procedure

Rats of 200–300 g weight are anaesthetized with thiopentone (50 mg/kg intraperitoneally). (Better perfusions are obtained in animals of 350–400 g wt.) The abdomen is opened in the mid-line in the standard way and the aorta, inferior vena cava, and renal vessels are exposed.

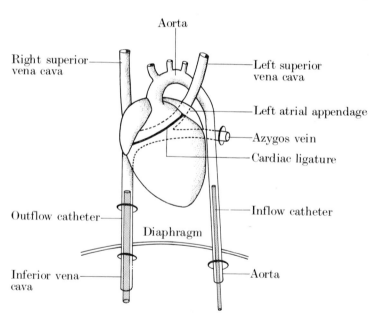

F IG. 8.4.    Perfusion of brain. Operative preparation of the rat head perfusion technique of Thompson *et al.* (1968). The ligatures required are illustrated and the operative technique is described in detail on p. 407.

Heparin (200 units intravenously) is given at this stage (the authors suggest that this should be into the left renal vein, which is then tied). Loose ligatures are placed individually around the inferior vena cava and the aorta just below the diaphragm. Each vessel is then cannulated from below and the cannulae are tied in place with the ligatures prepared. Perfusion begins at once. The thorax is opened and the heart exposed. A firm ligature is placed about the heart in the atrio-ventricular groove and, finally, the azygos vein (Fig. 8.4) is tied.

The outflow from the two superior venae cavae is therefore collected in the inferior vena cava, the remaining inflow being occluded by the ligature around the azygos vein. To restrict the area of arterial perfusion, clamps are placed at the base of each fore-limb and along the cut edges of the thorax.

*Characteristics of perfusion*

Flow, which starts at 16 ml/min, increases in the next 5 min, and is then apparently maintained for the remainder of a 3-h perfusion period. Perfusion pressure is determined by the pump, 120/90 mmHg. After 3 h, histology is apparently normal, apart from very limited areas of oedema.

More specifically, the electro-encephalogram shows an apparently normal pattern for the entire period and the regulatory functions of the brain on respiratory movement, constriction of the pupils, and a corneal reflex are preserved. No metabolic information is provided in this report and it is obviously such a complex preparation that it will be difficult to interpret the results in terms of cerebral metabolism.

**Method II. 'Isolated' perfused rat brain preparation. Method of Andjus et al.**

By cannulating the common carotid artery on each side and undertaking a prolonged operative procedure under hypothermia, these authors achieve a preparation which is more or less restricted to the skull and brain. With the semi-synthetic perfusion medium devised by Geiger *et al.* (see below) the brain continues to show electrical activity after 2 h of perfusion, and consumes glucose, with the formation of lactate at a linear rate over the same period.

*Apparatus*

A small-volume recirculation circuit (5–15 ml), using a roller-pump (Sigmamotor TM 20-2) and a simple, bubble oxygenator are used. The preparation is held in a funnel from which effluent medium drips to the oxygenator and, in addition, a filter (nylon mesh) is provided.

*Cannula.* A branched cannula, for insertion into the two carotid vessels has an end o.d. of 3 mm.

### Perfusion medium

Note that unlike the preparation of Thompson *et al.* (Method I above) in this preparation rat blood does not maintain electrical activity.

*Medium* I. The medium of Geiger *et al.* (1954) is used for most of the experiments reported. Canine red-cells (from citrated blood) up to 1 week old, are washed three times in 0·01 M phosphate buffer, pH 7·4, and finally in Krebs–Henseleit buffer. Cells are added to a solution of bovine serum albumin (Fraction V, Nutritional Biochemicals) to give a final haematocrit of 20–25 per cent and an albumin content of 7–8 per cent, pH 7·3. Details of preparation of the albumin (deionized) and of the medium are given on p. 39.

*Medium* II (Sloviter and Kamimoto 1967). A fluorocarbon dispersion (FX 80) in albumin and Krebs–Henseleit saline has been used with virtually identical results in e.e.g. and glucose uptake. Twenty per cent FX 80 in the albumin-buffer described in Medium I above and on p. 31 is prepared by ultra-sonication (IIOW power 20 s, Branson model W-185 C cell, with 0·5-in tip).

### Operative procedure

The special features of this preparation are the avoidance of anoxia and the use of hypothermia in place of chemical anaesthesia to avoid the effects of the latter on brain metabolism. Male rats of 250 g weight are used. A 'closed vessel' technique of hypothermia is applied until a rectal temperature of 16°C is achieved.

A mid-line incision in the ventral surface of the neck is made, exposing the common carotid vessel on each side. A tracheotomy is performed (p. 449) and the animal is ventilated artificially. Heparin is injected into one external jugular vein, which is then ligated, and then the external carotid and pterygopalatine vessels are located and tied (see Green (1963), Fig. 209). The bifid cannula is filled to the tip with medium and each limb is inserted into the common carotid artery on that side. Ligatures hold the cannulae in place and perfusion begins at once, at a low rate. The dissection continues with the removal of skin, mandible, and the muscles of the head, neck, and face.

The head is separated from the rest of the animal after tying a strong ligature around the neck at the level of the fourth cervical vertebra (C4). The vertebral canal must be blocked with cotton wool and with bone wax, and the preparation can then be transferred to the tray. The pump rate is increased to give a flow rate of 1·4 ml/min, and medium flows out through the cut jugular veins.

## Characteristics of perfusion

In a recirculation system of 15 ml a flow rate of 1·4 ml/min is maintained for 2 h or more. As tests of function, a normal e.e.g. and a linear rate of glucose uptake (2·1 $\mu$mol/min) and of lactate formation (0·83 $\mu$mol/min) continue for up to 2 h, when both metabolic and electrical activity appear to fail together.

## Method III. Isolated perfused brain of the cat. Method of Geiger and Magnes

Detailed accounts of this method have appeared in articles (Geiger, Magnes, Taylor, and Veralli 1954) and reviews (Geiger 1958), and it is outside the scope of this account to repeat them here. Method II (above) owes much to this method, in terms of perfusion medium and methods of assessment of the preparation. The work reported serves as a model for future metabolic studies in perfused brain preparations.

## Conclusions

Brain does not lend itself to simple and isolated perfusion, and all the preparations to date include more or less extraneural tissue. This heterogeneity, together with the many different tissues represented within the brain itself, makes perfusion a less valuable technique for metabolic studies than some of the existing *in vitro* or *in vivo* techniques. Nevertheless, it has proved possible to maintain a viable brain by this artificial means for several hours, and interesting metabolic studies are possible.

### RETICULOENDOTHELIAL SYSTEM: SPLEEN

Perfusion of the spleen in the rat is an established technique. Dornfest and Piliero (1966) studied reticuloendothelial function and elementary metabolic characteristics of the organ, while Ford and Gowans (1967) have studied its role in lymphocyte release. A perfused cat spleen preparation has been used to study the occurrence of 'transmitter overflows' following nerve stimulation (Blakeley, Brown, and Ferry 1963) and is of particular interest as a preparation in which pulsatile flow is considered to be essential (Blakeley 1965).

The two rat preparations differ in being recirculation (Ford and Gowans) and once-through systems (Dornfest and Piliero) respectively, but both these and the cat preparation are perfused with blood-based media. Ford's preparation will be described in detail, since the two methods do not differ significantly.

**Anatomy**

In the rat, the spleen lies in the left upper quadrant of the abdomen, largely covered by the stomach; the lower pole only is visible through a ventral abdominal incision. A long gastro-splenic ligament attaches the spleen to the dorsal curve of the stomach, while a narrow splenophrenic ligament attaches it to the under surface of the diaphragm. The dimensions of the rat spleen are $3\cdot5$–$4\cdot5 \times 0\cdot8$–$1\cdot0 \times 0\cdot5$–$0\cdot6$ cm, a long organ gently curved, concave medially. The approximate weight is 1 g.

*Arterial supply.* The splenic artery arises from the coeliac axis directly, runs over the surface of the pancreas, and its terminal branches enter the spleen in three groups of vessels (see p. 411 for details). No gastroepiploic branch is found in the rat, which simplifies the operative preparations.

*Venous drainage.* The splenic veins drain into a single vessel which empties into the portal vein (see Figs. 7.5 and 8.5).

*Microscopic anatomy.* The histology of the spleen is discussed by Cohrs and Schulz (1958), and by Bloom and Fawcett (1968). Of particular interest in perfusion studies is the distribution and route by which blood flows through the organ (Snook 1950). In a recent perfusion-fixation study (Pictet *et al.* 1969) the intermediate circulation of the spleen is nicely shown; 97 per cent of the normal blood flow traverses the sinuses, while only 3 per cent passes into the interstitium to come into contact with the macrophages.

The cellular composition of the spleen changes with the age of the rat; details are provided by Cohrs and Schulz (1958).

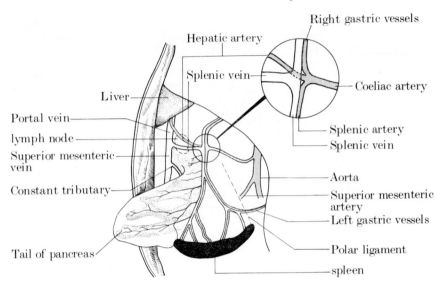

FIG. 8.5. Perfusion of spleen. Diagram of splenic blood supply of the rat as seen after cutting the avascular gastro-splenic ligament. The vessels supplying the spleen are actually embedded in pancreatic tissue. From Ford and Gowans (1967).

## Methods of perfusion

Two methods have been applied to the rat:

Method I. Recirculation-perfusion (Ford and Gowans 1967, Ford 1969).

Method II. 'Once-through' perfusion (Dornfest and Piliero 1966).

## Method I. Recirculation-perfusion. Method of Ford and Gowans and Ford

In the method of Ford and Gowans, the isolated spleen is perfused directly through the splenic artery with a blood-containing medium. Outflowing medium is collected from the splenic vein and recirculated by means of a roller pump. Such a preparation appears to be viable, in terms of lymphocyte uptake, for up to 10 h. The non-recirculating method of Dornfest and Piliero has been tested for up to 4 h.

### Apparatus

The apparatus originally used, with its more recent modifications is illustrated (Fig. 8.6). A direct pumping system, using a roller pump, and a somewhat complex spiral falling-film oxygenator is used. A nylon mesh filter is interposed between the pump and the splenic artery, with a manometer to record inflowing pressure, and a bubble trap. In more recent apparatus the falling-film oxygenator is replaced by a membrane oxygenator (Ford 1969), which consists of eight parallel channels of fine silicone rubber tubing (0·25 mm i.d., 0·5 mm o.d.). The reservoir

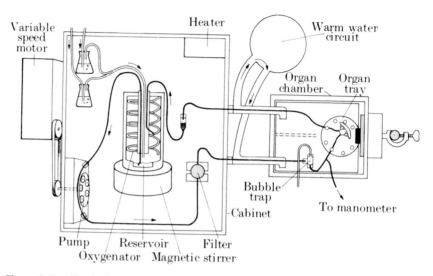

FIG. 8.6.   Perfusion of spleen. Apparatus for direct peristaltic perfusion of rat spleen (Ford and Gowans 1967, Ford 1969). The organ chamber is to the right, the pump-oxygenator to the left. Most of the apparatus is represented as seen from above.

too is modified to allow its volume to be altered during perfusion—a length of wide-bore tubing can be compressed with this result.

*Cannulae.* Bevelled polythene cannulae are recommended; i.d. 0·25 mm for the coeliac artery and i.d. 0·5 mm for the splenic vein.

### Perfusion medium

*Medium* I (Ford and Gowans): *modified rat plasma.* Rats are given heparin intravenously before bleeding by cardiac puncture. The heparinized blood is spun at 3000 $g$ for 30 min at 6°C to provide plasma, and the medium is prepared as follows: 7 ml whole heparinized rat blood+8·2 ml plasma (q.v.)+0·8 ml D-glucose (final conc. 2·5 mg/ml = 13·8 mM)+penicillin G (200 units/ml) and streptomycin 100 $\mu$g/ml). The final concentration of heparin is 35 units/ml; haematocrit 15–17 per cent; lymphocyte count 200/mm$^3$.

*Medium* II. For special studies, lymphocytes from thoracic lymph are added back to Medium I, resulting in lymphocyte counts of 4–10 × $10^3$ or $16 \times 10^3/\text{mm}^3$.

*Medium* III (Dornfest and Piliero). Whole, fresh heparinized rat blood is collected in heparinized syringes by aortic puncture, spun at 1500 rev/min for 20 min, and has the upper layer of white cells removed by aspiration. Glucose is added as 4 mg % solution to replace loss due to red-cell glycolysis, each hour during 4-h perfusions.

### Operative procedure

Rats of 230–260 g weight are anaesthetized with ether, and the abdomen is opened by a long median incision. This is extended by a horizontal incision towards the left flank. The spleen, which is handled as little as possible, is kept moist throughout the operation. The bloodless gastro-splenic ligament is divided and tributaries of the splenic vein and of the coeliac artery (other than those supplying the spleen) are divided between ligatures. The organ remains attached to the animal only by the coeliac artery and the splenic vein.

Heparin (500 units) is given intravenously; the femoral vein is a suitable route, and 10 min are allowed for its circulation. Then the aorta is clamped proximally to the origin of the coeliac artery; an incision is made in the aorta below the clamp to allow a cannula to be inserted into the coeliac artery and tied in place (Dornfest and Piliero describe the cannulation as extending into the splenic artery itself, but this is more difficult, and unnecessary if the hepatic and gastro-duodenal vessels have been tied as described above). The venous cannula is inserted into the portal vein (see p. 371) and advanced into the splenic vein before it is tied in place. Operation time is some 20 min, with 8–10 min of ischaemia.

### Characteristics of perfusion

Ford and Gowans describe an initial perfusion rate of 0·2 ml/min which is gradually increased to 1 ml/min as the result of a fall of resistance during the first hour of perfusion. The steady pressure required is 40 mmHg approximately. Dornfest and Piliero use flow rates of 0·1– 0·3 ml/min, at perfusion pressures of 50–80 mmHg, and comment that pressures over 125 mmHg result in the discharge of damaged cells from the spleen and this represents a 'failure' of perfusion.

Viability is assessed by these authors by the presence of an apparent difference in oxygenation of the inflowing and effluent medium (oxygen content and consumption have not been determined). Histology at the end of perfusion is normal, apart from some accumulation of iron pigment. A test of function might be contrived from the rate of removal of 'counts' from the medium to which labelled small lymphocytes have been added. This is linear for 4 h in the present preparation. To these criteria, the work of Dornfest and Piliero adds the rate of glucose uptake (15–21 mg/h, at flow rates 0·08–0·30 ml/min).

*Constancy of weight.* After 4 hours perfusion the spleen weights are the same as those of non-perfused controls. The spleen responds to vasoactive substances but, more specifically, the rate of release of leukocytes following the addition of adrenaline to the perfusate remains constant, despite a considerable reduction in flow rate. Further metabolic information on the perfused spleen preparation should be of considerable interest.

## Method II. 'Once-through' perfusion. Method of Dornfest and Piliero

The minor differences in medium and in operative technique between the two methods have already been discussed. In addition, these authors favour pentobarbitone anaesthesia. The main difference in the present preparation is the use of a 'once-through' system for perfusion. The apparatus is correspondingly simpler, but makes no special provision for oxygenation of the medium.

The results of the two methods can be compared only with difficulty. Method I gives perfusion of longer duration, 10 h against 3 or 4 h, and higher flow rates are obtained at lower perfusion pressure. In more recent studies (Dornfest, Cooper, and Gordon 1968) the preparation has been slightly modified; a Harvard peristaltic pump controls the flow at between 0·2–0·3 ml/min. The reservoir is gently shaken, not stirred, to prevent settling without damaging cells. Heating in this and in the earlier preparation is by means of a coil in a water bath at 37°C. Perfusions are continued for 4–5 h by the parameters already discussed.

*Splenic perfusion in the cat*

Although it is beyond the scope of this discussion to give details of the method of splenic perfusion in the cat, this is a standard preparation that has been used for the study of transmitter overflows after nervous stimulation (Blakeley *et al.* 1963, see Blakeley 1965 for refs). The method has some features in common with the rat spleen preparation of Ford (which is based upon it), but considerable attention is paid to the provision of a pulsatile perfusion pressure with all the details of the *in vivo* systolic pulse wave. Thus Blakeley (1965) describes a pump that provides a pulsatile wave form with a 'dicrotic notch', in immitation of the *in vivo* pulse. The possible relevance of such pressure wave form to perfusion for biochemical studies is discussed on p. 67.

### RETICULOENDOTHELIAL SYSTEM: THYMUS

Folkman, Winsey, Cole, and Hodes (1968) have applied a method of perfusion of the thymus gland, which was developed in the dog, to the rat. The apparatus has already been described in connection with perfusion of kidney and thyroid gland (see p. 342), and incorporates a roller pump, a filter, a warming coil, and a membrane oxygenator (p. 58). The perfusion medium was selected as one likely to preserve the gland for several days, and is basically Eagle's medium (see p. 27).

*Operative procedure*

Apparently the thymus is viable in the context of perfusion even after $2\frac{1}{2}$ h of anoxia, so that the following operative description, which involves 40 min of dissection under ischaemic conditions, may be acceptable. The thymus gland of the rat is intrathoracic (contrast that of the guinea-pig) and in the adult weighs approximately 0·5 g. It is perfused through the arch of the aorta, tying off the subclavian and common carotid vessels. Heparin (100 units intraperitoneally) is injected into the unanaesthetized rat, which is then killed, before beginning the dissection. (A modification, in which the animal is ventilated as described on p. 448 might allow the operation to be conducted with minimal ischaemia and result in a preparation more suited to biochemical use.)

The skin of the chest and neck is removed as a single flap. The chest wall is removed by lateral incisions and cut transversely at the level of the third interspace. The common carotid and subclavian vessels are tied distally and cut, leaving the subclavian vessels, from which the thymus supply is derived.

The arch of the aorta is tied distally and, in addition, the root of the aorta is tied and transected. The common carotid arteries are tied at their origin, taking care to avoid the subclavian vessels, from which arise the branches to the thymus. The distal aorta is cannulated (cannula = 3 mm o.d.), the remaining tissues are dissected free to allow the thymus to be removed from the thorax. Any leaking veins are ligated.

*Characteristics of the perfusion*

Direct pumping (pressure not given) results in a flow rate of 4–8 ml/min per g for several days. Viability is determined by the release of lymphocytes into the medium, while, in the dog but not yet reported in the rat, glucose uptake has been observed. For biochemical purposes, a less elaborate perfusion medium should be investigated.

## SALIVARY GLAND

There is, at present, no satisfactory perfusion of the rat salivary gland. For the cat, Douglas and Poisner (1962) describe a perfusion of the submaxillary gland which secretes saliva spontaneously and also responds with an increased saliva production to the addition of acetylcholine to the perfusion medium. Attempts to perfuse the submandibular gland of the rabbit, a species in which the atropine-resistant parasympathetic control of salivation is of particular interest, have been unsuccessful, as a result of oedema formation (L. H. Smaje, personal communication). The cat preparation will be described briefly as a technique that may have biochemical applications.

### Anatomy

The anatomy of the sublingual region of the cat is illustrated by Liddell and Sherrington (1929).

### Method of perfusion

Using Locke's solution as perfusion medium, and gassing with 95% $O_2$:5% $CO_2$ (pH 7·0), perfusion of both submaxillary salivary glands is performed through the external carotid artery on each side. Irrelevant branches of the artery are tied, leaving the external maxillary artery only. The venous effluent is not collected in this preparation, but drainage is assured by opening the chest and cutting the inferior vena cava. Saliva may be collected by means of a cannula in the submandibular duct.

*Characteristics of perfusion*

At a constant pressure of 90 mmHg, and at 23–5°C, experiments may be continued for 100 min. Saliva is produced in response to acetylcholine added to the perfusion medium. No metabolic parameters of function have been reported for this preparation, and the conditions required for a viable perfusion of salivary gland remain to be established.

## BONE

Preparations for the biochemical study of bone include tissue culture (Fell 1964), bone chips, bone slices, bone powder, and isolated cell preparations (see Vaughan (1970) for review). Enzyme extraction methods and histochemistry (Fullmer 1966) have contributed to an understanding of the metabolism of bone, but because of the hardness of the tissue this remains an area fraught with technical difficulty. Vaughan (1970) points out that different results are obtained in *in vivo* and *in vitro* studies of mineral binding by bone, and this may extend to other metabolic properties. An autoperfusion technique for bone has been introduced by MacIntyre, Parsons, and Robinson (1967), and later modified to allow perfusion with blood or with a semi-synthetic medium (Parsons and Robinson 1967). With this preparation it has been possible to demonstrate the expected release of $Ca^{2+}$ into the effluent blood in response to thyrocalcitonin added to the medium, and it may have other biochemical applications.

In the rat, the release of cells from bone marrow has been studied in two perfused preparations, the hind-limb perfusion (Kuna *et al.* 1959, Dornfest *et al.* 1962), and the isolated perfused femur (Dornfest *et al.* 1962). The first preparation is discussed in the context of perfusion of skeletal muscle (see p. 315), but the second belongs to this section as a preparation of isolated perfused bone that may have broader biochemical applications.

### Anatomy and blood supply

The skeletal system is said to receive 4–10 per cent of the resting cardiac output and, in the dog, most (70 per cent) of the diaphyseal blood supply is via the nutrient artery. Vaughan (1970) illustrates the various routes by which a long-bone receives its blood. Once within the bone, the bone canals (Haversian system) conduct capillaries and venous sinuses. Most of the flow is along the length of the bone and blood leaves by nutrient veins. This fact is the basis of the modification from nutrient artery perfusion to 'irrigation' introduced by Parsons and Robinson (1967) (see Method II, below).

MacIntyre, Parsons, and Robinson record that in the perfused tibia, infusion of barium sulphate at a time of normal blood flow still demonstrates inadequate perfusion of cortical bone. Since this is a perfusion via the nutrient artery, the modern views on blood supply to bone are relevant. The account by Brookes, Elkin, Harrison, and Heald (1961) and of Brookes and Harrison (1957) of the centrifugal direction of blood flow in bone is adopted by Vaughan (quoted above) and suggests that bone cortex is supplied outwards from within by blood that has already traversed the medulla. Blood from periosteal vessels is thought not to flow into the medulla except under special circumstances. However, this is still controversial. The conclusion that may be drawn from this view of the circulation through bone is that a marrow infusion (as suggested by Parsons and Robinson (1967)) is as likely to provide a 'physiological' blood supply to the bone cortex as is the more formal perfusion of the nutrient artery. Comparative studies are necessary to confirm this suggestion.

## Physiology

An excellent review of the subject, with the biochemical references is provided by Vaughan (1970).

## Methods of perfusion

Method I. Auto-perfusion, via the nutrient artery, of the cat tibia (MacIntyre *et al.* 1967).

Method II. Simplified 'irrigation' method for perfusion of cat tibia (Parsons and Robinson 1967).

Method III. Isolated perfused femur of rat (Dornfest *et al.* 1962).

## Method I. Auto-perfusion of cat tibia. Method of MacIntyre *et al.*

The tibia of the cat is perfused via the popliteal artery, from which arises its nutrient artery. This vessel is cannulated and provided with blood from the carotid artery of the same animal, with the introduction of a pump into the line between the two vessels to control the perfusion pressure and flow. Medium flows out of the bone around the distal end and is not recirculated, and such a preparation can be maintained for 3 h.

### Apparatus

A constant flow is provided by a prismatic roller pump (Bainbridge and Wright (1965)), and jacketed vinyl tubing is used.

*Arterial cannula.* Polythene tubing, o.d. 9 mm (NT/2) is suitable in the cat. No special oxygenator or filter is provided. The tibial preparation is contained in an organ bath, under paraffin.

*Perfusion medium*

Autologous blood. or blood direct from the carotid artery of a donor animal is used. No other procedure has been documented.

*Operative procedure*

The operative procedure recommended for the cat is described, with the suggestion that it may be equally applicable to smaller laboratory animals such as the rat.

Under chloralose and halothane anaesthesia, the popliteal artery is exposed behind the 'knee' by ligating and dividing both heads of gastrocnemius and all the thigh muscles near their tibial insertion. The branches of the popliteal artery, posterior tibial, medial and lateral geniculate, and the muscular branches, are thereby exposed and then tied. The popliteal vein is ligated and cut. To establish an 'auto-perfusion', the carotid artery is exposed and cannulated, its flow being directed into a cannula that is inserted into the popliteal artery. By interposing the pump, as previously described, the rate of flow can be maintained constant. Heparin (100 units) is injected intravenously.

Once perfusion is established, the tibia is isolated by ligating all the calf muscles at their origins and removing them. The knee and ankle joints are disarticulated and the tibia is mounted in the organ chamber. A rubber collar fixed round the tibial shaft prevents blood that has not traversed the bone from mixing with the outflow from the lower end of the tibia. This medium collects below paraffin and may be sampled.

*Characteristics of perfusion*

A pressure of 70–100 mmHg is maintained for up to 3 h and, with the assistance of the pump, a constant flow of 0·2 ml/min is maintained. Any fall in perfusion rate is attributed to the occurrence of micro-emboli. Plasma calcium, in the effluent medium remains constant for 1 hour but falls in the second hour; a definite effect of thyrocalcitonin in increasing the outflow of calcium has been demonstrated in this preparation.

## Method II. Simplified 'irrigation' technique. Method of Parsons and Robinson

Despite apparently satisfactory perfusion in Method I, its authors demonstrated, by means of barium injection after perfusion, that only a proportion of the cortex was perfused. The following method is simpler, and although not technically a vascular perfusion, it fulfils the same purpose, and has been used to demonstrate the thyrocalcitonin effect (q.v.). A tightly fitting tube is applied to the cut end of the tibia and medium is pumped into it under pressure. The bone is irrigated for the

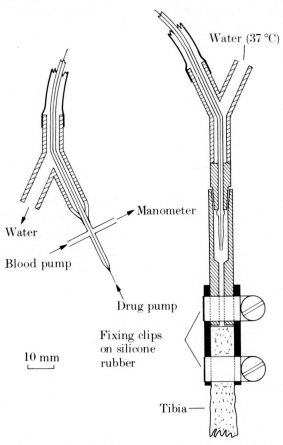

Fɪɢ. 8.7.  Perfusion of bone. Apparatus for perfusion of cat tibia (Parsons and Robinson 1967). The ends of a flexible tubing assembly for perfusion of isolated cat tibia are shown. A Perspex connector is fastened to the cut tibial shaft by a stout silicone rubber tube with clips, leaving a minimal dead-space, and the whole assembly is immersed in a bath of warm paraffin. Blood flowing out at the head of the bone (not shown) falls through the paraffin and is collected at intervals through a tap.

length of the shaft with blood or with a synthetic perfusion medium. Higher flow rates are obtained than with Method I (0·5 ml/min) although the authors suggest that only 40 per cent of this flow passes through bone capillaries. The apparatus is illustrated in Fig. 8.7 and is self-explanatory.

**Method III. Isolated perfused femur of rat. Method of Dornfest** *et al.*

By direct pumping, or under gravity, perfusion medium is introduced into the proximal end of the isolated rat femur by bathing the entrances to the nutrient foramina, and emerges through the distal nutrient foramina where it can be collected. The efficacy of such an approach is based upon electron-microscopic data from Zamboni and Pease (1961). Perfused (or 'irrigated') in this way, bone may be maintained for up to 20 h, as determined by discharge of marrow cells, or for 4 h in routine experiments. The operative preparation is brief and the apparatus simple, which makes this an attractive technique for possible bio-chemical use.

*Apparatus*

A non-recirculation apparatus with Tygon tubing and siliconed glass is recommended by the authors. A rotary-action pump provides the perfusion pressure, but later experiments report satisfactory perfusion under gravity from a simple reservoir. No formal oxygenator is included; temperature control is by means of a coil of Tygon tubing immersed in a water bath at 37 °C, and the reservoir is constantly stirred to prevent settling.

*Cannulae.* No cannulae as such are required in this technique. A length of rubber (silicone rubber would be preferable) is slipped over each end of the bone and tied firmly about the diaphysis (shaft) of the bone. Tubing of internal diam. 0·5 cm is suitable, with appropriate adaptors to the assembled apparatus.

*Perfusion medium*

*Medium* I. Fresh heparinized rat blood from animals below 400 g body weight is recommended. Blood from older animals is said to lead to increasing vascular resistance during perfusion. The blood is collected under ether anaesthesia, by aortic puncture (see p. 33).

*Medium* II (see also Medium III for spleen). Leukocyte-depleted medium is prepared by centrifuging Medium I at 2000 rev/min for 20 min, aspirating the 'buffy layer'. The resulting medium has 50–150 cells/mm$^3$. Both Media I and II are filtered through stainless steel gauze before use.

*Characteristics of perfusion*

Perfusion, without recirculation of medium, is maintained at 3–8 ml/h. Initially, a perfusion pressure of 10–30 mmHg (14–40 cm $H_2O$) is required, but with Medium I the perfusion pressure is increased each hour to maintain this rate of flow, up to 100 mmHg. Higher pressures are marked by failure of the perfusion. Apparently, with cell-free medium (Medium II) no such increase in resistance is noted. However, nothing is known of the metabolism of the preparation and whether or not cells are required for some other aspect of bone metabolism cannot be ascertained. Nor is there any information concerning the oxygen requirements of bone under these conditions. Nevertheless, this preparation appears to maintain bone in a viable state for several hours and might be further investigated.

## PLACENTA

Attempts to maintain the placenta *in vitro* by perfusion are confronted by the problems posed by its very special vascular arrangement. No preparation to date meets the requirements of stability and reproducibility required for biochemical experiment. Simply stated, the placenta represents a membrane dividing two circulations: the foetal, which is formed of a distinct capillary bed, supplied and drained by the umbilical vessels, and the maternal circulation which is derived in a 'semi-parasitic' way from the vessels supplying the uterine wall. On this side, there are no formal vessels; vascular sinusoids bathe the maternal surface of the placenta, and the act of separating the placenta from the uterine wall disrupts the fine syncytial arrangement.

Three approaches are possible:

(1) The foetal circulation alone may be perfused, which would allow studies of metabolism but not of placental transfer of materials.

(2) Perfusion of the foetal circulation, with an attempt to mimic the maternal circulation by a multiple injection method (Panigel, Pascaud, and Brun 1967). This more complex preparation would allow transport studies to be undertaken. Such a preparation is available for the human placenta.

(3) Nixon's suggestion (1962) was to perfuse the uterine vessels of the pregnant sheep, thereby maintaining the maternal circulation. For the biochemist, such a preparation has the disadvantages of a complex operative procedure and a heterogeneity of tissue perfused.

In the event of interest in perfusion of the placenta in small animals, Panigel describes the vascular anatomy of the rat and guinea-pig placentae (Panigel 1959). However, for the present, biochemical studies on the placenta might best be conducted in more standard incubation techniques using slices and homogenates (Szabo and Grimaldi 1970).

## MAMMARY GLAND

Mammary gland perfusion in larger animals, the dog, cat, and the goat have been conducted by Linzell (1954 and 1959). These experiments were conducted with a pump assisting the circulation, and in series with the experimental animal, so that the mammary gland is not particularly accessible to biochemical study. A detailed description is given by Hardwick and Linzell (1960). Isolated preparations are described by Petersen, Shaw, and Visscher (1941) and by Peeters and Massart (1947), but are again in large animals. It should be possible to devise a similar preparation for the rat, but at present most biochemical investigations on mammary tissue are conducted in incubations, either of the whole gland (Folley 1956) or of fragments (see Kon and Cowie (1961), for review).

## PERFUSION OF BLOOD-VESSELS

The biochemistry of blood-vessels is of great interest, and has till now been studied in homogenates (for example, by Böttger *et al.* 1968) or extracts (for example, by Briggs, Chernik, and Chaikoff 1949). These studies concerned lipid synthesis and $^{14}C$ acetate incorporation into fats, with interest in the aetiology of atherosclerosis. This has been reviewed by Wahl and Sanwald (1969). More recently, Hoff and Hayes (1969) have suggested that the technique of recirculation-perfusion might be applied to the study of the metabolism of blood-vessels and describe an apparatus that would be suitable for such a purpose. Several other preparations of arterial perfusion in small animals have already been described; the mesenteric artery preparation of McGregor (1965), a renal artery perfusion (Hrdina, Bonaccorsi, and Garattini 1967), a perfused ductus arteriosus of guinea-pig (Kovalčík 1963), a perfusion of rat-tail artery (Hinke and Wilson 1962), and a preparation perfusing the entire tail of the rat (Wade and Beilin 1970), with particular interest in each case, in the vasomotor responses of the preparation.

It is not improbable that each of these preparations will find some biochemical application, and that such methods may be extended to include other types of artery (all the existing preparations deal with 'muscular' arteries), and of veins. In the vessels at present under consideration, intima, a layer of cells of endothelial origin, and the media, of which smooth muscle is the principle component, are the tissues perfused. However, it should be borne in mind, that blood-vessels have their own vasculature, the vasa vasorum. These are seen in the outer layer of the vessel wall, the adventitia. There is no information about how much substrate reaches the layers of the vessel wall from the lumen and from the vasa vasorum and in practice, luminal perfusion has been effective in maintaining at least smooth muscle activity in blood-vessels.

Two preparations, that of Hinke and Wilson (1962) and the newer apparatus of Hoff and Hayes (1969) aim to irrigate both the lumen and the outer surface of the isolated arterial segment. This is an approach which has been productive in biochemical studies of intestine (Fisher and Parsons and others; see Intestine, Chapter 7) but it differs in principle from 'perfusion' as defined in Chapter 1. It seems appropriate to include such methods as are available to study the biochemistry of blood-vessels since, apart from their intrinsic value, every organ perfusion technique itself incorporates the perfusion of the vascular bed of that organ.

### Anatomy

Relevant details of vascular anatomy are described in the appropriate section below. Microscopically, the walls of arteries are in three layers: an endothelium (intima), a muscular (or elastic) layer, the media, and an outer layer, the adventitia.

Veins are also composed of three layers, but these may be less well-defined than in arteries. The larger veins contain considerable quantities of smooth muscle in the medial layer. In the larger arteries (over 1 mm diameter), blood-vessels of blood-vessels, vasa vasorum, supply the outer layers, i.e. adventitia and external media, and this complicates the concept of vascular perfusion in such vessels. For the present purpose it may be assumed that muscular arteries, those most frequently studied in perfusion, have no vasa vasorum.

*Nerves and lymphatics.* Veins as well as arteries have a copious nerve supply, mainly of sympathetic origin, and perivascular lymphatics are prominent.

The methods to be described have been introduced for another purpose than biochemical study, namely for the investigation of vasomotor responses. The anatomical and methodological details, given here might help a worker to select a preparation from which a 'biochemical' perfusion could be developed (see Chapter 1). In addition, more formal attempts at perfusion of blood-vessels for general experimental purposes are described on p. 431.

## Methods of perfusion of blood-vessels *per se*

Method I. Mesenteric artery preparation of McGregor (1965). Modification of Northover (1968).

Method II. Miscellaneous arterial perfusions:

(a) Renal artery: Hrdina *et al.* (1967).
(b) Ductus arteriosus: Kovalčík (1963).
(c) Middle cerebral artery: Uchida and Bohr (1969).

Method III. Perfused tail of rat: Wade and Beilin (1970).

Formal perfusion studies:

(a) Isolated perfused rat-tail artery preparation: Hinke and Wilson (1962).
(b) Pulsatile-circulatory artery perfusion system: Hoff and Hayes (1969).

## Method I. Mesenteric artery preparation of McGregor

The preparation most commonly used in vasomotor studies is that developed by McGregor (1965). The superior mesenteric artery is perfused with direct pulsatile pressure; side branches are tied, with the exception of four terminal ileal branches through which outflow occurs. Northover's modification leaves only one ileal artery as outflow. The outflow is not cannulated, and medium is not normally collected in this preparation, which is detached from the intestine and transferred to an organ bath.

### Apparatus

A water-jacketed organ chamber houses the preparation, and perfusion is by means of a roller pump, for example, Harvard 1210. This provides a flow of 25 ml medium/min at a pressure of 60 mmHg. (Note that the resistance of the cannula in this system is sufficient to give a perfusion pressure recording of 60 mmHg, and adding the vessel increases this to 85 mmHg.)

*Cannula.* A glass cannula of external diameter, *c.* 1 mm is suitable.

*Perfusion medium*

*Medium* I. McGregor recommends Tyrode solution (see p. 27), gassed with 95% $O_2$:5% $CO_2$ and Northover used a similar solution with satisfactory vasomotor responses. viz. NaCl 138 mM; KCl 2·74; $NaHCO_3$ 10·1; $MgCl_2$ 1·06; $CaCl_2$ 0·582; $NaH_2PO_4$ 0·416 mM; plus glucose 5·69 mM. This medium was said to be 'aerated' but the pH is not reported; flows of 4–10 ml/min were obtained with perfusion pressures considerably below that employed by McGregor (22 mmHg = 30 cm $H_2O$).

*Medium* II. A depolarizing medium (Northover 1968) consisting of $K_2SO_4$ (92 mM) or KCl (138 mM) to replace NaCl in Medium I gave satisfactory perfusions and might have application to studies in other organs.

*Medium* III. Whole heparinized rat blood (from a donor animal by recirculation, in the original description of MacGregor (1965)) gives satisfactory perfusion at a rate of 1 ml/min.

*Operative procedure*

Under ether anaesthesia, the abdomen of the rat is opened in the usual way. The superior mesenteric artery is identified at its origin from the aorta, level with the right renal artery, by sweeping the intestine to the animal's left. Its major branches are easily seen when the mesentery of the bowel is held up to the light. All identifiable branches (e.g. pancreatic, jejunal, colic) are tied, near to the main trunk, leaving the four main branches which supply the ileum. (In the modification suggested by Northover (1968) and Uchida and Bohr (1969), a single ileal vessel is preserved.) Loose ligatures are placed about the origin of the superior mesenteric artery and it is cannulated as described in Chapter 7 (p. 365). The cannula is tied in place and the perfusion begins when the distal ileal vessels are cut, close to the intestine. The remainder of the mesentery is cut to free the preparation, which may now be transferred to the organ chamber.

*Characteristics of perfusion*

With Medium I, flow rates of 25 ml/min are obtained at perfusion pressures about 60–85 mmHg; and in this once-through perfusion, the rate of flow and the vasoconstrictor response to noradrenaline are preserved for 4 h. Experiments have been conducted at either 22°C or

37°C (38°C) without significant difference, but there is no metabolic information available, nor is the effluent medium collected in the present system.

**Method II. Renal artery preparation (Hrdina *et al.*); Ductus arteriosus preparation (Kovalčík); Middle cerebral artery preparation (Uchida and Bohr)**

These preparations have in common the establishment of conditions for long-term 'once-through' perfusion with at least evidence of survival of vasomotor response. In one case, the preparation of Uchida and Bohr, such evidence was available of survival of the preparation for 13 h and this is encouraging to the use of such a preparation for metabolic studies.

*Apparatus*

A variety of apparatus has been used; the segment of vessel is generally removed from the animal after cannulation and suspended in a temperature-controlled bath of buffered saline. Perfusion is by means of a roller pump and oxygenation when provided, is by means of a simple bubbler (see p. 44).

*Perfusion medium*

(a) *Renal artery*. The medium is said to be 'Krebs' saline, and is in fact a balanced salt medium with 15 mM $NaHCO_3$, gassed with 95% $O_2$: 5% $CO_2$ (pH 7·2).

(b) *Ductus arteriosus*. Tyrode solution, gassed with 95% $N_2$:5% $CO_2$ (this is a special situation in which oxygen induces total constriction of the vessel). Again, the pH must lie around 7·1–7·2, rather than 7·4 as suggested.

(c) *Middle cerebral artery*. A variant of Tyrode solution is used to obtain very prolonged perfusions, viz. NaCl 119 mM, KCl 4·7, $KH_2PO_4$ 1·18, $MgSO_4.7H_2O$ 1·17, $CaCl_2$ 1·6, $NaHCO_3$ 14·9, D-glucose 5·5 mM and sucrose, 50 mM. The medium is prepared in demineralized water, with 0·026 mM $CaNa_2$ EDTA.

*Operative procedures*

(a) *Renal artery*. The left renal artery is prepared for perfusion by excision from the rat, before mounting it on a stainless steel cannula. The authors recommend beginning perfusion within 5 min of its removal from the animal, but comment that the vessel retains its reactivity even if stored for 24 h at 4°C. The supra-renal branch is then tied, and the perfusing vessel is immersed in buffer. Medium flows freely from the distal, cut end of the artery which is perfused with a flow rate of 6–7 ml/min at 45–50 mmHg pressure. Noradrenaline and ATP are reported to produce reversible and reproducible increases in perfusion pressure, but the survival time of the preparation is not discussed. Under certain circumstances, for example, for biochemical experimentation, it may be preferable to perfuse the vessel without an anoxic interval. For this purpose, the right renal artery may be prepared according to the method discussed for kidney perfusion (Method IA, p. 234) by cannulation from the superior mesenteric artery.

(b) *Ductus arteriosus* (guinea-pig). This preparation is of interest to physiologists as a vessel that closes completely when the newborn animal breathes. It is sensitive to oxygen tension and has been successfully perfused with a medium gassed with $N_2$ (q.v.). Equally, it may have interesting biochemical properties appropriate to this unique vessel. The great vessels are cut from the animal (foetus), a procedure which requires some skill. The left and right pulmonary arteries are ligated and a cannula (o.d. = 1 mm) is tied into the pulmonary trunk. Medium then flows freely from the cut ends of the aorta. (Alternatively, one end could be ligated and the other cannulated for collection of effluent medium.) No information is available about the viability of this preparation beyond the demonstration that it becomes occluded when the bathing medium is gassed with 95% $O_2$:5% $CO_2$.

(c) *Other arteries.* No special description is required of the operative preparation of other blood-vessels that have been perfused. Vessels of particular interest are the femoral artery (of dog) (Carrier and Holland 1965), iliac artery of rat (Uchida and Bohr 1969), the central vessel of the ear (of rabbit) (De la Lande and Rand 1965), and the umbilical vessels of the newborn (Yasargil 1960), descriptions of which are available as cited.

### Method III. Perfused rat-tail preparation. (Wade and Beilin)

The rat tail provides a vascular bed of convenient size, with a distribution to the skin, bone, and tendon. It has, as a result been used in studies of the control of vasomotor tone, and to assay vasomotor hormones, with a minimum of interference from the surrounding tissues (Friedman, Nakashima, and Friedman 1963; Wade and Beilin 1970). No metabolic studies have yet been undertaken with this preparation, and it may indeed be too heterogeneous for ready interpretation of the results of such experiments. Nevertheless, the preparation to be described has been well characterized and is highly reproducible (Wade and Beilin 1970) and may have applications in the study of biochemistry of blood-vessels. The main artery of the rat tail, the 'ileo-lumbar' artery, is cannulated *in situ* as it lies on the ventral surface just beneath the skin. When the artificial circulation, of albumin in Krebs–Henseleit saline, pumped from a reservoir directly into the artery is established, the tail is removed from the rat and transferred to the specially designed trough. The existing apparatus allows the effluent medium to be sampled, although there are multiple veins and cannulation is impracticable.

*Anatomy*

A single artery supplies all the tissues of the tail, and drains through a capillary bed into four veins that empty into the iliac veins.

*Apparatus*

Fig. 8.8 illustrates the apparatus used; a Watson–Marlow pump supplies medium from the reservoir (oxygenated with 95% $O_2$:5%

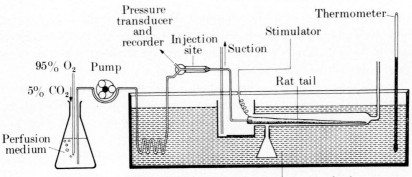

FIG. 8.8. Perfusion of blood vessels *per se*. Apparatus for perfusion of the isolated rat tail (Wade and Beilin 1970). The medium flask is constantly gassed with a mixture of oxygen 95% and carbon dioxide 5%. Perfusion at a known constant rate is maintained by the roller pump acting on a short segment of silicone rubber tubing joined in turn to the coil of nylon tubing enclosed in the covered constant temperature water bath. Venous outflow is removed by a suction lead.

$CO_2$), by means of a simple bubbler), via a 'damping' tube to the tail artery. A pulse wave of *c.* 8 mmHg persists, despite the damping. The pressure is recorded by a transducer attached to a side-arm of the main flow, and a second side-arm allows the circuit to be broken periodically (every 30 min is recommended) to release bubbles that interfere with the pressure recording. The primitive oxygenator is responsible for this problem, which may be solved by means of a bubble trap, or by the use of a more elaborate oxygenator (see Chapter 1). The tail-support is of special design (see Fig. 8.8) and is held in a water bath at 34°C.

*Cannula.* A 1·02 mm (o.d.) polythene catheter with Luer fitting, and drawn out slightly at the tip to correspond to the arterial diameter, is used. No venous cannulation is attempted in the present preparation.

## Perfusion medium

The final composition of the medium is: NaCl 132·8 mM, KCl 4·8, $NaHCO_3$ 15, $CaCl_2$ 2·5, $MgCl_2$ 0·1, D-glucose 0·69 mM and bovine serum albumin 4·0 g %. It is prepared as follows: Tyrode solution, containing in addition 12 g % albumin and 0·69 mM glucose is dialysed against Tyrode plus glucose for 36 h and then diluted to give the final concentration of 4·0 g albumin per cent.

## Operative procedure

Rats, ideally above 350 g in weight (tail weight 9–10 g) are anaesthetized with intra-peritoneal Nembutal. Heparin (100 units) is injected into the femoral vein in the usual way (see Chapter 3), and the animal is laid on its back. Tapes hold the hind limbs, and an additional transverse tape holds fur and the perineum away from the operating area. The ventral aspect of the tail is cleaned with spirit, and shaved over its first 2 cm. A mid-ventral incision is made through the skin, and underlying fascia over a length of about 1 cm starting at the base of the tail. The artery lies immediately below the surface at this point. A pair of curved forceps is inserted, closed, into the distal end of the fibrous tunnel thus exposed, and protect the artery whilst the incision is extended a further centimetre. The vessel is now cleared over a length of 1 cm and two ligatures passed behind it, with a curved forceps. The upper ligature is tied, the cannula is inserted through a small incision made with scissors, into the ventral surface of the artery, advanced about 1 cm, withdrawn 2–3 mm and finally tied in place with the lower ligature. Blood flows back freely in the event of successful cannulation and the tail can then be transferred to the organ tray.

The tail is removed from the rat by an incision with a scalpel through the next inter-vertebral space above the site of cannulation, weighed and attached to the inflow catheter of the apparatus, taking care to avoid the entry of bubbles. Perfusion begins at once, although some adjustment of the tail and the cannula may be necessary to achieve the expected flow rate and perfusion pressure (see below).

Total operation time is about 10–15 min, with anoxia for 1–2 min. On the basis of the response to noradrenaline, this period of anoxia appears not to be critical.

*Characteristics of preparation*

Perfusions of 2–3 h duration are feasible in this preparation, with a steady flow rate (0·5 ml/min) obtained at perfusion pressures about 25 mmHg. Even with the present albumin concentration, there is a gain in weight by the tail of some 10 per cent attributable to oedema. It occurs early (cf. kidney) and does not thereafter increase significantly. The best criterion of function in this preparation is the pressor respones to various constrictor hormones. No studies have been undertaken on the effluent medium, but it is possible to collect this and the range of possible experiment is thereby increased.

## Formal perfusion studies

Two systems exist which may be more directly of interest in bio-chemical experimentation on perfused blood-vessels. In the method of Hinke and Wilson (1962), the isolated segment of tail artery is cannula-ted at both ends and perfused, both within the lumen and in the circuit that bathes the outer surface of the vessels. More specifically, Hoff and Hayes (1969) have introduced an apparatus in which segments of blood-vessels may be perfused with pulsatile pressure under standard conditions, and in which facilities exist for examination of medium from both lumen side and outer side. Its authors suggest that the apparatus may be of use in metabolic studies.

### (a) Isolated perfused rat-tail artery

A length of 3 cm of artery is removed from the tail and perfused under hydrostatic pressure with Krebs–Henseleit buffer, containing glucose (11 mM). The vessel is cannulated at both ends for luminal perfusion and, in addition, the bathing fluid may be recirculated.

*Apparatus*

Fig. 8.9 shows the quite simple perfusion apparatus required. No pump or oxygenator is used, instead, gas pressure from a cylinder of 95% $O_2$:5% $CO_2$ drives medium from the reservoir (cf. Leichtweiss, Kidney, Chapter 4).

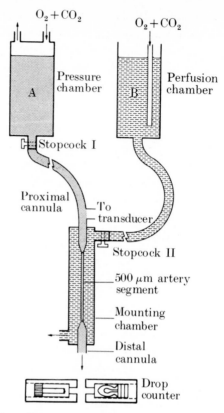

FIG. 8.9.    Perfusion of blood-vessels *per se*. Block diagram of Hinke and Wilson apparatus (1962) for perfusion of a 500-$\mu$m segment of rat-tail artery.

*Cannulae.* Glass cannulae of 1 mm inside diameter are used; the distal cannula is turned downwards to facilitate collection in the once-through system. Temperature control is by water-jacketing.

### Perfusion medium

Standard Krebs–Henseleit buffer (see p. 23) is used, with the following additions; D-glucose 11 mM, Ca disodium versenate 0·026 mM (to prevent autoxidation of noradrenaline used in these experiments). No albumin is used (contrast Wade and Beilin method).

### Operative procedure

In rats anaesthetized with ether, a longitudinal ventral incision exposes the artery of the tail. A length of 3 cm is excised and mounted on the cannulae within the perfusion chamber. Alternatively, to avoid

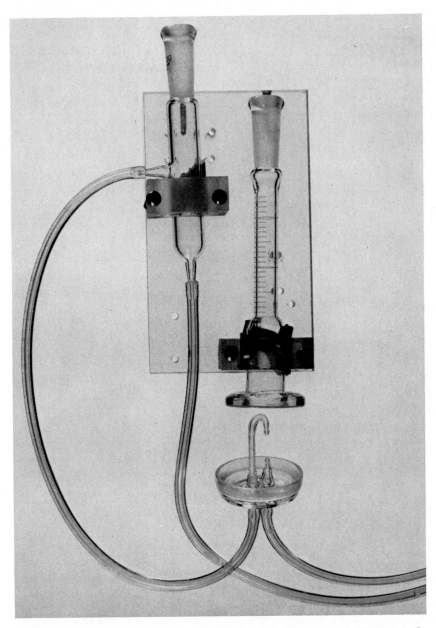

FIG. 8.10. Perfusion of blood-vessels *per se*. Apparatus for recirculation-perfusion of a blood vessel (Hoff and Hayes 1969).

anoxia, and the entry of air bubbles, the vessel could be cannulated in *situ* (cf. previous method).

*Characteristics*

These investigators pre-perfused at 40–60 mmHg pressure for $1\frac{1}{2}$–2 h. The test of viability is a reduction of flow rate of perfusion medium (e.g. from 2 drops/s to 0·19 drops/s) following the addition of 5 ml K–H buffer containing 0·5 $\mu$g/ml noradrenaline. By this criterion, successful perfusion may be continued for 5 h, and the preparation has been used to study pressure: flow relationships and the effects of adrenaline, noradrenaline, vasopressin, and angiotensin II, as well as the effect of variation in the ionic composition of the medium. There have been no biochemical studies to date.

### (b) Pulsatile-circulatory artery perfusion system (Hoff and Hayes)

The system suggested by Hoff and Hayes differs from the preceding in providing pulsatile rather than steady flow, and the external medium although it can be sampled, is not circulated. The apparatus has a major advantage in being 'closed' to allow recirculation, and it is both simple and compact. The apparatus illustrated (Fig. 8.10) is of glass, with Tygon tubing. It contains 5–6 ml of medium and is suitable for a 1-cm length of artery. No proposals are made concerning perfusion medium or flow rate, but some of the information in the preceding methods might be applicable.

REFERENCES

*General*

ANDJUS, R. U., SUHARA, K., and SLOVITER, H. (1967) An isolated, perfused rat brain preparation, its spontaneous and stimulated activity. *J. appl. Physiol.* **22**, 1033.

HO, R. J. and MENG, H. C. (1964) A technique for the cannulation and perfusion of isolated rat epididymal fat pad. *J. Lipid Res.* **5**, 203.

ROBERT, A. and SCOW, R. O. (1963) Perfusion of rat adipose tissue. *Am. J. Physiol.* **205**, 405.

*Adipose tissue*

LONG, J. A. and LYONS, W. R. (1954) A small perfusion apparatus for the study of surviving, isolated organs. *J. Lab. clin. med.* **44**, 614.

MORGAN, H. E., HENDERSON, M. J., REGEN, D. M., and PARK, C. R. (1961) Regulation of glucose uptake in muscle. I. The effect of insulin and anoxia on glucose transport and phosphorylation in the isolated, perfused heart of normal rats. *J. biol. Chem.* **236**, 253.

NISHIITSUTSUJI–UWO, J. M., ROSS, B. D., and KREBS, H. A. (1967) Metabolic activities of the isolated, perfused rat kidney. *Biochem. J.* **103**, 852.

Renold, A. G. and Cahill, G. F. (1965) Metabolism of isolated adipose tissue: a summary. *Handbook of physiology*, p. 483. American Physiological Society, Washington.

Rodbell, M. (1964) Metabolism of isolated fat cells. 1. Effects of hormones in glucose metabolism and lipolysis. *J. biol. Chem.* **239**, 375.

—— and Scow, R. O. (1965) Chylomicron metabolism; uptake and metabolism by perfused adipose tissue. *Handbook of physiology*, Adipose tissue, Chapter 49. American Physiological Society, Washington.

Salaman, M. R. and Robinson, D. S. (1966) Clearing-factor lipase in adipose tissue—A medium in which the enzyme activity of tissue from starved rats increases *in vitro. Biochem. J.* **99**, 640.

Scow, R. O. (1965) Perfusion of isolated adipose tissue: FFA release and blood-flow in rat parametrial fat body. *Handbook of physiology*, Chapter 45, p. 437; American Physiological Society, Washington.

Sheldon, H. (1965) Morphology of adipose tissue. *Handbook of physiology*, p. 125.

Wassermann, F. (1965) The development of adipose tissue. *Handbook of physiology* (eds. A. E. Renold and G. F. Cahill, Jr.), p. 87. American Physiological Society, Washington.

*The brain*

Bloom, W. and Fawcett, D. C. (1968) *A textbook of histology*, Chapter 12. Saunders, Philadelphia.

Geiger, A. (1958) Correlation of brain metabolism and function by the use of a brain perfusion method *in situ. Physiol. Rev.* **38**, 1.

—— and Magnes, J. (1947) The isolation of the cerebral circulation and the perfusion of the brain in the living cat. *Am. J. Physiol.* **149**, 517.

—— —— Taylor, R. M., and Veralli, M. (1954) Effect of blood constituents on uptake of glucose and on metabolic rate of the brain in perfusion experiments. *Am. J. Physiol.* **177** 138.

Gerhard, L., Meesen, H., and Veith, G. (1958) Nervensystem. *Pathologie der Laboratoriumstiere* (eds. P. Cohrs, R. Jaffé, and H. Meesen), p. 698. Springer-Verlag, Berlin.

Green, G. C. (1963) *Anatomy of the rat*. Trans. Am. phil. Soc. Hafner, New York (3rd printing).

Miller, L. L., Bly, C. G., Watson, M. L., and Bale, W. F. (1951) The dominant role of the liver in plasma protein synthesis. *J. exp. Med.* **94**, 431.

Palay, S. L., McGee-Russell, S. M., Gordon, S., and Grillo, M. A. (1962) Fixation of neural tissues for electron microscopy by perfusion with solutions of osmium tetroxide. *J. Cell Biol.* **12**, 385.

Rusznyák, I., Foldi, M., and Szabo, G. (1967) *Lymphatics and lymph circulation*, 2nd edn. Pergamon Press, New York.

Scremin, O. U. (1968) Sodium fluorescein penetration from the perfused cerebral ventricles into different areas of the rat hypothalamus. *J. Physiol., Lond.* **194**, 42p.

Sloviter, H. A. and Kamimoto, T. (1967) Erythrocyte substitute for perfusion of brain. *Nature, Lond.* **216**, 458.

Thompson, A. M., Cavert, H. M., and Lifson, N. (1958) Kinetics of distribution of $D_2O$ and antipyrine in isolated, perfused rat liver. *Am. J. Physiol.* **192**, 531.

—— Robertson, R. C. and Bauer, T. A. (1968) A rat head-perfusion technique developed for the study of brain uptake of materials. *J. appl. Physiol.* **24**, 407.

TORACK, R. M. (1969) Sodium demonstration in rat cerebrum following perfusion with hydroxyadipaldehyde—antimonate. *Acta neuropathol.* **12**, 173.

*Reticuloendothelial system: (a) Spleen*

BLAKELEY, A. G. H., BROWN, G. L., and FERRY, C. B. (1963) Pharmacological experiments on the release of the sympathetic transmitter. *J. Physiol., Lond.* **167**, 505.

—— (1965) Neuro-effector mechanisms of the autonomic nervous system. D.Phil. Thesis, University of Oxford.

BLOOM, W. and FAWCETT, D. W. (1968) The spleen. *A textbook of histology*, 9th edn., p. 403. Saunders, Philadelphia.

COHRS, P. and SCHULZ, L-U. (1958) *Pathologie der Laboratoriumstiere*, Vol. 1, p. 330. Springer-Verlag, Berlin.

DORNFEST, B. S., COOPER, G. W., and GORDON, A. S. (1968) Effect of erythropoietin on reticulocyte release from the isolated perfused rat spleen. *Ann. N.Y. Acad. Sci.* **149**, 249.

—— and PILIERO, S. J. (1966) Mechanism of leukocyte production and release. VII. Factors influencing leukocyte release from isolated perfused rat spleen with special reference to vaso-active agents. *J. Lab. clin. Med.* **68**, 577.

FORD, W. L. (1969*a, b*) The kinetics of lymphocyte recirculation within the rat spleen. *Cell Tiss. Kinet.* **2**, 171; The immunological and migratory properties of the lymphocytes recirculating through the rat spleen. *Br. J. exp. Path.* **50**, 257.

—— and GOWANS, J. L. (1967) The role of lymphocytes in antibody formation. II. The influence of lymphocyte migration on the initiation of antibody formation in the isolated, perfused spleen. *Proc. R. Soc.* **B168**, 244.

PICTET, R., ORCI, L., FORSSMANN, W. G., and GIRARDIER, L. (1969) An electron microscope study of the perfusion—fixed spleen. 1. The splenic circulation and the RES concept. *Z. Zellforsch. mikrosk. Anat.* **96**, 372.

SNOOK, T. (1950) A comparative study of the vascular arrangements in mammalian spleens. *Am. J. Anat.* **87**, 31.

*(b) Thymus*

FOLKMAN, J., WINSEY, S., COLE, P., and HODES, R. (1968) Isolated perfusion of thymus. *Expl. Cell Res.* **53**, 205.

*Salivary gland*

DOUGLAS, W. W. and POISNER, A. M. (1963) The influence of calcium on the secretory response of the submaxillary gland to acetylcholine or to noradrenaline. *J. Physiol., Lond.* **165**, 528.

LIDDELL, G. T. and SHERRINGTON, C. (1929) *Mammalian physiology*, 2nd edn, p. 78. Clarendon Press, Oxford.

*Bone*

BAINBRIDGE, D. R. and WRIGHT, B. M. (1965) A multi-channel long-term infusion system for small animals. *J. Physiol., Lond.* **177**, 6P.

BROOKES, M., ELKIN, A. C., HARRISON, R. G., and HEALD, C. B. (1961) A new concept of capillary circulation in bone cortex. *Lancet* (i), 1078.

—— and HARRISON, R. G. (1957) The vascularisation of the rabbit femur and tibofibula. *J. Anat.* **91**, 61.

DORNFEST, B. S., LoBUE, J., HANDLER, E. S., GORDON, A. S., and QUASTLER, H. (1962) Mechanisms of leukocyte production and release. II. Factors influencing leukocyte release from isolated perfused rat legs. *J. Lab. clin. Med.* **60**, 777.

FELL, H. B. (1964) Some factors in the regulation of cell physiology in skeletal tissues. *Bone biodynamics* (ed. H. M. Frost), p. 189. Churchill, London.

FULLMER, H. M. (1966) Enzymes in mineralised tissues. *Clin. Orthop.* **48**, 285.

KUNA, S., GORDON, A. S., MORSE, B. S., LANE, F. B., and CHARRIPER, H. A. (1959) Bone marrow function in perfused, isolated hind-legs of rats. *Am. J. Physiol.* **196**, 769.

MacINTYRE, I., PARSONS, J. A., and ROBINSON, C. J. (1967) The effect of thyro-calcitonin on blood-bone calcium equilibrium in the perfused tibia of the cat. *J. Physiol., Lond.* **191**, 393.

PARSONS, J. A. and ROBINSON, C. J. (1967) A simplified perfusion method for the study of calcium equilibrium in isolated bone. *J. Physiol., Lond.* **194**, 59P.

VAUGHAN, J. M. (1970) *The physiology of bone.* Clarendon Press, Oxford.

ZAMBONI, L. and PEASE, D. C. (1961) The vascular bed of red bone marrow. *J. Ultrastruct. Res.* **5**, 65.

*Placenta*

NIXON, D. A. (1962) The transplacental passage of fructose, urea and meso-inositol in the direction from foetus to mother, as demonstrated by perfusion studies in the sheep. *J. Physiol., Lond.* **166**, 351.

PANIGEL, M. (1959) Anatomie vasculaire du placenta chez des rongeurs. *C. r. Ass. Anat.* **46**, 565.

—— PASCAUD, M., and BRUN, J. L. (1967) Une nouvelle technique de perfusion de l'espace intervilleux dans le placenta humain isolé. *Path, Biol., Paris* **15**, 821.

SZABO, R. J. and GRIMALDI, R. D. (1970) The metabolism of the placenta. *Adv. met. dis.* **4**, 185.

*Mammary gland*

FOLLEY, S. J. (1956) *The physiology and biochemistry of lactation.* Oliver and Boyd, Edinburgh and London.

HARDWICK, D. C. and LINZELL, J. L. (1960) Some factors affecting milk secretion by the isolated, perfused mammary gland. *J. Physiol., Lond.* **154**, 547.

KON, S. K. and COWIE, A. T. (1961) *Milk: the mammary gland and its secretion.* Academic Press, New York.

LINZELL, J. L. (1954) Some observations on the use of the perfused lactating mammary gland. *Rev. Can. Biol.* **13**, 291.

—— (1959) Physiology of the mammary glands. *Physiol. Rev.* **39**, 534.

PEETERS, G. and MASSART, L. (1947) La perfusion de la glande mammaire isolée. *Archs int. Pharmacodyn. Thér.* **74**, 83.

PETERSEN, W. E., SHAW, J. C., and VISSCHER, M. B. (1941) A technique for perfusing exercised bovine mammary glands. *J. Dairy Sci.* **24**, 139.

*Blood-vessels* per se

BÖTTGER, I., DEUTICKE, U., EVERTZ-PRÜSSE, E., ROSS, B. D., and WIELAND, O. (1968) Über das Verhalten von freien Acetat beim Miniaturschwein. *Z. ges. exp. Med.* **145**, 346.

BRIGGS, F. N., CHERNIK, S., and CHAIKOFF, I. L. (1949) The metabolism of arterial tissue. 1. Respiration of rat thoracic aorta. *J. biol. Chem.* **179**, 103.

CARRIER, O. and HOLLAND, W. C. (1965) Supersensitivity in perfused isolated arteries after reserpine. *J. Pharmac. exp. Ther.* **149**, 212.

DE LA LANDE, I. S. and RAND, M. J. (1965) A simple isolated nerve-blood vessel preparation. *Aust. J. exp. Biol. Med. Sci.* **43**, 639.

FRIEDMAN, S. M., NAKASHIMA, M., and FRIEDMAN, C. L. (1963) Sodium, potassium and peripheral resistance in the rat tail. *Circulation Res.* **13**, 223.

HINKE, J. A. M. and WILSON, M. L. (1962) A study of the elastic properties of a 550 $\mu$ artery *in vitro*. *Am. J. Physiol.* **203**, 1153; Effects of electrolytes on contractility of artery segments *in vitro*. ibid. **203**, 1161.

HOFF, H. F. and HAYES, T. L. (1969) Description of a pulsatile-circulatory artery perfusion system. *Pflügers Arch. ges. Physiol.* **305**, 292.

HRDINA, P., BONACCORSI, A., and GARATTINI, S. (1967) Pharmacological studies on isolated and perfused rat renal arteries. *Eur. J. Pharmacol.* **1**, 99.

KOVALČÍK, V. (1963) The response of the isolated ductus arteriosus to oxygen and anoxia. *J. Physiol., Lond.* **169**, 185.

MCGREGOR, D. D. (1965) The effect of sympathetic nerve stimulation on vasoconstrictor responses in perfused mesenteric blood vessels of the rat. *J. Physiol., Lond.* **177**, 21.

NORTHOVER, B. J. (1968) The effect of drugs on the constriction of isolated, depolarised blood vessels, in response to calcium or barium. *Br. J. Pharmacol.* **34**, 417.

UCHIDA, E. and BOHR, D. F. (1969) Myogenic tone in isolated perfused resistance vessels from rats. *Am. J. Physiol.* **216**, 1343.

WADE, D. N. and BEILIN, L. J. (1970) Vascular resistance in the perfused isolated rat tail. *Br. J. Pharmacol.* **38**, 20.

WAHL, P. and SANWALD, R. (1969) *Atherosclerosis* (eds. F. G. Schettler, and G. S. Boyd), p. 141. Elsevier, Amsterdam.

YASARGIL, G. M. (1960) Die antagonistische Wirkung von Kalium und Calciumionen an *in vitro* durchströmten menschlichen Nabelschnurarterien. *Helv. physiol. pharmac. Acta* **18**, 491.

# 9. Lung

THE respiratory function of lung, gas exchange, has been exhaustively studied in whole animals and in perfused lungs of larger experimental animals since the introduction of the heart–lung preparation by Knowlton and Starling (1912). More recently, however, the non-respiratory functions of the lung have been recognized, and studies with isolated perfused lung have been undertaken to elucidate some of them (see Heinemann and Fishman 1969).

The biochemistry and intermediary metabolism of the lung have scarcely been considered, and there is clearly scope for further metabolic studies. Homogenates, slices (see Heinemann and Fishman for references), and a tissue-dicing method which yields homogeneous tissue fragments that can be pipetted or weighed have been used (Mongar and Schild 1953). More recently, a variety of rat-lung perfusion techniques has been devised, some or all of which may find applications in biochemistry. The perfused lung has special advantages in that:

(i) the complex anatomical relationships are maintained;
(ii) the preparation is isolated from other metabolically active tissues and organs;
(iii) the lung has a particularly high blood-flow which includes, unchanged, all blood leaving the liver, and the arrival of substrates may be important in establishing the metabolic function of the lung;
(iv) excretory and detoxicating functions of the lung have been described, and a recirculating isolated system such as perfusion lends itself particularly to such studies (see Chapter 1).

Finally, recirculation of a small volume of perfusion medium circumvents the problem of establishing arterio–venous differences in an organ with very high blood-flow. Information about the metabolic activities of the lung is fragmentary, having been obtained in a non-systematic way, often as a by-product of research primarily concerned with something other than the metabolic performance of lung. Oxygen consumption and substrate incorporation into fatty acids and thence into phospholipids have been studied in a preparation of perfused rabbit lung (Popjak and Beeckmans 1950). Acetate incorporation into lung lipids has been demonstrated in a similar preparation, and fatty acid oxidation, lipolysis, and pentose-shunt activity are known to be present in the lung of rats and rabbits and therefore presumably of other species (references include Lands (1958), Uspenskü (1965), and Felts (1964)). Metabolic adaptation to diet has been confirmed in the same experiments with perfused rabbit lung in which the incorporation of acetate into fatty acids is increased in starvation.

Experiments of two basic types have been considered but in neither case have intensive investigations been carried out. The first type is concerned with the handling of substrates added to the pulmonary circulation. Till now, glucose is the only substrate to have been studied in the rat (Levey and Gast 1966), and its conversion to lactate and pyruvate has been demonstrated. The second type of investigation is based on the observation that the lung, as the only major tissue interposed in the right-heart circulation, may have detoxicating or activating effects on plasma-borne substrates passing through it. Bakhle, Reynard, and Vane (1969) and Leary and Ledingham (1969) have used the perfused rat lung to establish that angiotensin I is converted to angiotensin II and that angiotensin II is not destroyed in the right-heart circuit.

Sperling and Marcus (1968) have used yet another isolated ventilated perfused lung preparation to differentiate between the effect of a drug on the tracheobronchial system and one on the alveoli. Airborne drugs in a perfused ventilated lung reach the tracheobronchial system first, while intravascular addition of a drug should bring it first into contact with the alveoli (Sperling, Marcus, and Delaunois 1968). This last approach to perfused lung has not yet been applied to the study of its metabolism.

### Anatomy

*Lobes*

The anatomy of the rat lung differs slightly from the standard description of mammalian lung in textbooks on human anatomy. Green (1963) describes briefly and Lauche (1958) more extensively and accurately, the division of the right lung into four or possibly five lobes, while the left has but one. Consequently the right lung is much larger. The lobes on the right are the superior, middle, and inferior (the latter is sometimes further divided by a deep fissure, allowing this to be considered as two lobes) and the deeper median lobe, which is in relation to the heart. The latter lobe is sometimes referred to as the post-caval lobe, since it lies deep and posterior to the superior vena cava. The lobes may be of importance in perfusion studies for reasons similar to those described for the liver. Thus, individual lobes may be sampled for intermediates or enzymes, without interrupting the perfusion of the remainder of the organ.

## Airways

The branches of the airways of the lung correspond to the overlying lobar divisions of the lung; trachea, two bronchi, and smaller bronchioli terminate in the air-sacs or alveoli where gas exchange occurs. As in man, the trachea and bronchi have cartilaginous rings supporting their circumference, while their main tissue components are smooth muscle, connective tissue, glands that secrete mucus, and an epithelial lining that is extensively ciliated.

## Circulation

The pulmonary arterial circuit is similar to that described for man but some differences between man and the rat occur in the bronchial arterial system. The major circulation of the lungs is via the pulmonary arteries and pulmonary veins. This brings deoxygenated blood from the right ventricle to the lung and delivers it, after exchange of $CO_2$ and $O_2$, to the left atrium. The entire cardiac output must traverse this circuit; the vessels are large, and the flow rate of blood through them is correspondingly high. The secondary circulation, via the bronchial arteries, supplies the bronchi and main air passages down to the respiratory bronchioles (Allison, de Burgh Daly, and Waaler 1961) with oxygenated blood from the left heart. In man and dog, and probably in other species, the bronchial supply accounts for no more than 1 per cent of the total blood reaching the lungs. No figures are available for the rat, and as yet no attempt has been made to maintain this secondary circulation in the isolated perfused rat-lung preparation. (In an isolated perfused lobe of lung from the dog (Allison *et al.* 1961) the bronchial arterial supply has been shown to be essential to maintain the vasomotor reflexes.)

## Note on the bronchial arteries of the rat

The bronchial arteries differ from the corresponding vessels in man both in their origin and in their mode of distribution within the lung (Verloop 1949). Instead of arising directly from the aorta, as in man, the left bronchial artery of the rat arises from the left internal mammary artery, via its pericardiaco-mediastinal branch, and the right bronchial artery from the costocervical trunk (Green 1963, see Fig. 225). This makes them very difficult to perfuse. Within the lung, in man, there are reports of arteriolar anastomoses between the bronchial and pulmonary circuits (von Hayek 1960), while in the rat this is not the case. Lauche (1958) quotes Verloop (1949) (and there have been no subsequent reports), who injected four rat aortas with dye and reported no case in which arterioles of the pulmonary artery became filled. Only capillaries are so affected and the conclusion is that there is no anastomosis of consequence between the two circulations in the rat lung.

*Nerve supply*

The trachea is supplied by the vagus nerve (recurrent laryngeal branch) and the remaining pulmonary system is supplied copiously from the parasympathetic nervous system via the vagus and from the sympathetic nervous system, via its major thoracic plexuses. In larger animals electrodes at the hilum have been used to stimulate nerves to the lung, but no information on such experiments is available for the rat. The function of these nerves is in any case, obscure.

A rich lymphatic supply is found in rat lung and becomes particularly evident during inflammatory processes (Lauche 1958). It may be of some importance in lung perfusion *in situ*, as in the perfusion of other organs *in situ*, in that it offers a possible unaccountable route of fluid loss.

### METHODS OF LUNG PERFUSION

A variety of methods has been proposed for perfusion of the rat lung and these have been simply classified as follows:

(i) *in situ* or isolated,
(ii) with ventilation or without ventilation.

The advantages of isolation of the lung are:

(1) that all medium entering the lung can be recovered; Leary (1969) comments that in the *in situ* preparation some loss of medium through 'collateral vessels' supplying the chest wall was unavoidable, and that this was not recirculated; exudation from the lung surface and formation of lymph are other sources of fluid loss;
(2) sampling from lung lobes during perfusion, with continued perfusion of the remaining lung may have some applications in biochemical experiments and this is clearly more easily accomplished in the isolated lung (by analogy with the isolated liver, see Chapter 3).

The advantage of ventilation is in offering the physiological route of gas exchange. However, it may not be essential in studies concerned primarily with lung biochemistry.

There is a further effect of ventilation on the distribution of 'blood' within the lung, and this may apply in the perfused organ (see p. 70). The methods available are described under the following headings.

I. *In situ* perfusion (without ventilation): Leary and Ledingham 1969.
II. *In situ* perfusion (with ventilation): modification of Method I.
III. Isolated lung perfusion (without ventilation): Bakhle, Reynard, and Vane (1969).
IV. Isolated lung perfusion (with positive-pressure ventilation):
   A. Pulsatile perfusion: Hauge (1968).
   B. Hydrostatic perfusion: Levey and Gast (1966).
V. Isolated lung perfusion (with negative-pressure ventilation): Delaunois (1964).

Of these preparations, only that of Levey and Gast has been applied to biochemical problems, and there have been no comparative studies upon which to base an assessment of the various methods of perfusion available. All the methods to be described may have some feature of value in studies of lung biochemistry.

**Method I.** *In situ* **lung perfusion without ventilation**

This, the simplest method, has been published by Leary and Ledingham (1969) and described in a little more detail by Leary (1969). Medium is perfused into the right ventricle by means of a cannula inserted through the right atrium, enters the lung via the pulmonary artery, and leaves the lung to enter the left side of the heart. It drains into a cannula inserted into the left ventricle, through the ascending aorta. The lung and heart are left *in situ*.

This preparation has been used for only very brief perfusions; 6 min in the experiments of which the results were published by Leary and Ledingham (1969), and apart from determining flow rate, the removal of angiotensin is the only activity that has been studied as yet.

*Apparatus*

Perfusion is under gravity from a multibulb oxygenator, using the glassware, pump, and temperature-controlled cabinet described in Chapter 3.

*Cannulae.* Polythene cannulae of 3-mm external diameter are used for both inflow and outflow. The end should be flared to hold the cannula in place when the ligatures are tied.

## Perfusion medium

Leary and Ledingham modified the medium of Hems *et al.* (1966) (see Chapter 1) by omitting the washed human red cells. This was done to avoid interference from angiotensinases present in red cells or blood. For a similar reason, the first 5–10 ml of medium were discarded at the start of perfusion to wash the remaining rat blood from the circuit.

## Operative procedure

Well-fed rats, male or female, of 250–350 g weight, are anaesthetized with intraperitoneal Nembutal. Heparin, 0·1 ml (100 int. units), is injected intravenously into the femoral vein at the start of the operation. With the animal on the operating tray (see Fig. 1.13), the skin is incised in the mid-line from xiphisternum to manubrium sterni and the rib cage is carefully opened with a point of the scissors, first horizontally just above the diaphragm and then, by two lateral and vertical incisions, the anterior chest wall is removed. A large clamp is put across the upper ribs before completing the removal by a transverse incision. Great care must be taken not to puncture the lung during this procedure.

The lungs collapse, since no ventilation is instituted, and the period of partial anoxia begins (i.e. circulation continues but there is no oxygenation). The thorax is packed with crushed ice (in theory, to minimize metabolic changes due to anoxia; this may be omitted without significant effect on the subsequent perfusion) and the ligatures shown in Fig. 9.1 are placed (A) behind the inferior vena cava just above the diaphragm, (C) round the right atrio-ventricular groove, and (E) round the root of the aorta, carefully avoiding the pulmonary artery. (E) is the only ligature that may present some difficulty; it should be positioned first, while the circulation is still vigorous and the aorta clearly visible. A curved forceps is used first to clear connective tissue parallel to the first quarter of a centimetre of the ascending aorta, separating it from the pulmonary artery, which lies to its left in this situation. The same curved forceps is used to pass a ligature carefully behind the vessel. Ligature (A) round the inferior vena cava just above the diaphragm presents no difficulty. Ligature (C) is passed with curved forceps behind the superior vena cava and the tip of the right auricle.

Unless care is taken, the left superior vena cava, with its azygoz branch, is omitted from this ligature. The ligature illustrated by Leary as passing round the atrio-ventricular groove is particularly difficult to position. The following procedure may be adopted. A ligature is passed round the right superior vena cava but is not tied. A curved forceps is now passed behind the inferior vena cava at the point of its

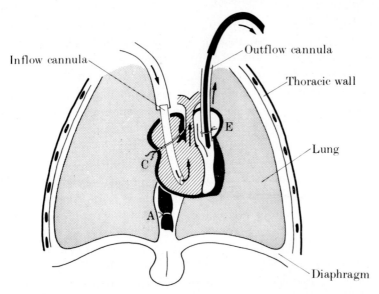

FIG. 9.1.   Operative preparation for *in situ* perfusion of lung (method of Leary and Ledingham 1969). For details of positioning of ligatures A, C, and E see p. 444.

entry to the right atrium and the caudal (outer) free end of the preceding ligature grasped to be taken behind the inferior vena cava. This ligature is now loosely tied and lies approximately over the atrial ventricular groove.

Nevertheless it has been found necessary (personal observation) to ligate the left superior vena cava separately (Ligature (F), not shown in Fig. 9.1). In positioning each of these ligatures extreme care is required to avoid including either of the arteries, or one of their subdivisions, in the ligature.

*Cannulation.* Inflow cannula (B) (Fig. 9.1) is filled to the tip with medium, the right auricle is grasped in a curved forceps (as for inferior vena cava cannulation in liver perfusion; see Fig. 3.11), and the cannula pushed through the auricle. Blood flows back up the cannula and a small tissue-paper plug at the end prevents excessive blood loss. Ligature (C) is tightened and tied again round the cannula when its tip has been advanced into the right ventricle.

Outflow cannula (D) is inserted into the ascending aorta, the vessel is grasped in curved forceps, and a small incision is made with pointed scissors 2 mm from the heart. Blood loss is minimal, since the supply to the left heart has already been considerably reduced by ligation of the venae cavae. The cannula is advanced against some resistance beyond the aortic valve and, when in position, ligature (E) is tightened. It is important to include the root of the aorta, with the origins of the two coronary arteries, in this ligature, to avoid loss of medium from the circuit into cardiac muscle with the additional complication of the cardiac metabolism superimposed on that of the lungs.

Finally, ligature (A) around the inferior vena cava is tied and circulation through the pulmonary artery, lung capillary bed, and pulmonary vein is begun by attaching the right ventricular cannula to the oxygenator outflow. Flow begins immediately. In the experiments described by Leary and Ledingham the first 5–10 ml of effluent medium were discarded.

### Characteristics of perfusion

This preparation has till now been used only for very brief experiments—6 min is the maximum reported by Leary and Ledingham (1969). Flow rate was constant (over 30 ml/min) for the time tested. Angiotensin II added to the medium was not removed by lung tissue; no other parameter of function is recorded.

### Method II. *In situ* lung perfusion with ventilation

Preliminary studies have been conducted with a preparation identical to the preceding one, but in which a tracheotomy permits positive pressure ventilation of the lungs (see p. 449 for details).

Perfusions have been conducted for from 30 min to 1 h, with steadily falling flow rate (initial flow 30–40 ml/min).

### Method III. Isolated lung perfusion, without ventilation (lungs inflated)

This method, described briefly by Bakhle, Reynard, and Vane (1969), was used in conjunction with a simple recirculation circuit to assay vasoactive-hormone destruction in a single passage through the lungs. The method is suitable for both rat and guinea-pig.

*Apparatus*

No description of the apparatus is given.

*Perfusion medium*

Krebs–Ringer buffer as described by Umbreit (see p. 26) was used at 37°C, but no details of gassing mixture or the final pH are given and no protein or other oncotic substances were added.

*Operative procedure*

Under Nembutal anaesthesia tracheotomy is performed (see p. 449), a cannula tied into the trachea, and the chest is opened as described in Method I. Direct cannulation of the pulmonary artery is carried out (no details are given), and then the left heart is cut away. After starting a flow of medium the lungs are dissected free; the period of anoxia is therefore shorter than in Method I. The isolated lung is suspended from its two cannulae (tracheal and arterial) with effluent dripping from the cut pulmonary veins. The final step is to partially inflate the lungs via the tracheal cannulae and then to clamp the cannula. This has the effect of restoring the normal lung architecture but cannot contribute substantially to the process of gas exchange.

*Characteristics of the preparation*

No information is given by the authors apart from the ability of the isolated lung to degrade or inactivate some polypeptide hormones, and in its present form this preparation is unsuited to controlled biochemical experimentation.

## Method IV. Isolated lung perfusion with positive pressure-ventilation

Two preparations are available, one using pulsatile flow and devised to study vasomotor responses to hypoxia and temperature change (Hauge 1968), the other using steady flow under hydrostatic pressure and devised to study metabolic activity in the isolated lung (Levey and Gast 1966).

## Method IVA. Perfusion with direct pulsatile flow: method of Hauge

The isolated rat lung is perfused by means of a pulsatile flow pump through an inflow cannula inserted into the pulmonary artery through the right ventricle. Outflow is through a cannula lying in the left atrium or in the left ventricle and the lung is subject to rhythmic positive-pressure ventilation through a cannula in the trachea. The whole preparation is housed in a water-jacketed organ chamber and can be perfused for 3 h with either whole rat blood or rat plasma.

### Apparatus

The apparatus consists of a water-jacketed organ chamber in which the lungs are suspended from the three cannulae already described. The circulating water is kept at 36–38°C. The inflow cannula is attached directly to a pump with a side-arm manometer to record inflow pressure. Medium flows out into a small open reservoir whence it is collected by the pump. No oxygenator is described, since, in this system, oxygen is supplied by ventilation of the lungs.

*Pump.* The piston pump of Iversen and Steen (see p. 63) is used and produces a pulsatile flow. Its output can be adjusted and it is set to give a mean pressure of 18·8 mmHg (25·5 cm $H_2O$), the figure given for the normal pulmonary artery pressure of the rat (Smith and Bennett 1934).

*Ventilator.* The Starling ('Ideal') pump is used with an expiratory overflow arrangement to prevent over-inflation. A simple device (Konzett and Rössler 1940) uses a side-arm beneath a water-seal for this purpose. The rate can be varied, as can the pressure or volume of gas delivered to the lungs. For the present purpose the authors use 34 strokes per min, an inspiratory pressure of 10 cm $H_2O$, and an expiratory pressure of 1·5–2·0 cm $H_2O$. The gas mixture used for ventilation can also be varied at will—in these experiments 8 per cent $CO_2$ in air was used.

### Perfusion medium

*Medium* I. Whole fresh heparinized blood (100 i.u. Heparin/10 ml blood), obtained by heart puncture from three or more donor rats. The medium sufficient to prime this apparatus is 18–22 ml.

*Medium* II. Rat plasma, obtained by centrifuging Medium I, is reported to be equally effective. However, no comparative studies are presented.

### Operative procedure

Rats, male or female, of 180–200 g weight were used; or of 350–400 g in the later experiments. Anaesthesia is with intraperitoneal pentobarbitone (see p. 85). A tracheotomy is performed, and positive-pressure ventilation, at the pressure, volume, and rate given in the previous paragraph is initiated.

*Tracheotomy.* With the rat anaesthetized and lying on its back, an incision is made, with scissors, through the skin of the throat. A further incision across the superficial muscles, exposes the trachea, and a curved forceps may be passed behind it. Keeping to the mid-line permits this to be done without damage to major blood-vessels.

A ligature is drawn behind the trachea and tied loosely. To cannulate the trachea, an incision is made across half its diameter, about 1 cm above the point of entry into the thorax. The cannula (polythene or glass, o.d. 3–4 mm) is inserted well down the trachea and tied in place; ventilation should be begun when the chest wall is opened.

The operation takes less than 1 min and may be used in techniques of perfusion apart from the lung, for example, in liver (Powis, see Chapter 3), large intestine (Powis, see Chapter 7) to reduce anoxia during the surgical preparation.

The chest is opened as described on p. 444. The trachea, lungs, and the great vessels are dissected free of connective tissue, especially posteriorly and in relation to the diaphragm. Loose ligatures are placed around the superior and inferior venae cavae; heparin is injected (in this description directly into the left ventricle ($0.5$ ml = 100 int. units of Nova heparin)) and the venae cavae are then tied. Alternatively, heparin may easily be injected into the femoral vein. Further loose ligatures are placed round the ventricles and round the base of the ascending aorta. Before cannulation of the pulmonary artery and aorta, the authors recommend stopping ventilation. This introduces a period of anoxia and with practice it is possible to introduce the cannulae satisfactorily without this precaution. Stainless steel cannulae are used.

The inflow cannula, filled to the tip with medium, is inserted through the apex of the right ventricle (a small incision helps) and is advanced into the pulmonary artery. The outflow cannula is similarly inserted in the left ventricle and it is advised that the tip should not be advanced beyond the mitral valve. This is because it is thought that the cannula tip in the left atrium might impede venous drainage from the lungs. Tightening the ligature round the ventricles now fixes both cannulae in place. Finally, the aorta is occluded by tightening the ligature already positioned. The lungs, trachea, and heart are now lifted free from the thorax and suspended from their cannulae in the organ chamber.

Flow of medium begins and ventilation, if stopped during the operation, is now recommenced. The operation is said to take 15 min, of which 10–11 min are relatively anoxic.

*Characteristics of perfusion*

*Flow.* A steady perfusion rate is reported for 3 h at 7–8·5 ml/min. This flow rate is low, since the physiological cardiac output (i.e. pulmonary blood flow) is nearer to 30 ml/min. Pressure is said to be constant in this preparation; but the pump is basically of 'constant pressure' design and this observation is of limited significance.

*Effluent oxygen tension.* An increase in oxygen content of the effluent, compared with inflowing medium, confirms that gas exchange occurs in the ventilated lung. No information on oxygen consumption of the lung itself can be derived from this data.

*pH*. The authors refer to a steady pH of 7·38–7·48 for about 30 min but this then falls. No details are given but two factors may be involved. Equilibration of rat blood ($HCO_3 = 25$ mM) with 8% $CO_2$ would give a pH of approximately 7·1 (Henderson–Hasselbach equation). Alternatively, glycolysis, first by the red cells of the medium and possibly by the lung (see Levey and Gast), would also tend to lower pH if the buffer capacity of 25 ml medium were exceeded.

## Method IVB. Perfusion with constant hydrostatic pressure. Method of Levey and Gast

In this method a hydrostatic perfusion apparatus (Miller, see Chapter 3) is used. A roller pump delivers medium to the top of a multibulb oxygenator, whence it flows into a reservoir with an overflow which determines the pressure at which medium enters the pulmonary artery. An outflow cannula drains medium to a reservoir which in turn feeds the pump and the whole apparatus is enclosed in a heated cabinet at 37°C.

*Apparatus*

The cabinet and glassware used by Levey and Gast are similar to those described by Miller *et al.* (1951) and commercially available from Metalloglass Inc. (see Chapter 1). The cabinet perfusion system of Hems *et al.* (1966) could be used with equal effect. Levey and Gast introduced a second, identical perfusion apparatus within the same cabinet, omitting the rat lung, in order to follow the rate of glycolysis by the erythrocytes of the perfusion medium.

*Pump.* A standard roller pump is suitable, for example, that of Watson Marlow (see p. 66).

*Cannulae.* An inflow cannula of external diameter 1·8 mm (wall thickness 0·4 mm) and an outflow cannula of external diameter 2·8 mm (wall thickness 0·5 mm), both in polythene, are recommended by the authors.

*Organ chamber.* The authors give no detailed description of the organ chamber other than to report that the lung is mounted so as to be in the natural anatomical position, i.e. trachea horizontal, with the 'anterior' lung surfaces lying ventrally. (This is the reverse of the position of the lungs in the *in situ* technique described above.)

*Perfusion medium*

The medium of Miller (see Chapter 1) is used. Fresh heparinized rat blood, pooled from several fed animals, is filtered through nylon mesh ($50 \times 40$/cm). Saline (0·9%) is added to give an haematocrit of 20- 25 per cent, i.e. diluted 1:1, and glucose is added to a final concentration of 150 mg % (9 mM approximately). In order to fill the double perfusion apparatus described, 110–130 ml medium is required so that the lungs, 3·2–4·0 g weight, are perfused with a volume of 55–65 ml medium. The medium is gassed with 95% $O_2$:5% $CO_2$, which may be expected to result in pH 7·1–7·2.

*Operative procedure*

The operative procedure of Hauge is followed with the following differences.

(1) The trachea is not cannulated but a wide-bore needle is inserted directly and tied in place.
(2) The right ventricle is excised, allowing the inflow cannula to be inserted into the pulmonary artery under direct vision and tied in place.
(3) The outflow cannula is described as lying in the left atrium but may be inserted through the wall of the left atrium as already described by Hauge (see Method IVA).

*Ventilation.* Intermittent positive pressure-ventilation with air is continued throughout the perfusion at 100 strokes/min. This is nearer to the normal respiratory rate of the rat and some three times the rate used by Hauge.

## Characteristics of perfusion

Flow rate is 4–11 ml/min, in the same range as that reported by Hauge. No other comparable data is available so that the relative effect of pulsatile and non-pulsatile perfusion upon the lungs cannot be assessed.

*Metabolic function.* Glucose uptake with a formation of lactate and pyruvate has been demonstrated with this preparation and appears to occur at a constant rate throughout the 3-h experiments. The rate of glucose removal is approximately 0·25 $\mu$mol/min per g wet weight of lung (assuming 4 g of lung perfused). Glycolysis by red cells in the medium is more rapid and there may be some advantage in applying the medium of Hems *et al.* (see Chapter 1) to studies of this sort. The great advantage of this preparation appears to be its relatively long survival time.

## Method V. Isolated lung perfusion with negative-pressure ventilation. Method of Delaunois

The method and apparatus described by Delaunois are too complex to be of ready application in biochemical experiments. It has been used by Sperling and Marcus (see p. 440) to differentiate the effects of drugs added to the ventilating gas or to the blood stream.

Rats or guinea pigs of 250–400 g weight may be used, and the isolated lungs are ventilated at a maximum negative pressure of 16 mmHg for the guinea-pig (8 mmHg for the rat). The perfusion medium used by Delaunois was Tyrode solution and the flow rate achieved, only 0·2–0·3 ml/min, is very much lower than that found *in vivo*, or in other preparations.

For negative-pressure ventilation, the lungs must be enclosed in a sealed chamber. They are therefore inaccessible, a disadvantage in some metabolic experiments, and the preparation cannot at present be recommended for biochemical studies.

SUMMARY

Techniques of lung perfusion for biochemical studies are at an early stage of development. The preparation of Levey and Gast (1966)

(Method IVB) deserves further attention, especially in the choice of a defined perfusion medium. The preparation of Leary and Ledingham (Methods I and II) is simpler, and, by analogy with the *in situ* perfusion of the liver, may prove to be satisfactory. However, oedema formation and falling perfusion rate are problems that remain to be solved in this preparation.

For the lung, as for other organs, it may be of value to apply metabolic tests of function to the assessment of a method of perfusion.

## REFERENCES

ALLISON, P. R., DE BURGH DALY, I., and WAALER, B. A. (1961) Bronchial circulation and pulmonary vasomotor nerve responses in isolated perfused lungs. *J. Physiol., Lond.* **157**, 462.

BAKHLE, T. S., REYNARD, A. M., and VANE, J. R. (1969) Metabolism of the angiotensins in isolated perfused tissues. *Nature, Lond.*, **222**, 956.

DELAUNOIS, A. L. (1964) A new isolated lung apparatus. *Archs int. Pharmacodyn. Ther.* **148**, 597.

FELTS, J. M. (1964) Biochemistry of the lung. *Health Phys.* **10**, 973.

GREEN, E. C. (1963) *Anatomy of the rat.* Hafner, New York.

HAUGE, A. (1968) Conditions governing the pressor response to ventilation hypoxia in isolated perfused rat lungs. *Acta physiol. scand.* **72**, 33.

HEINEMANN, H. O. and FISHMAN, A. P. (1969) Non-respiratory functions of mammalian lung. *Physiol. Rev.* **49**, 1.

HEMS, R., ROSS, B. D., BERRY, M. N., and KREBS, H. A. (1966) Gluconeogenesis in the perfused rat liver. *Biochem. J.* **101**, 284.

KNOWLTON, F. P. and STARLING, E. H. (1912) The influence of variations in temperature and blood pressure on the performance of the isolated mammalian heart ('Heart-Lung' preparation of later authors). *J. Physiol., Lond.* **44**, 206.

KONZETT, H. and RÖSSLER, R. (1940) Versuchsanordnung zu Untersuchungen an der Bronchialmuskulatur. *Naunyn-Schmiedebergs Arch. exp. Path. Pharmak.* **195**, 73.

LANDS, W. E. M. (1958) Metabolism of glycerolipides: a comparison of lecithin and triglyceride synthesis. *J. biol. Chem.* **231**, 883.

LAUCHE, A. (1958) Trachea, Bronchien, Lungen und Pleura. *Pathologie der Laboratoriumstiere*, Vol. 1, p. 28. Springer-Verlag, Berlin.

LEARY, W. P. P. (1969) D.Phil. Thesis, University of Oxford.

—— and LEDINGHAM, J. G. (1969) Removal of angiotensin by isolated perfused organs of the rat. *Nature, Lond.* **222**, 959.

LEVEY, S. and GAST, R. (1966) Isolated perfused rat lung preparation. *J. appl. Physiol.* **21**, 313.

MILLER, L. L., BLY, C. G., WATSON, M. L., and BALE, W. F. (1951) The dominant role of the liver in plasma protein analysis. *J. exp. Med.* **94**, 431.

MONGAR, J. L. and SCHILD, H. O. (1953) Quantitative measurement of the histamine-releasing activity of a series of mono-alkyl-amines using minced guinea-pig lung. *Br. J. Pharmac.* **8**, 103.

POPJAK, G. and BEECKMANS, M. (1950) Extra-hepatic lipid synthesis. *Biochem. J.* **47**, 233.

SMITH, F. J. C. and BENNETT, G. A. (1934) The pulmonary arterial pressure in normal albino rats and the effect thereon of epinephrine. *J. exp. Med.* **59**, 173.

SPERLING, F. and MARCUS, W. L. (1968) Turpentine-induced histological changes in isolated rat and guinea-pig lungs. *Archs int. Pharmacodyn. Thér.* **175**, 330.
—— —— and DELAUNOIS, A. L. (1968) Responses of isolated, perfused rat and guinea-pig lungs to barbiturates. *Archs int. Pharmacodyn. Thér.* **176**, 309.
USPENSKǓ, V. I. (1965) Lipolytic activity of liver and lungs in experimental alloxan diabetes. *Fedn Proc. Fedn Am. Socs. exp. Biol.* **24**, T.43.
VERLOOP, M. C. (1949) On the arteriae bronchiales and their anastomosing with the arteria pulmonalis in some rodents. A micro-anatomical study. *Acta anat.* **7**, 1.
VON HAYEK, H. (1960) *The human lung.* Hafner, New York.

# Appendix I

## Working diagram for a perfusion cabinet

Details are given of the perfusion cabinet of Hems *et al.* (1966), which is discussed in connection with perfusion of liver, kidney, hind-limb, and lung. The dimensions and materials described as indicated are suitable for a cabinet containing two perfusion circuits, maintained at 37°C. Included is an electrical wiring diagram for the perfusion cabinet.

Duplicate and independant controls are provided for two simultaneous perfusions. For further details see p. 76, and the relevant chapters.

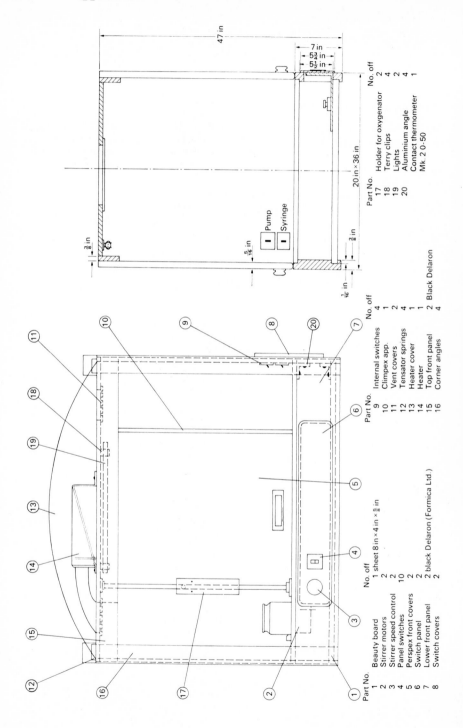

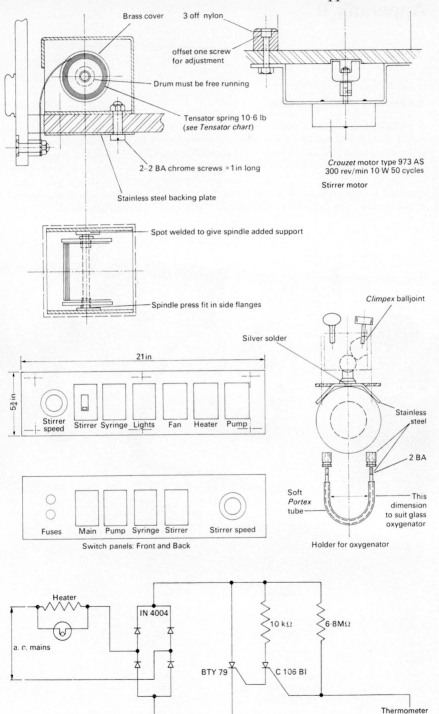

Brass cover

3 off nylon

offset one screw for adjustment

Drum must be free running

Tensator spring 10·6 lb (*see Tensator chart*)

2–2 BA chrome screws ×1 in long

Stainless steel backing plate

*Crouzet* motor type 973 AS 300 rev/min 10 W 50 cycles

Stirrer motor

Spot welded to give spindle added support

Spindle press fit in side flanges

*Climpex* balljoint

Silver solder

Stainless steel

21 in

5¾ in

Stirrer speed   Stirrer   Syringe   Lights   Fan   Heater   Pump

2 BA

Soft *Portex* tube

This dimension to suit glass oxygenator

Fuses   Main   Pump   Syringe   Stirrer   Stirrer speed

Switch panels: Front and Back

Holder for oxygenator

Heater

IN 4004

a. c. mains

10 kΩ   6·8 MΩ

BTY 79   C 106 BI

Thermometer

# Appendix II

## Ambec perfusion apparatus

A perfusion apparatus available from MX International Inc., Denver, Colorado, U.S.A. is that discussed as suitable for perfusion of pancreas (p. 333) or small intestine (p. 379). The water-heated apparatus is described in Chapter I, p. 72. The photograph illustrates the rotary pump (top left), the organ tray, with cannula holders, and inlet and outlet ports (centre), together with the rotating mesh oxygenator, described in more detail in the Text (p. 55). Technical specifications are available from the manufacturers, and in the reference given in the text.

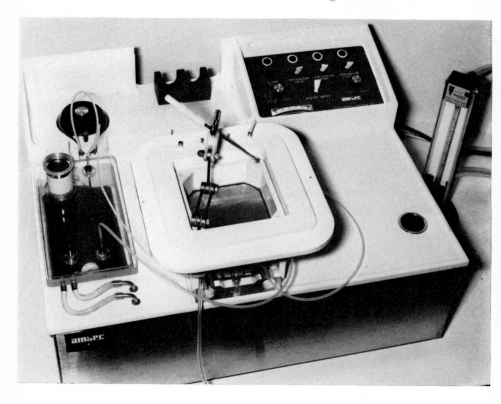

# Author Index

# Subject Index

acetate, 250; *see also* buffer; gluconeogenesis
incorporation into steroids, 350
incorporation into lipid in lung, 439
acetyl COA and liver, 113, 114
adenine nucleotides
effect on perfusion pressure, 428
in kidney, 237–8
in liver, 162, 163, 174, 203
in muscle, 314
ATP/ADP ratio, 174, 237–8
adenosine monophosphate in kidney, 237–
238
adenosine triphosphate requirement, 192
adipose tissue, 395 *et seq.*
anatomy, 395–6, 397
adipose tissue perfusion, 396
effect of annoxia, 403
methods of perfusion, 396
tests of function, 398, 399; adrenaline
effect, 398; flow, 398, 399, 402; free
fatty-acid release, 398, 403; oedema
and weight gain, 398, 403; oxygen
consumption, 401, 403; pressure,
398, 402
adrenal gland perfusion, 346–8
tests of function, 347, 348; flow, 346,
347; 11-hydroxylation of steroids,
347; noradrenaline release, 347,
348; oxygen consumption, 347
alanine in Schnitger medium, 22
release from liver, 165
albumin, 38–40, 209, 232, 327; *see also*
plasma expanders
binding of bicarbonate, 25, 39
bovine, fraction V, 20, 21, 28, 29, 30, 31,
37, 254
concentration, ideal, 38
defatting, 39, 209; *see also* free fatty acid
denaturation, 44
dialysis, 21, 209, 232, 254
human, fraction V, 38, 328
in heart perfusion medium, 294
oxygen solubility, effect on, 39
precipitation, 39, 107–8
preparations, 20, 21
problems with, 38
amino acids, addition to perfusion medium,
189, 209
release from perfused liver, 165
*see also under individual acids*

amylase, 197
synthesis by perfused liver, 197, 202,
207; by perfused pancreas, 332
anaesthesia and anaesthetics, 85–7, 212
amytal, 212
carbon dioxide in air, 86
chloralose, 86, 212, 419
ether, 85–6, 345, 413; effect on intestine,
364, 368; open, 85, 86, 87
'Evipan', 161
for liver perfusion, 212
halothane, 86, 419
hexobarbitone, 212
hypothermia, in place of, 408
'Inactin', 253
insulin, release by, 86–7, 212
pentobarbitone (Nembutal), 86, 87, 170,
212, 310, 444; effect on intestine,
364, 368
phenobarbitone, 86, 212
thiopentone, 406
urethane, 212
animal, choice of experimental, 16–17; *see
also individual species*
anoxia, 111
effect on heart, 270
effect on kidney, 237
effect on metabolism of liver, 205
in perfused intestine, 381, 391
*see also* flow rate
antibiotics in perfusion medium, 21, 83,
160
actinomycin D, 207
effect on liver metabolism, 206–7
mycostatin, 30, 343
neomycin, 30, 343
penicillin, 28, 29, 177, 207, 316, 343, 351,
387, 397, 412; clearance of, 157
puromycin, 207
streptomycin, 28, 29, 30, 177, 207, 316,
351, 387, 397, 412
tetracycline, 20, 21, 30, 207
anticoagulation, 9, 37, 136
defibrination, 36, 189, 364, 397
effect of liver metabolism, 37, 207–8
with ACD, 10, 37
with citrate, 10, 37
with heparin, 10, 37, 136, 208, 246, 251,
270, 316, 334, 364, 381, 383, 389,
412, 413, 444
with alternative agents, 37